WATER ENGINEERING AND MANAGEMENT THROUGH TIME – LEARNING FROM HISTORY

Water Engineering and Management through Time – Learning from History

Editors

Enrique Cabrera & Francisco Arregui
ITA, Universidad Politécnica de Valencia, Spain

CRC Press
Taylor & Francis Group
Boca Raton London New York Leiden

CRC Press is an imprint of the
Taylor & Francis Group, an **informa** business
A BALKEMA BOOK

Published by:
CRC Press/Balkema
P.O. Box 447, 2300 AK Leiden, The Netherlands
e-mail: Pub.NL@taylorandfrancis.com
www.crcpress.com – www.taylorandfrancis.com

ISBN 13: 978-0-415-48002-4 (hbk)

Visit the Taylor & Francis Web site at
http://www.taylorandfrancis.com

and the CRC Press Web site at
http://www.crcpress.com

Typeset by MPS Ltd. (A Macmillan Company), Chennai, India

Table of Contents

Part D – Conclusions

Foreword

Historically nobody has doubt about the importance of water as a fundamental resource, necessary for the human being but also for the proper economic and social development of cultures and civilizations.

But it is in the last years where the public awareness about water has gained bigger importance. The increasing needs of water for human supply and agricultural use, together with a less availability of the resources has make the water be a permanent matter of attention, and its management, an authentic challenge for the companies involved in that task, to whom they arise constant needs of providing innovative and sustainable solutions of the management pattern of water's integral cycle.

Therefore it has a huge value to look back and observe what our predecessors has done in this hard and noble task of putting the water at the citizens disposition, which difficulties have they had and how they find the solutions in order to learn the lessons that water management history through the pass of time has left us, to try to face with the biggest success the future challenge of the management of a limited and essential resource like water.

The book that you have in your hands just exactly deals about this and it is a great pleasure for aqualia to collaborate in this line with the university world, trying once more to combine the academic knowledge and the daily practice, hence to be more useful to the whole society.

With actions like this we will try to approach to all the people and show them that behind water's enjoyment in quality and quantity there is a very complex process that has to be managed by qualified and skilled professionals, experts in all the phases of water's integral cycle. With our participation in publications like this we will like to contribute a little bit more in the popularization and knowledge of this sector.

Therefore I invite you to use the information contained in every chapter of the book and enjoy the reading, learn and thought it over.

Fernando Moreno García
General Manager of aqualia Gestión Integral del Agua

Preface

The challenges water policy has to face this 21st century are enormous. Among others, it is worth to mention in first place the need to guarantee access to drinking water and to a decent hygiene level for all the inhabitants of a planet that has almost tripled its population in the last six decades. The second issue to be mentioned is a growing contamination that must be dramatically reduced. In tune with the growth of mankind, during last decade pollution has increased at an unsustainable pace. Last, water policy must ensure to cover, with scarcer resources, not only the human, industrial and agricultural needs, but those required by the ecosystems as well. It is worth to underline that in the last few decades have supported a deep deterioration. This is a rather complex task because in some decades, climate change threatens will reduce available water resources in dry areas a very significant amount (up to 40%).

This book focuses on these and other issues of to the future's water policy. Nothing new under the sun, since rivers of ink have been spent, are spend and will be spent trying to identify not only the actions that are convenient to ensure a more sustainable future than the present is, but also the great difficulties to overcome to put these actions in practice. The novelty lies, we believe, in the approach to perform the analysis. It is inspired in the great historian Edward Gibbon who, while walking around Rome's ruins, wondered how such an impressive culture had fallen so low. The answer can be found in his famous book, "The History of the Decline and Fall of the Roman Empire". To some extent, the Mediterranean' water culture has lived a similar history.

In fact, water engineering history has written its most glorious pages in many countries in which actually water is poorly managed. In most of them, current water policies are simply unsustainable. And history repeats itself. Brilliant solutions of the past – though in another context – claim for an adaptation to present day. And this is not an easy task. In his conclusions Gibbon states that *what does not evolve, is decadent*. After all, it is the immobilism what encumbers policies valid until some few decades ago, now unsustainable. If the present work contributes to unblock what is now blocked, mainly in countries lying on the Mediterranean' shores, the effort put on a book of complex genesis will be worth. Its root, papers presented at an international seminar which was held under the same name at the University of Alicante (Spain) in mid-2006. But, because the final objective of this publication was to become a book rather than the proceedings of a meeting, a later analysis of their contents evidenced some weakness to overcome. This is the reason why this publication includes five contributions not scheduled initially. By the other hand, most of papers presented at the seminar have been updated by the authors.

The result is a book of fifteen chapters organised in four sections. The introduction includes just a chapter that provides the general framework. The second section, *Water across time*, gather five lessons corresponding to periods in which Water Engineering has written some of its most brilliant pages. The third section, under the title *Great challenges of water in the 21st century*, is integrated by seven chapters that review some of the more relevant problems of present-day water policy. Last, a shortest section includes some conclusions and summarises the contents of the preceding chapters.

Arrived to this point of this prologue, must be recognised the obvious. There are many periods of this history and many relevant cultures that are not described in the book and, for sure, some actual serious concerns are not discussed. The reason is evident. A wider analysis would require much more time, making unfeasible this work. In fact, Gibbon devoted nearly twenty years of his life to his book. Nevertheless the contents as it is should be enough to achieve the aim we initially set to ourselves, to identify the way of the future. And for such purpose it is necessary to gain some historical perspective otherwise, we will not be able to see wood for the trees.

This book presents a singular water engineering history. Singular because it has mostly been written by engineers, with the inevitable advantages and disadvantages involved. After all, most of the improvements that water management has witnessed over time have been developed by engineers. However, they have not anticipated the strong environmental impacts caused by many of the solutions they conceived. Because of that, the analysis of the best and the worst of these solutions is the right way to learn from history. We presently live in a new era, the 21st century, which requires solutions able to integrate many different points of view in a world – that of great hydraulic public works – whose scale has grown considerably in the last decades. After all, what has multiplied the dimensions of the problem is precisely this change of scale.

Nowadays water policy must be analysed from many different perspectives. Here lies its grandeur and its complexity at the same time. It is fascinating because it involves jurists, biologists, historians, geographers, engineers, economists, chemists, geologists, sociologists and, last but not least, politicians and the society as a whole. All of these groups have a specific opinion about it. Moreover, it will also deeply concern professionals and citizens of forthcoming generations. After all, today's decisions affect them much more than it will influence those who are now adopting them. Although they are key players in this process they will never have the chance to participate and take decisions in the crucial issues. Water policy must harmonise many opinions and interests, most of them not directly represented. The main objective of this book is to show water policy integration from an engineering and historical perspective. You, as a reader, will judge to what extent we have succeeded in our objective.

Last, we must mention those who have made possible this book. First and foremost, thanks to the authors, excellent professionals but, above all, friends. Secondly, our thankfulness goes to the University of Alicante, represented by Professor Concepción Bru, co-author of one of the chapters. After all, that University housed the embryo of this book, the seminar previously mentioned that was supported as well by Iberdrola, CAM, and Aguas de Alicante. Thirdly, we want to thank AQUALIA. Its sponsorship, has covered the costs generated by the preparation and printing of this book. And last, it would be unfair to close this list without mentioning Janjaap Blom and CRC Press/Balkema – Taylor and Francis Group. Their patience for the meticulous and careful edition of the book is very much appreciated. They all have our most sincere gratitude.

Valencia, April 2010

Enrique Cabrera and Francisco Arregui,
ITA
Universidad Politécnica de Valencia
Spain

Part A
Introduction

CHAPTER 1

Engineering and water management over time. Learning from history

Enrique Cabrera & Francisco Arregui
ITA, Universidad Politécnica de Valencia, Spain

ABSTRACT: If there is an activity in which human beings have displayed all their ingenuity, it is water management. The need for water both as a means of support and as an essential sustenance made the first irrigation systems appear already in the earliest civilisations. The present book reviews the inseparable binomial 'human ingenuity-water management', a harmonic relationship until the early 20th century. Everybody did the right thing in each historical period until then. But the beginning of last century brought a number of vertiginous changes which were going to alter the harmonic relationship that had always existed. These changes became actually faster over the years, to such an extent that the traditional harmonic relationship has finally ceased to exist during the last decades. The problem lies in the fact that the dramatic technological and social changes have not been accompanied by the institutional and cultural changes required to ensure that the spectacular economic growth was also sustainable. The reflection that follows – a prelude of the historical review of water engineering carried out in this book – tries to show how those vertiginous changes have not had the necessary counterweights, which has caused clear imbalances. The imbalances are so serious that water now forms part of the politicians' agenda in every country and not only in arid countries, as was the case until very recently. And this is happening increasingly often. Our ultimate aim is therefore to provide the reader with a perspective that is broad enough to have a better understanding of the tremendous challenge that the current generation has to face. After all, only an exhaustive knowledge of the problem will guarantee success at its resolution.

1 INTRODUCTION

We are living in a period during which the magnitude of the changes that occur, and the speed at which they succeed each other, are so significant that, from this perspective, one of the current decades would be equivalent to a century for those who preceded us. Indeed, the world left by the present-day generations has nothing to do with the world that they knew during their childhood. This is the differential fact which characterises the time we are living now as opposed to the one that our ancestors lived through. Until just over a century ago, it hardly mattered from any point of view (economic, social or cultural) to have been born one hundred years earlier or later. It was the same to live in the 11th century or in the 12th century, for instance. But this does not apply to us, who were born in the 20th century, and it will not apply either to those who have just arrived, or who still have to arrive, during the present 21st century. It is obvious that the improvement experienced in nearly all the aspects that form the broad concept that we know as 'quality of life' has been spectacular. However, that huge improvement has had a clear loser, the natural environment where we live, the essential ingredient of which is water – the central topic in this book.

The aforementioned changes summarise the transformation of a largely rural population, that of the early 20th century, into an urban population, the one that is typical of the 21st century. The demographic growth experienced in the last few decades and its concentration in very small spaces (an issue that this book is going to treat in greater detail in the chapter specifically devoted to water and the city), has generated a number of dramatic environmental impacts that, since they are unavoidable, it will be advisable to minimise. This is certainly a hard task, as more often than not there are conflicting interests at stake. What is convenient in the short term (a rapid economic growth) is not the best choice from the long-term perspective: to respect the natural environment. Therefore, reaching that balance point which can reconcile both perspectives is not a straightforward, immediate task.

After all, the culture that prevails today is based on the short term, if not on immediacy. The Latin philosophy of *carpe diem* is in the DNA of 21st-century's society, and the natural environment is the main loser in that obsession with obtaining immediate results. The term 'sustainability' is permanently found in the politician's discourse, simply because almost nothing is sustainable nowadays, which grants full validity to the Latin expression *Excusatio non petita, accusatio manifesta.* People speak about ecological agriculture as a different way of cultivating the land, when just a few centuries ago, the term did not exist, simply because all agriculture was ecological. We have coined the concept of 'environmental impact' to quantify the extent to which a specific anthropic action affects the natural environment – another new term which was unnecessary before the 20th century. And finally, it was the enormous environmental impacts generated by the great projects which, after arousing deep social concern, catalysed one of the most socially relevant initiatives undertaken by the United Nations. First, with the Brundtland report entitled "Our Common Future" (CMMAD, 1988), which laid the foundations of sustainable development, and shortly after, with the Rio Summit of 1992, where the ideas materialised in specific plans and road maps.

The water policy of the last decades represents a paradigmatic example of the far-reaching transformation which took place during the 20th century. Within the context of the massive hydraulic development that characterised the first half of the last century, man thinks that it is possible to satisfy the ancient wish to transport water from where it is abundant to where it is scarce. And states, as they always did, assume the costs associated with a set of impressive infrastructures which are built enthusiastically because they are the 'banners' of modernity. Nobody raises any objection whatsoever. Nobody expresses their opposition to them. Nobody contemplates the possibility of carrying out a cost-benefit analysis that can justify them. And because their environmental impacts (the clear collateral damages caused by these great infrastructures) are still unknown, euphoria runs wild. We must wait until the second half of the 20th century to see society starting to question the construction of so many works, an unrest that will culminate, when the end of the century is near, in the abovementioned report elaborated by the Brundtland Commission.

Technological development entails the disproportionate self-esteem of human beings, who even believe that they will be able to dominate Nature. So much so that society enthrones those who plan these works. This is proved by the statement of Rouse, one of the most remarkable civil engineers of the 20th century: *Hydraulic engineers are human too* (Rouse, 1.987), which shows the enormous prestige that civil engineers had in mid-twentieth-century society. But this comment is made when the zenith of the great hydraulic work has already been reached, which can be easily associated with the construction of the Aswan dam, right in the middle of the 1960s. Curiously enough, that zenith or peak of the massive hydraulic development policy is going to pronounce the death sentence of the most mythical delta in the world, that of the river Nile. With a capacity to store five times as much water volume as the Hoover dam, the most emblematic one in the United States (it is worth remembering that this dam changed the 'face' of Las Vegas desert), Aswan was 'sold' to the society as *The barrier against famine in Egypt*, a slogan that time has eventually placed in its right context. The reality is summarised with great mastery by Kerisel, a brilliant French civil engineer, in his book *The Nile, the hope and the anger. From wisdom to lack of moderation* (Kerisel, 1999).

And, of course, the most favourable context for this culture to take full root is represented by the areas where those desires for water have always existed. In other words, it is on the shores of the Mediterranean, as is going to be seen in the following chapters, that the history of water engineering has written its most brilliant pages. It is a wish that will slow down the changes that the new 'man-natural environment' dialogue is going to demand in its new context. Because it is undoubtedly in these regions that the weight of history is most influential and the inertia is stronger. And while technology and society evolved so slowly between the dawn of civilisation and the late 19th century, that water policy did not have much trouble to adapt to the successive changes which took place;

this harmony is broken with the arrival of the 20th century. The desirable thing would have been to match the speed at which changes took place with an agile response that could adapt the culture and management structures to the new framework. But the reality has been quite different. Culture and vested interests have encumbered changes to a greater extent precisely in those countries which most badly needed them. Amongst others, all those bathed by the Mediterranean, where water has always been scarce, especially during the frequent drought episodes. This scarcity has generated a culture which still remains intact today.

In short, as far as the relationship between man and natural environment is concerned, the changes occurred in the last one hundred years have exceeded by far the variations seen during several millennia. These changes were the materialisation of the immense possibilities offered by modern technology. And the speed of change contrasts with the inertia and culture of a society that had always been able to manage water wisely, until just a few decades ago. For this reason, the challenge that present-day society has to face now is to match up to its ancestors: to give the adequate response to the moment in which that society is living.

2 ASPECTS IN THE MAN-WATER INTERRELATIONSHIP WITHOUT BACKGROUND CHANGES

The first human settlements were established on the banks of springs and rivers, simply because there is no life without water. But man soon learns to transport water across the distance, which is going to allow him to occupy new territories. And also very soon, man observes that irrigation multiplies crops, which justifies why the history of water linked to irrigation is as old as fascinating. That is not the case for the third conventional use, the industrial one, which will have to wait until the eighteenth-century industrial revolution to start competing with the traditional uses that had prevailed until then: the human use and the agricultural one. What has been said above explains that the history of the water-man relationship is the history of mankind itself which, packed with nuances and anecdotes, has of course been kept through time as it should.

Although their dimensions are quite different, many of the aspects in the ancient man-water relationship have hardly changed. This is so because, though the actions of human beings on water have become more aggressive with the passing of time, that technological development which permits to attack water also contributes somehow to laminate it. And when the necessary changes have benefited everyone (with all-win solutions), they have been introduced in a relatively easy way, facilitating a harmonic relationship. The problems arise when there are conflicting interests which hinder the adaptation measures required, i.e. those which can help us minimise impacts.

This section is going to list the main aspects in the man-water relationship – ten in all – which the passing of time has not significantly altered, though the dimensions of that interrelationship are of course completely different. We will later analyse other interrelationships which either develop in a context that has nothing to do with that of antiquity or are simply new relationships that have proved to be unsustainable over time.

The importance of civil engineering in the world of water. The next chapters provide a detailed description of some of the infrastructures that man has constructed through the centuries seeking to achieve a better use and management of water. Dams, canals, aqueducts, tunnels, and pluvial water collection facilities, thousand-year-old works that still amaze us. In any case, the discovery of reinforced concrete during the second half of the 19th century substantially changed the scale of a relationship that had been much friendlier until then. Large dams are going to multiply the advantages and the disadvantages, which is why they are one of the specific issues that will be discussed later on, separating them from the general set of civil works.

Water and extreme events. Human beings have always been concerned not only about rises in river levels and floods but also about droughts. Chapter 41 of the Genesis refers to the droughts that Egypt periodically suffered. It is shown during the episode in which Joseph interprets the Pharaoh's dream. There is also evidence of periodical overflowings of the Nile which contributed to increase the fertility of the lands situated near its banks. More or less frequently, there is no geographical area on Earth that is unaware of some extreme events that climate change threatens to boost. In the Mediterranean, however, people have always coexisted with them.

The large dams which started to be built in the early 20th century largely alter – for the better, in this case – the consequences of these extreme phenomena. On the one hand, they permit to laminate the floods (and therefore the overflowings) and, on the other hand, they permit to increase the volume of water stored, thanks to which a better management of droughts is possible. The negative consequences entailed by their construction will be reviewed later on.

Conflicts over water. It is worth remembering that the word *rival* comes from the Latin *rivalis* – those who are on the banks of the river (*riva*). And because, especially in those places where water is scarce, man has always wanted to control this natural resource competing with whoever it was necessary, the term *rival* has been extended to any kind of dispute. However, it is also necessary to underline that those disputes have seldom led to wars (Wolf and col., 2005). A completely different matter is the use of water during a war, e.g. the cutting (or poisoning) of the supply sources of a city as a strategic weapon. The next section – water and wars – will deal with this issue.

In recent years, the conflicts associated with water have deserved a lot of attention, above all in the United States (Gleick, 1998; Beach *et al.*, 2000; Pryor, 2006) and all the analyses draw the same conclusion: water has nearly always been a catalyst of peace rather than a cause of war (Asmal, 2000). And occasions for discrepancy are abundant. After all, nearly 300 basins are shared between different countries throughout the world. As a matter of fact, there were 214 in 1978 but, after the dismembering of the Soviet Union and Yugoslavia (completed in 2005 in the second case), there are nearly fifty more now (263). And we can also find frequent internal conflicts between different regions of the same country. Spain is one of the countries where these conflicts are becoming increasingly frequent (Cabezas and col., 2010). There are even cases of conflicts inside the same region where the different uses (generally the growing urban demand as opposed to the traditional agricultural use) compete with one another (Molle and Berkoff, 2006). A particularly complex case is that of the capital of Mexico, to which we will refer later on.

The problem lies in the fact that, whereas rivalry was confined to lands situated near the banks or shores in the ancient times, now technology has made it possible to transport water as far as we want, as a result of which disputes are arising increasingly often between regions which are hundreds of kilometres away from each other. Two web pages offer a detailed list of the numerous conflicts that have taken place. One of these pages (www.transboundarywaters.orst.edu) corresponds to the University of Oregon, specialised in these matters, as shown by the fact that it imparts a Programme on Conflict Management [in the context of] water policy. Also the Pacific Institute specifies the chronology for many of these disputes (www.worldwater.org/conflict.html), while at the same time it makes an invitation to add items to a list that will become significantly longer during the 21st century. Not in vain, these conflicts are intrinsic to human condition and, of course, to human needs. How else can we understand sentences like that of Mark Twain (he lived in California at the end of the 19th century): *"In the west, whisky to drink and water to fight"*? Or the one which has formed part of the cultural heritage of the fertile regions of Valencia for many centuries *"Water makes you more drunk than wine"*.

Water and wars. Because water was needed to survive, human beings have always tried to inhabit places where water supply was guaranteed, even when towns were besieged. All the necessary works were undertaken for that purpose. After all, the fastest way to make a town surrender was to cut its water supply. Bonnin describes some of the infrastructures that were developed in order to ensure water supply (Bonnin, 1984), which sometimes included the construction of large subterranean galleries which provided access to nearby inconspicuous water sources always situated outside the walled town. Amongst other cases, Bonnin describes the gallery that King David constructed in Jerusalem three thousand years ago in order to gain access to the springs in Gihon.

The literature offers countless examples of besieged towns to which water supply was cut, this being always the first action of those who were attacking it. Even the Romans, who used this strategy on numerous occasions, suffered it in the city of Rome itself. It was in 537 A.D. – when the Roman Empire was already falling into decline – when the Ostrogoth Vitiges cut the 14 aqueducts that fed it during the siege to which he subjected it (Dembskey, 2009). The eternal city resisted thanks to its wells and, above all, to the Tiber.

This strategy of cutting or poisoning the water supply to towns has sadly returned to the foreground in some countries after the attacks against the Twin Towers on September 11, 2001. It is the case of the United States or Israel. The situation is so serious that the Journal of Water Resources Planning and Management, ASCE, devoted a whole monographic issue to it in 2006. Its editorial (Ostfeld, 2006) summarises the state of the art in this field.

Water and laws. Due to the common disputes provoked by water, as soon as the earliest social communities were established, one of the first issues that they subjected to regulations was the right to – and the use of – water. One of the earliest pieces of evidence can be found in the code of Hammurabi (Fig. 1a), which dedicated seven articles to the regulation of these issues already four thousand years ago (Bonnin, 1984). The thousand-year-old *Tribunal de las Aguas* [Water Court] of Valencia still remains active (Fig. 1b). Of Arab provenance, it was created by Abderrahman III and its origins date back to the 10th century (Giner Boira, 1997).

Water legislation is one of the most complex issues in civil law nowadays. The coexistence of historical rights – strongly consolidated from the legal point of view – with the more modern legislation required to deal with present-day problems – such as the contamination to which the whole Water Framework Directive (UE, 2000) has been dedicated – makes water legislation become more and more complicated each day. This is especially true in countries with a long legislative tradition, without a doubt those where water has always been a scarce resource. However, if the difficulties derived from the new environmental framework were not enough, the current trend to political decentralisation ends up in new federal – or similar – structures which increase complexity even more in many countries (Embid and Hölling, 2009). It is the case of Spain. And it all without forgetting the international legislation that has to deal with the problems inherent to cross-border rivers (Phelps, 2007). In any case, the current legal difficulties must have the same order of magnitude as the ones that our ancestors had to face, with the distance imposed by the time elapsed, of course.

Figure 1a. The code of Hammurabi (Louvre Museum).

Figure 1b. The *Tribunal de las Aguas* in Valencia today.

Figure 2. The Albolafia today.

Water as a source of renewable energy. The kinetic energy of rivers was very soon used to drive waterwheels which permitted to raise water. According to Rouse, wheels were used for this purpose at least one thousand years before Christ in Egypt, Mesopotamia and China (Rouse and Ince, 1963), though other authors date the appearance of these wheels 500 years later (Bonnin, 1984). The use of waterwheels in Spain was above all spread by the Arabs, and it is even possible to visit some of these wheels, like the Albolafia (Fig. 2) in Cordova. Built in the 9th century by Abderrahman II, it raised the water from the river Guadalquivir to the Emirs' Palace – now the Episcopal Palace. It is reported to have been functioning until the late 15th century when Queen Isabel – who was staying at the Alcázar in 1492, a few months before Columbus' first departure toward America – had it dismantled because the squeaking of the buckets moving around the wheel did not allow her to sleep.

Not only waterwheels and wheels but also many other hydraulic machines were used in ancient times. Amongst others, stand out the Archimedean screw (also known as Archimedes' screw) or Ctesibius' piston pump. It is particularly interesting to have a look at Bonnin's chronological table of the raising machines used in antiquity which additionally includes their specific hydraulic capabilities (Bonnin, 1984).

As far as the modern hydraulic turbines are concerned, we have to wait until the mid-eighteenth century when Euler first describes jet turbines (Rouse and Ince, 1963). However, these machines would still have to wait two more centuries – when the great dams of the 20th century were built – to reach all their splendour. Their presence creates spectacular slopes and they make it possible to take huge volumes of flow through the turbines. The rise and development of hydroelectricity throughout the 20th century is impressive. Viollet wrote a brilliant chronicle about this story not long ago (Viollet, 2005).

Figure 3. The Itaipu hydroelectric power station at the border between Brazil and Paraguay.

Itaipu (see Fig. 2) – at the border between Brazil and Paraguay – stands out among the greatest hydroelectric exploitations. It started operating in 1982 and, when three decades have gone by, it is still the world's largest hydroelectric power station with its 14,000-Mw power, though it will lose that status as soon as the hydroelectric power station built next to the Three Gorges dam in China starts functioning.

The Three Gorges dam serves to clearly highlight the inconveniences and advantages of works that have made possible man's old wish: to dominate the natural environment in order to put it at the service of his interests. This dam is going to house the largest hydroelectric power station in the world. Its 22,500 Mw can be at work shortly (about 2011) and will exceed by 50% Itaipu's current record. The dam permits to regulate the floodings of the river and generate an enormous amount of clean electricity for China, the country which emits the most greenhouse effect gases. Its environmental and social cost is inestimable, though. The ecosystems in the surrounding environment have been irreversibly affected and its construction entailed the displacement of more than a million people.

Regarding the water-energy binomial, it thus seems evident that human beings are taking full advantage of nature's hydroelectric wealth. And if they not exploit that wealth even more, it is not so much due to the respect for the natural environment but, above all, because the cost-benefit ratio of the infrastructures that still have to be planned does not justify it. This is why, at this stage, it is advisable to ask oneself whether all these actions are sustainable over time or they will take its toll sooner or later. Obviously, we are by no means questioning the end sought: to obtain the clean, renewable energy that contributes to such an extent to reduce the emission of greenhouse effect gases. What can be debated upon is the way to achieve it: the dam. But, of course, man has always aspired to taking as much advantage as possible of nature. One way or other, only time will tell if we have perhaps gone too far.

Water and communications. When speed does not matter too much, fluvial transport has been more advantageous than land transport for heavy and sizeable objects. And, of course, since time did not matter too much in antiquity, maritime and fluvial transport acquired great importance. In

Figure 4. Floodgates in the Panama Canal between the Atlantic and the Pacific.

fact, the Egyptians used the Nile more than 4,000 years ago as the means of transport for the large stone blocks with which they built their pyramids and obelisks. There is even evidence (Bonnin, 1984) of the possibility that existed to navigate from the Mediterranean to the Red Sea 3,500 years before the construction of the current Suez Canal promoted by Fernando Lesseps. This is recorded, amongst others, by the great historian Herodotus of Halicarnassus. Navigation mostly took place in one arm of the Nile.

With the passing of time, the importance of this transport has never stopped growing and civil engineering has indeed played a beneficial role from any point of view in this field. It has made possible to turn non-navigable stretches into navigable ones and, with the help of floodgates, it has permitted to solve the problem posed by the slopes that dams generate in rivers, or, as in the case of the Panama Canal, by the slopes existing between two oceans (Fig. 4).

The river Danube constitutes one of the most remarkable examples of fluvial navigation in the world. It is worth highlighting that it is the second longest river in Europe (2,850 Km) and its basin is shared by 17 countries. It is, therefore, a unique case (Wolf and col., 2005) that acquired great relevance in antiquity, both because in the times of the Roman Empire its course formed a border and because it was the main connection link with the Asian regions. At present, it is the only fluvial corridor in the European Union (Fig. 5) and, using the canal that links the Danube with the rivers Rhine and Main, it permits to navigate from the Black Sea to the port of Rotterdam, already in the Atlantic.

Nevertheless, from a global perspective, maritime transport has lost some of the importance that it used to have in ancient times, especially after the irruption of railway and sea transport. However, it is the most sustainable of all environmentally speaking and its cost by unit of weight is approximately seven times lower than that of road transport.

Water and measurement. Man has always felt the need to measure the flow of water that circulated through rivers and canals. But it took him a long time to establish the ratio between the useful passage section and the speed, despite the fact that Heron of Alexandria had correctly formulated

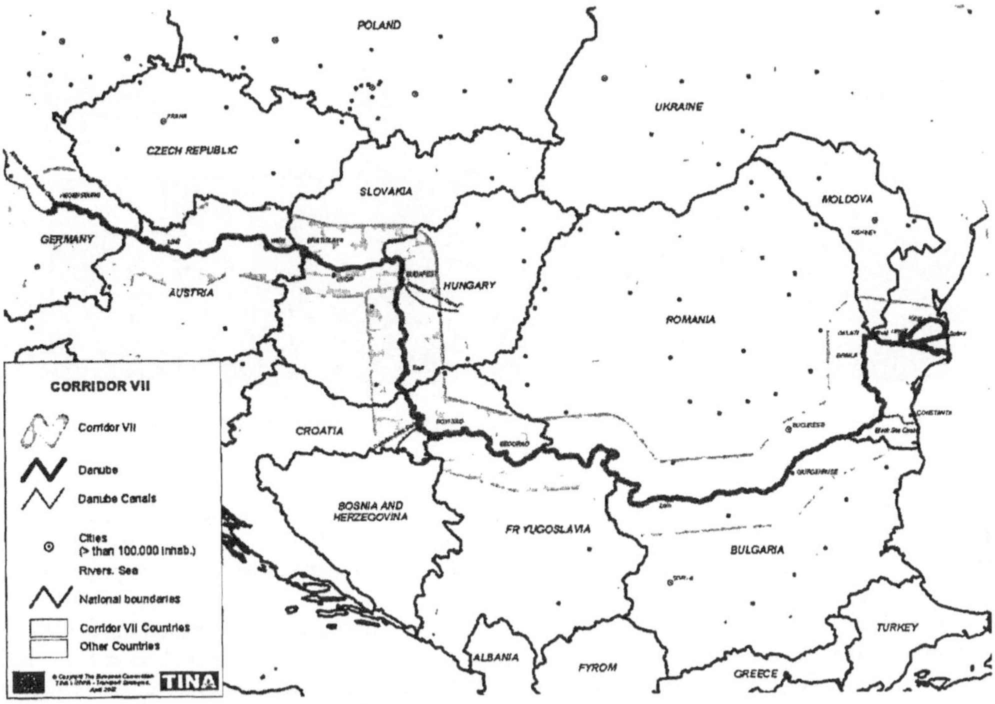

Figure 5. The Danube, one of the ten Pan-European transport corridors.

the continuity equation already in the 2nd century before Christ. Centuries later, everything seems to suggest that the Romans were not aware of the ratio existing between speed and flow (Rouse and Ince, 1963). Due to all this, the measurement consisted in monitoring water level in ancient times. It is worth highlighting among all these measuring instruments the well-known nilometers (Viollet, 2000), the most famous ones being those which can still be visited on the Elephantine island, very near to the Aswan dam. The level-measuring instruments have been used across the centuries and, in fact, they permitted to divide or distribute the water for irrigation among the different farmers' communities in the Middle Ages. Hence the name of '*partidores*' (dividers) that they have in the fertile regions of Valencia. The sentence pronounced by the Count of Ribagorza about the distribution of the waters from the river Mijares in 1347 is another example of this (García, 1997).

And while the water in rivers and canals was measured in limnimeters, the consumption of pressurised water was monitored from the very first moment with calibrated tubes known as 'calix' (Bonnin, 1984). Made of bronze (and not of lead, in order to prevent deformation), their diameter and length were perfectly defined, which permitted to control the flow supplied for a specific pressure. This system is still used today. In the case of Spain, it was used until the installation of water meters became widespread. However, in those countries where it is not obligatory to measure, the system is still at work.

In fact, it is necessary to wait until Leonardo reformulated the continuity equation at the beginning of Renaissance (Barbera, 1983), through it is Castelli that will first establish it formally in 1628, more than one hundred years after Leonardo's death. Therefore, most of the measuring instruments used nowadays (with the exception of volumetric instruments and limnimeters) determine the circulating volume from the flow speed at different points of a specific passage section (Arregui and col., 2007).

Figure 6. Thermae of Caracalla ruins in Rome (beginning of the 3rd century A.D.).

In any case, water flow measuring has been a permanent concern for human beings, a concern that is being aggravated as time goes by, due both to the scarcity of this resource and to the costs that its sustainable management entails. After all, it is essential to determine its consumption so that each party can assume their corresponding expenses.

Water, health and leisure. The modern SPA (*Salus per Aquam*) facilities, which have become habitual in many higher-range hotels, have inherited not only the tradition of Roman baths but even their name. They reached their maximum splendour during the Roman Empire – as is visible from the ruins that have survived to the present day (Fig. 6) – but they were already common much earlier, associated with Greek gymnasiums. Actually, the first baths about which there is a written record are those of the Knossos palace in Crete, already nearly four thousand years ago (Bonnin, 1984).

As for the water-health-leisure trinomial, things have changed very little, or rather have not changed at all, with the passing of time. Whereas in most of the preceding comparisons, even though the essence was kept, man's action has quite different dimensions, the same thing cannot be said about the thermae. The rooms where those hot baths were located two thousand years ago were decorated with wonderful statues, frescos and mosaics. They could easily stand alongside the best facilities of this kind available today.

Water and beliefs. In nearly all sets of beliefs, water has a spiritual value that any other natural resource lacks, no matter if it is a precious metal like gold or a precious stone like the diamond or the emerald. It is particularly relevant in this respect to remember the declaration that faith groups made in 2006 within the framework of the Water World Forum held in Mexico (FMA, 2006). It literally says that *For Judaism and Christianity, water is essential at the beginning of rituals. Letting the clean, fresh and living water fall symbolises God's spirit and makes possible the manifestation of a new spiritual world. For Islamism, the character of cleanness and the power of water are vital. For Muslims, cleanness becomes a rite before approaching God in their prayers. For Hinduism, water also occupies a special place due to the spiritual cleanness powers, as Hindus strive to reach physical and spiritual purity. For the native peoples, water is sacred; it is an offer of life and connection to everything that exists within a broad unity that is celebrated through rituals of cleanness and gratitude.*

And if the water-beliefs relationship has so much relevance nowadays, you can imagine how important it was in ancient times when man's inability to understand natural phenomena immediately suggested him associating extreme events (droughts, heavy rains or floods) with supernatural causes. Thus, many rivers were considered divinities (in the case of Egypt, for instance, the Nile was the second deity after the Sun God) while purification rites with water were present in nearly every culture. Consequently, one can hardly expect water to lose that halo of spirituality which has always accompanied it.

3 THE NEW FRAMEWORK IN THE MAN-WATER INTERRELATIONSHIP

While man's anthropic action was compatible with nature because the impact caused by the engineering works carried out was negligible, the interrelationship between man and water was sustainable. But in the 20th century, massive hydraulic development along with pollution start to break the balance to such an extent that the side effects are clearly noticeable after a few decades. This generates the social unrest that precedes any innovative policy, which will effectively break the inertia of the past. It will culminate in 1983 with the assignment made by the UN General Assembly to a Commission specifically created for that purpose, the World Commission on Environment and Development, which would be presided by the Norwegian Gro Harlem Brundtland. The assignment consisted in preparing *"A global programme for change"* with very specific aims:

- To propose environmental strategies to reach sustainable development in 2000
- To materialise the concern about the environment in a higher level of international collaboration
- To explore the most suitable strategies to deal with environmental problems
- To define common environmental sensitivities

In the light of the facts, we have not only failed to achieve these aims, but have in effect moved gradually away from them, which is the reason why hydraulic engineering – the brilliant history and evolution of which is going to be reviewed in the following chapters – must rigorously reflect once again on the role that it played in the 20th century. It has now become clear that the idea is not to subjugate nature – as it was initially believed – but rather to act in tune with it. We must consequently reconcile development and the improvements in the quality of life standards of society with nature conservation: that is what sustainable development means. In fact, few years after the publication of the Brundtland report, the Task Committee on Hydraulic Engineering Research Advocacy (ASCE, 1996) carried out a deep self-criticism exercise admitting that:

- Research and education have not been articulated properly.
- Researchers do not adequately connect with the real needs of society.
- Hydraulic training has not been adapted to the needs of the labour market.
- Hydraulic engineers have to think more broadly and with greater foresight.

It is evident that civil engineering has played an essential role in everything that regards the management of water resources, so much so that the 20th century is known as the last Golden Age of hydraulics (Rouse, 1987 and Plate, 1987) and because he made ancient dreams come true, the hydraulic engineer achieved the maximum social prestige during those decades. It has already been said that Rouse found it necessary to state that hydraulic engineers were not gods but human beings (Rouse, 1987). However, everything has a limit and, very soon, the crisis of the massive hydraulic development policy is going to show that water policy needs to be designed from different – and simultaneously complementary – perspectives.

Nevertheless, the civil engineer's role in water management is irreplaceable and will always have the maximum relevance. It cannot be forgotten that the solutions have come, are coming and will inevitably come from the field of engineering. For this reason, it does not seem logical to apply the pendulum law either. And something like this happened when, during the third World Water Forum of Kyoto in 2003, Profesor Stephenson, in his condition as representative of the IAHR (International Association of Hydraulic and Engineering Research) felt that *"in the Forum, Hydraulic Engineering was only a drop inside an ocean"* (IAHR, 2003). In this increasingly transversal and interdisciplinary world, the engineer cannot be left out of the decision-making bodies. That is why more and more engineers are defending the need for them to have a more active participation in the decisions adopted by politicians (Sheer, 2010).

Therefore, after reviewing the aspects of the man-water relationship – the essence of water has not been significantly altered – and following the analysis of the causes that start to make visible the exhaustion of the relationship as it had always been understood, it is convenient to examine the actual changes operated. First, we review the aspects in that relationship which, due to the

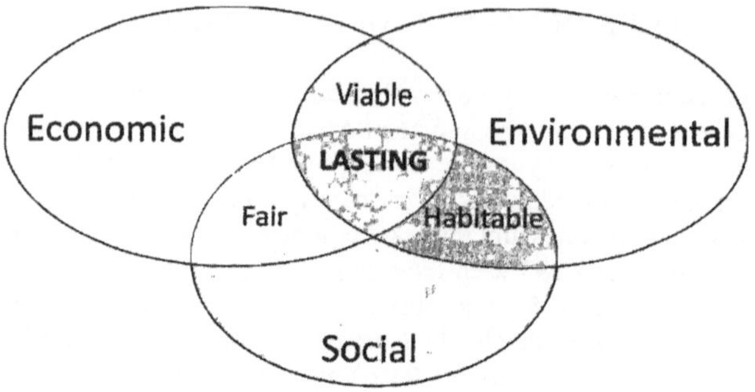

Figure 7. Framework for water policy development in the 21st century.

spectacular technological progress, has been altered to a large extent throughout the 20th century, and especially during its second half. We have organised these aspects in two blocks. The first block includes those in which there is a balance between advantages and disadvantages, whereas the second one contains those in which the ratio has eventually become worse.

3.1 *Significant changes occurred in the man-water relationship during the 20th century with positive and negative aspects*

3.1.1 *A new framework for the man-water interrelationship*

The litany that is often used to highlight the importance of water and which essentially focuses on emphasising that it is fundamental both for the life of human beings and to keep all the ecosystems alive has not lost and will never lose the slightest bit of truth. Water becomes much more important every day, because it is needed by sectors such as tourism, industry and leisure. Therefore, we must add to its traditionally acknowledged social character of water its status as an economic good, without forgetting its environmental character either, of course. Water has always had this character but it went unnoticed until overexploitation and contamination highlighted the need to take it into account, above all if we do not want to further jeopardise the future of the coming generations. Figure 7 shows that new framework which now houses the water policy.

Since what is more convenient for one axis goes against the interests of the other two in most cases, the new framework is far more complex than the simpler one in which the water policy developed until the last decades of the 20th century. The large hydraulic infrastructures – which were built under a dogma, that of general interest, which nobody questioned – were not even subjected to an elementary cost-benefit analysis, and their potential environmental impact was simply ignored. At present, though, works can only be undertaken in any developed country if they successfully go through the filter represented by each one of the three axes.

It is obvious that water policy in the early 20th century did nothing but follow the inertia of history. Of course, the modest magnitude of the actions carried out until then (compared to the dimensions of the large infrastructures that reinforced concrete will permit to build) did not alter the natural environment. On the other hand, the absence of alternatives to the traditional (urban and agricultural) uses and the impossibility to transport large flows over long distances guaranteed a very slight pressure on water resources. And the impossibility to transport large volumes of water across long distances also prevented the territorial conflicts that are so well-known to us today. Summing up, the greater or lesser degree of exploitation of water resources carried out in each historical period depended on the technological possibilities of the moment.

The three dimensions in the new framework are directly related to the following sections, as they shape the difference between the traditional water policy and the policy that it is necessary to implement if we want to guarantee the survival of future generations. We are referring to the economy of water, to the competition between uses and the significant problems it entails and, finally, to the environmental problems generated by the large hydraulic works.

3.1.2 *The economy of water*

Many reasons advise us to pass on consumers all the costs associated with their utilisation of water. Two of them stand out from the rest. The first one is to guarantee both the efficient management of the water distribution company and the rational use of those who consume it. In effect, the efficiency of the system largely depends on the variable cost of water. It is explained by the concept of optimum level of leaks within a water network; the value is determined minimising the sum of the variable cost that can be attributed to the escaped water and the expenses required to maintain the network with a specific level of losses (Cabrera and col., 2004). And from the consumers' point of view, the price that they pay for water not only conditions their demand but, above all, encourages them to be more efficient in its utilisation. Thus, for example, the investment required to reuse grey waters or take advantage of rain waters in a dwelling will be repaid within a short period of time if all the costs are recovered. If water is subsidised, the user has no motivation to bet on this type of facilities, which save so much water.

The second reason lies in the economic sustainability of hydraulic infrastructures. At present, every large investment demands to carry out a rigorous cost-benefit analysis that can justify it. And it must additionally be demonstrated that the large infrastructure in question is the best solution among all the possible alternatives. Apart from being highly indebted, the governments that used to subsidise these works now have to face the growing social expenses associated with a population whose life expectancy and needs grow over time.

In Europe, the importance of rigorously applying the principle of cost recovery appears in all the documents published by the European Union in relation to water. From the Water Framework Directive, which specifically dedicates article 9 to it (UE, 2000) until the more recent "Facing the challenge of water scarcity and droughts" (CEC, 2007), where section 2.1 recommends that the price of water should take into account all the costs derived from its sustainable use.

Nevertheless, irrigation has always been and is still highly subsidised in countries with an agricultural tradition. Regardless of the fact that, if subsidies exist, they should encourage saving (EEA, 2009) – because this is actually not the case in the subsidies applied at present – recent studies commissioned by the European Union have shown that many of them not only do not encourage saving but also contribute to deteriorate the environment, which is much worse (IEEP, 2009). One of the examples proposed in these analyses is precisely the subsidy to irrigation in Spanish agriculture.

In conclusion, the economy of water – which was practically a marginal issue until a few decades ago – is now going to become a key tool in the water policy of the 21st century, with all likelihood the most important one.

3.1.3 *Competition between uses*

As said above, the massive concentration of population in urban areas, the deep changes occurred and, finally, the technological development of the last decades has favoured the appearance of a new scenario completely different from the one seen by the preceding generations. It is a scenario that has made previously unthinkable conflicts come to the surface. Many others are going to be described in what follows. Among them, we could highlight two specific cases: the disputes in the Jucar basin between traditional farmers and the new crops on irrigated land, and the social conflict generated by the enormous water needs of the Mexican capital city.

The example of the river Jucar is particularly appealing. The traditional farmers with thousand-year-old historical rights over its waters work on lands near the coast where the mild climate has always permitted to grow profitable products. At present, traditional farmers compete with new irrigators who sow lands that, mainly for climate-related reasons, nobody had thought of cultivating until a few decades ago. The European Union's agricultural policies have done the rest. Subsidising crops with dubious profitability, they distort what has been dictated by nature's climate. It is not a minor issue, as all the farmers involved are situated on the banks of the Jucar (the new ones on the

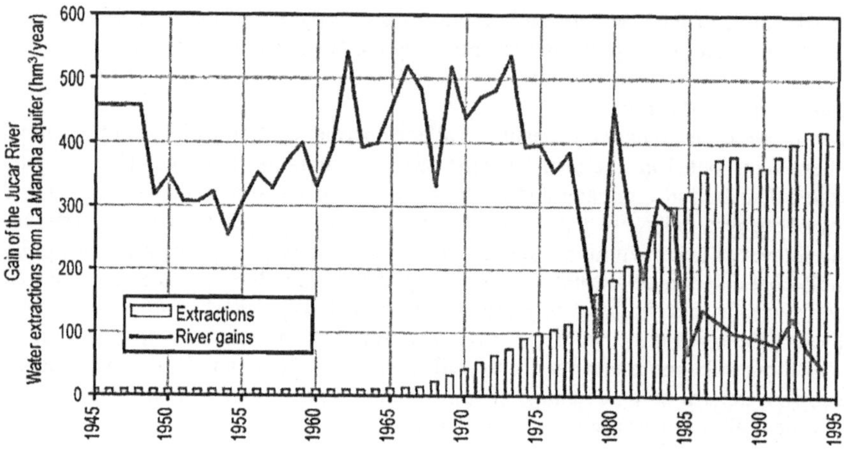

Figure 8. Gain of the river Júcar – water volume pumped in La Mancha aquifer (MIMAM, 2000).

upper stretch and the traditional ones on the lower stretch) and all of them have the right to use its waters.

The solution adopted has been to encourage saving in the traditional irrigated lands and to release part of the old concessions (the historical rights, 1000 Hm^3/year, were reduced to 350 Hm^3/year in 1999, a value that is still generous considering the irrigated surface area). But the problem has not disappeared because, without any controls, irrigators extract water from the aquifer that feeds the river Júcar, which has seen how its volume of water has diminished alarmingly (MIMAM, 2000). This can be seen in Figure 8, which relates the pumpings of La Mancha aquifer with the water gain of the river in the associated stretch. The natural underground contributions have fallen at the same pace as the water volumes raised.

The second example – that of Mexico City – is well-known. Due to its spectacular growth during the last decades, the aquifers which have always supplied water to the city are now insufficient (their current contribution is situated about 65%). They soon had to resort to neighbouring basins, the first one of them, the Lerma basin in the 1950s but, as the demand continued to grow, they had to use the Cutzamala basin in 1982, planning a water transfer of more than 100 kilometres, apart from other remarkable complementary works (eight new dams and some pumping stations to overcome slopes of more that 1,000 metres). However, as the demand does not seem to have a limit, they are thinking of boosting this transfer, which requires building a new dam, in the river Temascaltepec this time.

We are talking about a huge social problem (Perló and González, 2005), because they cannot leave part of a city like the capital of Mexico without water supply. But, on the other hand, the native communities of the granting basins are witnessing their economic as well as social and environmental problems multiply because of the endless drain into which Mexico City has been transformed. It should consequently not surprise us to see how the opposition to new transfers is bigger each day. This problem is really difficult to solve. On the one hand (Delgado, 2007), because the natural limits of basins do not coincide with the administrative ones, an increasingly frequent difficulty as we have highlighted in the section dedicated to water and laws. These situations could never arise in the past because the technology available did not permit to move so much water across such long distances, additionally overcoming spectacular topographical obstacles. Another important difficulty has been highlighted (Delgado, 2007), namely the fact that the administration with competences is fragmented, this being a problem to which Spain should find a solution too (Cabrera and García-Serra, 1997).

We thus find ourselves before a scenario which was not contemplated by the individuals who established the current rules of the game in the past. Consequently, there is an urgent need to design new rules that permit to resolve all these conflicts in a rational way, leaving passion for other less important issues. Time will only multiply problems. We must also urgently rethink and redesign water administration so that it can become more efficient, logical and endowed with a greater decision-making capacity. All this leads us once more to underline the main message behind this reflection. The changes occurred at a vertiginous speed during the 20th century while the response that they needed was strongly encumbered by the weight of history.

3.1.4 *The great civil works and the associated environmental impacts*

Although a reflection about the advantages and disadvantages of great civil works has already been made in an indirect way, it is time to refer specifically to the advantages and disadvantages of dams, those impressive engineering works which have revolutionised water management during the last one hundred years. And it is appropriate because they undoubtedly bring together and summarise the essence of the pros and cons associated with the technological development that has taken place.

Questioning the advantages that the possibility of storing and regulating large water volumes entails from the operational point of view seems ridiculous. Having water available when rainfall is scarce and being able to laminate the floods that follow a heavy rain period represents an improvement that ancient civilisations would have loved to use. After all, it was always one of the greatest wishes, as shown by the fact that, already six thousand years ago, one of the first kings of the Menes dynasty ordered the construction in Memphis – the capital of Egypt at that time (it is about 20 kilometres away from Cairo) – of the first documented dam (Rouse and Ince, 1963); and it all without forgetting the renewable energy that can be generated through them, an issue that we have referred to above.

However, the great benefits associated with reservoirs cannot hide the enormous impacts caused by the presence of dams in the dynamics of rivers. Indeed, any river constitutes a complex ecological system and its functioning is affected to a great extent by the presence of these artificial barriers. The natural regime of water flows, the transport of solids, the dynamics of nutrients, the temperature regime and, ultimately, water quality, all of it is altered, especially in the dry periods that are so frequent in those geographical where dams are significantly abundant. It is worth remembering that climate irregularity actually constitutes the main reason for their construction.

At this stage, and since dams are simply essential for many countries in the world, there are only three possible action lines. The first one, despite being aware of the fact that it is impossible to bring fluvial spaces back to its original condition, would be to manage them as sustainably as possible (Armengol and col., 2008). The second one would be to use water in the most efficient possible way to interfere with the natural environment as little as possible. Dams are the last solution and not, as it happened during a large part of the 20th century, the first one. And the third line – when the reasons justifying their construction vanish into thin air – is to demolish them in order to bring the fluvial space back to its original state. This is what has been done lately in the United States (Wildman and col., 2008).

3.2 *Significant negative changes occurred in the man-water relationship during the 20th century*

In the course of the last few decades, society has become fully aware that water in particular and natural resources in general require a more sustainable management. However, the problem not only continues but is even becoming worse because the solutions and measures that are being adopted, despite being numerous, are still insufficient to counteract man's anthropic action. The following subsections highlight some of the most relevant problems directly or indirectly related to water.

3.2.1 *The growing increase of contamination*

The utilisation of water degrades its quality, but the impact of spillages of used waters on the natural environment until the mid-twentieth century was non-existent in the medium-long term because the natural depurative process sufficed to return its original quality to water. However, halfway through the 20th century, the contamination generated by human activity provoked much more unrest because the natural environment is unable to assimilate the spillages that it receives, those coming from the city both because their water volumes increase (urban growth has been spectacular from 1950 onwards) and because they additionally include non-biodegradable products like those which are present in detergents.

Widespread agricultural contamination is not too far back in time. The irruption of agrochemicals in the fields – the synthesis of new products that will be used as pesticides and herbicides (Pascual, 2009) – was spectacular after the discovery of DDT in 1939. Its discoverer, Muller, received the Nobel Prize for this achievement. Farmers used aggressive and hardly selective chemical products which not only finished with the characteristic plagues of crops but also attacked all the ancillary fauna and altered the system's ecological balance. And it is also during this period that nitrogenous fertilisers start to be administered generously. Due to the lack of knowledge about the processes, a large proportion of the fertilisers did not reach the plants. Instead, they ended up contaminating aquifers after being dragged by the irrigation water and together with agrochemicals.

And finally, we must refer to the most worrying contamination, the industrial one. The economic and technological development that followed World War II – the third industrial revolution – indicates the beginning of the globalisation of the economy. International borders are opened and competition on a global scale, and with it the need to increase competitiveness, becomes the differential fact. The main victim of this globalisation is most probably going to be the natural environment in general and the water environment in particular. Many industrial processes require water. It will receive a contaminant load (metals included) during its utilisation. Initially, that water would be spilt with no treatment whatsoever, which is why developed countries were going to react soon before the evident deterioration of the receiving masses. This has not been the case in many developing countries, where industrial spillages are not subjected to any type of treatment yet. That is why contamination is the most serious problem that 21st-century water policy has to face. Especially in countries like China (Gleick, 2009), which has based its spectacular economic growth during the last decade on the minimisation of its production costs, unattainable for the rest of industrialised countries, amongst other reasons, because they are sparing themselves the environmental costs, among which stands out the one associated with giving its initial quality back to water.

The importance has been widely acknowledged since antiquity. Bonnin tells us a number of episodes in which the springs that gave supply to a population nucleus were poisoned. It became a key strategy at war times (Bonnin, 1984). The poisoning provoked by the lead of the pipes which transported water was also common in Rome, a problem that exists still today. And, finally, also in Rome, they built the aqueducts that brought the pure water from the Appenines in order not to drink the water from the Tiber, the course that received the flow of the city sewers. In any case, the dimension of those isolated and transitory contamination problems suffered by those who preceded us have nothing to do with the current ones.

Present-day contamination is consequently an extremely negative differential fact, and it is most probably the main problem that water policy will have to face in the 21st century. This is shown by the Water Framework Directive (UE, 2000), the aim of which is simply to recover and protect all waters (continental surface, transition, coastal and ground waters). Neither should it come as a surprise that the motto chosen for the World Water Day in this year 2010 was "Clean water for a healthy world".

3.2.2 *The complex access to water and hygiene for millions of inhabitants*

To ensure that all the planet inhabitants can drink good-quality water and enjoy a minimum basic level of hygiene is one the greatest challenges that Society has to face. For this reason, one of the chapters in this book is specifically dedicated to the compliance of the millennium challenges. In any case, we now summarise the state of the art taking into account the last World Health Organisation report (WHO, 2010). In particular, regarding hygiene, there is a significant delay with respect to the millennium goals. Whereas the objective for 2015 was that only 23% of the world's population would lack such a basic service, 36% of those who inhabit this planet (2.7 billion people) will still have this problem in that year. Luckily, access to drinking water is going somewhat better than expected. Only 672 million people (9%) will not have it at their disposal, though many more, 3.5 billion people (47%) will not have an easy access to it, i.e. through a tap at their homes. Finally, we should not forget either that 4,000 children die every day due to the absence of these services, a figure that may be irrelevant in relative terms, but heartbreaking in absolute values.

3.2.3 *The overexploitation of surface resources*

Man's wish to exploit all surface waters reaches its peak in Spain with the figure of the illustrious regenerationist politician Joaquín Costa who, after the collapse of the colonial dream in 1898 after the loss of the last enclave in America, the island of Cuba, arrived to say that "Spain will not leave behind its backward state while rivers lose one drop of water in the sea" (Costa, 1911). And indeed, today, in the 21st century, there are many Spanish Mediterranean rivers (Mijares, Turia and Segura, amongst others) which do not get to the sea. And these cases are not exclusive to the east of Spain. The same happens to one of the most emblematic rivers in the United States, the river Colorado. Overregulation has turned that wild river excavated by the world's most famous canyon into a different river which languishes and dies before reaching the Gulf of California.

Therefore, the wish of those who lived centuries ago in areas where water was scarce has come true with its advantages and disadvantages, making desalination play an increasingly prominent role in areas near the coastline where rivers were already exhausted. This was not at zero cost, though. Its high energy consumption (with all the emission of greenhouse effect gases that it entails) and its high production cost, especially compared to the almost non-existent cost associated with the surface water of traditional rivers, limit its use to isolated cases for the time being.

3.2.4 *The overexploitation of ground resources*

Taking into account the essential role that ground waters play at present, it may well be stated that they were not widely used in the ancient times despite the fact that man became aware of their existence at a very early stage. Most probably, the first one was a chance contact (Bonnin, 1984). Needing to drill the ground looking for shelter, a hiding place or simply to bury the dead, he must have found water at few metres' depth. Thus, the first documented well was going to be built more than 6,000 years ago about 10 km away from Belgrade.

Nevertheless, the difficulty involved in drilling the ground with the means available at that time, and especially the impossibility to raise water in significant amounts, made human beings excavate galleries originating in the natural springs through which water came to the surface – except in not very deep phreatic strata. After all, constructing horizontal galleries is much easier than drilling the land. The earliest documented ones, situated in Armenia, date back to the 8th century before Christ.

Because galleries permitted considerable water flows to rise naturally (which was absolutely impossible with wells), many lands were irrigated with water coming from these galleries. Very frequent in the South-East – like in any other territory where surface waters were scarce and the phreatic stratum was not deep – they were constructed until the early 20th century, the moment in which the technology that allowed human beings to raise water from considerable depths became widespread (Hermosilla, 2006). The difficulty to raise water from those great depths had been the greatest limiting factor until then.

The history of intensive exploitation of ground waters is therefore little more than a century long. Because water collecting points were situated next to its utilisation place, the final costs were reasonable and could be directly assumed by the developers. This absence of subsidies favoured a very efficient use of this water. The higher supply guarantee during dry periods always contributed to its implementation (Sahuquillo and col., 2005).

But precisely some of the abovementioned advantages have caused the main problems that this kind of exploitation is facing nowadays. As the use of these waters was driven by private initiative, the administration has hardly controlled the drillings made and even less the volume of water raised. In Spain, most of them are illegal and many aquifers are overexploited because the water volume extracted exceeds the natural recharge capacity nearly every year. More specifically, Figure 9 shows one of the most overexploited aquifers in the Alicante province, that of Carche-Salinas. It is

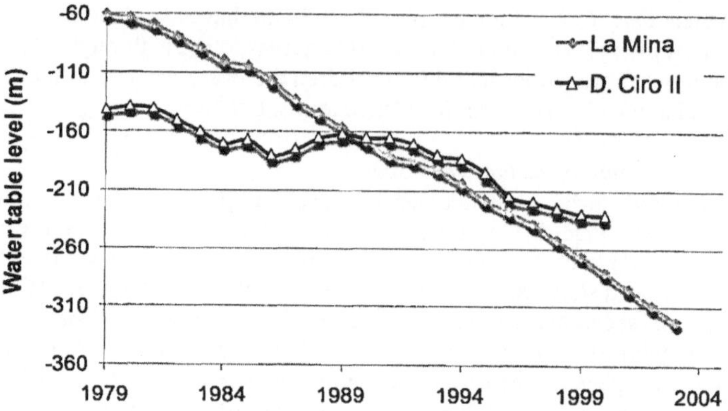

Figure 9. Evolution in the water raising depth at Carche-Salinas aquifer (Gil and Rico, 2006).

certainly worrying to see that in only 26 years (between 1979 and 2005) the water table has gone down 250 m (from 60 m to 310 m), with an annual average of 10 metres descent. And in dry years, that descent can reach 40 m on some specific occasions (Gil and Rico, 2006).

The short history of some aquifers is thus being extremely turbulent. Only a few generations will have sufficed to squander a tremendously important natural heritage. We should not forget either that the overexploited aquifers situated near the coastline end up becoming salinised. The conclusion is clear: this situation has to be changed as soon as possible. We must impose order where there is lack of order and make available all the necessary technical and human means so that these highly strategic water reserves can be managed in a sustainable manner.

3.2.5 *The loss of biodiversity*

Until little more than a hundred years ago, the planet's ecosystems had at their disposal practically all the water in the planet. It is with the massive hydraulic development of the early 20th century that man starts to interfere in the centuries-old water-ecosystems balance, diverting more and more water each year for its use to the detriment of biodiversity. The worrying current state is described in detail by a recent European Union report (EC, 2010) which admits – and its environment ministers have just certified it precisely in the International Year of biodiversity – that they have failed in the attempt to stop its progressive deterioration by 2010, a goal that they had set themselves some years before. The figures are actually very worrying. 60% of the ecosystems are degraded and biodiversity losses exceed (between 100 and 1,000 times) the normal rate. And what is worse, it is known that over one third of the species evaluated are on the verge of extinction.

The loss of biodiversity is closely linked to climate change, to which we are going to refer next. These are global problems that go beyond borders and one could even say that they are the two sides of the same coin. That is why their resolution demands a joint treatment, though biodiversity has been the poor brother so far. This was not seen as a real problem but rather as a question of solidarity with the different life forms existing in the planet. However, it is actually more, much more than that (Worm and col., 2006; NAAA, 2009), because the loss of biodiversity means an economic – so far underestimated – cost of 50 billion euros a year for Europe. And unless the trend is reversed, the bill will go up to 1.1 quintillion euros per year by 2050, 4% of its gross domestic product. Therefore, we must act at once, which is why 2020 is the new deadline that the European Union has set itself to start reversing the situation once and for all (EU, 2010).

3.2.6 *The climate change*

The climate change-water policy relationship is more than evident. According to most of the prediction models used by the IPCC (Milly and col., 2008), halfway through the 21st century the majority of arid or semiarid areas in the world will see how their water availability is reduced to a

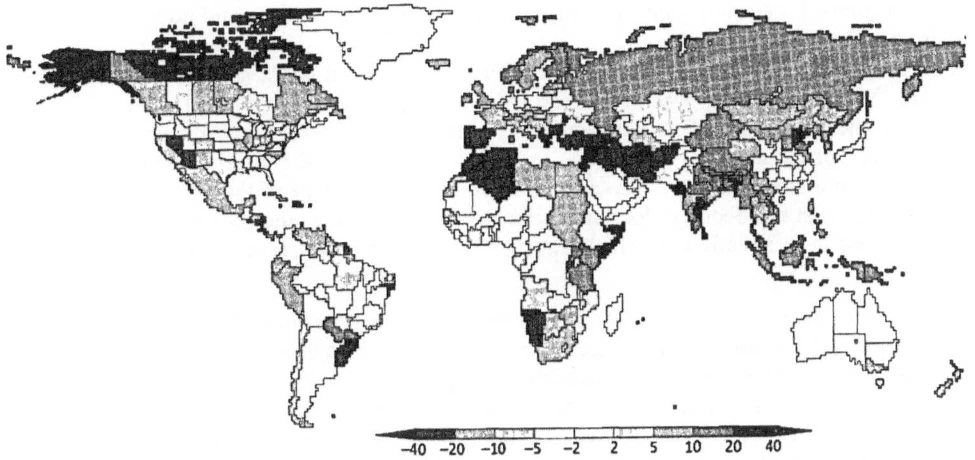

Human influences. Dramatic changes in runoff volume from ice-free land are projected in many parts of the world by the middle of the 21st century (relative to historical conditions from the 1900 to 1970 period). Color denotes percentage change (median value from 12 climate models). Where a country or smaller political unit is colored, 8 or more of 12 models agreed on the direction (increase versus decrease) of runoff change under the Intergovernmental Panel on Climate Change's "SRES A1B" emissions scenario.

Figure 10. Forecast for the variation in the water resources available halfway through the 21st century (Milly and col., 2010).

very significant extent, up to 40% (Fig. 10), which means that the hydrological planning carried out until now will need to go through an in-depth revision.

It seems consequently obvious that we must encourage water saving policies as much as possible, and not only because of the lower availability (which is important too, of course) but also because the sustainable management of water consumes a lot of energy, about 19% of the total in California (CEC, 2005), which means that saving water is equivalent to significantly reducing the emission of greenhouse effect gases and, therefore, to mitigating the effects of climate change.

4 THE CHANGE OF PARADIGM

In the light of the explanations above and regarding water policy, it is crystal clear that this generation must inescapably succeed in overcoming formidable challenges during the next few decades. Only if they cope successfully with these challenges will they be able to leave a habitable planet for the coming generations. However, the current policies need to change to a great extent if we want to succeed, especially in relation to time scales. Nowadays, nearly all the decisions are focused on immediacy, or at best on the short term. However, what we really need is generosity and foresight.

It is not an easy change. Democracies elect their decision-makers for short periods of time. Terms of office – generally between four and six years – represent very brief periods if we measure them with respect to the time scale that applies to the natural environment. And since politicians have to justify what they have done and accredit their good moves or decisions with specific results at the end of their term of office, one can hardly expect them to adopt decisions in which results will only be visible in the medium-long term – unless citizens, with a solid environmental education, can understand the convenience of measures that are as unpopular as necessary. Therefore, it is vital to make the general public aware of the serious risk we are running, and not only us but especially the coming generations. As is going to be explained in greater depth later in another chapter of this book, this task is more complex and necessary in semi-arid countries like the Mediterranean ones. The brilliant history that we have just outlined, which is going to be shown in more detail through the following chapters, is full of realisations and wishes. After several millennia, many of them came true during the 20th century. It is not easy, therefore, to explain that what was valid across so many centuries – since it really was valid then – has stopped being valid now. The change of paradigm is thus complex.

This change of paradigm has come to be known as *Nueva Cultura del Agua* [New Water Culture] in Spain (Martínez, 1997), whereas in the USA, Gleick (Palaniappan M., Gleick P.H., 2009) has coined the expression "*Soft Path for Water*" for it. It is consequently a counterpoint to the predominance of the large civil work that prevailed during the last century, a trend which could be given the alternative name of "*Hard Path for Water*". The essence of this new paradigm can be guessed from what has been put forward so far because, somehow, the six basic pillars on which it is supported have already come to the surface in a natural way. More precisely, they are the following:

1. To guarantee the water required to cover the needs of the whole population
2. To guarantee the water needed to ensure the survival of ecosystems
3. To adapt the quality of water to the use that is made of it
4. To adapt the scale of facilities and infrastructures to that of the needs
5. To promote and encourage the involvement of citizens in the water policy
6. To implement rating systems that favour fairness and efficiency

None of the six preceding items requires a specific clarification because their importance has already been highlighted, even that of the central ones (items 3 and 4) through indirectly in this case. After all, they highlight how relevant it is to reuse water (grey water in dwellings and treated urban ones for other uses, such as irrigation), to take advantage of rainwater or, ultimately, to decentralise draining as much as possible (Sieker, 2008).

Finally, and within this change of paradigm, it is worth mentioning two new terms that have acquired great popularity in recent years. We are referring to 'virtual water' and 'water print', two very didactic and interrelated concepts which, despite not solving anything themselves, do provide valuable information. The first one (Allan, 2003) is the result of counting the water needed to produce a good – generally food – though the water required to produce an industrial good is also counted. Thus, for example, if a plantation of orange trees of one hectare irrigated with 5,000 m^3/year of water produces 40,000 Kg of fruit: the unitary consumption of this citrus fruit per unit of weight and in these specific conditions is 120 l/kg. Obviously, this value is only an order of magnitude because it can vary to a great extent from one year to another. Pluviometry and productivity, amongst other factors, have an influence on its value. The second term refers to a unit of consumption, whether it is a person, a group of people or, ultimately, a nation. Thus, a person's water print would be the sum of the water that he/she uses directly from the supply network (say, about 125 l/day) plus the one consumed indirectly with the food and drinks which the person in question ingests.

Within a globalised world, the preceding concepts permit to convert the food trade into imaginary water transfers. Therefore, it seems reasonable for a semi-arid country to encourage the production of food that requires little water and to import those foodstuffs whose production requires large volumes. Thus, the country where water is in short supply is importing virtual water from countries where there is plenty. Some authors (Hoekstra and Hung, 2002) have made calculations for the commerce of virtual water between countries. On the other hand, the water print corresponding to one unit of consumption, for example, one country (Chapagain and Hoekstra, 2004) makes it possible to evaluate the extent to which the water resources that it owns permit its self-supply and, at the same time, to value the policies that can contribute to raise the supply guarantee. In short, it is information of considerable interest on the path that leads to a more rational and sustainable use of water.

5 THE CHALLENGES FOR THE FUTURE

The preceding analysis shows that the main problem for the current water policy lies in its inability to evolve at the same speed as the events that have succeeded each other during the last century, especially since the third industrial revolution which took place more than sixty years ago. It is worth remembering – and this is only an example – that in such a short period of time (measured, of course, with respect to the time scale for the history of mankind) human beings have overexploited and contaminated strategic aquifers, a water source which was practically untouched one hundred years ago and which has played a strategic role during that period of time. As far as the natural environment is concerned, man has undoubtedly gone too far in recent years, above all in those countries where, due to their water scarcity, history is packed with memorable milestones. The marvellous thousand-year-old culture of water has a strong inertia – too much to respond to the pace demanded by the vertiginous changes occurred during the last decades. On the contrary, Northern Europe – with less inertia and more flexibility – has found better responses to the challenges posed by the future. This issue is thoroughly examined in one of the following chapters.

It is extremely complex for Spain in particular and for Mediterranean countries in general to change the current *status quo* in order to walk gradually toward the *Soft Path for Water.* In order to be able to do it, it is previously necessary to introduce deep structural changes, starting with water administration itself. And there are many interests which hinder it at very different levels. Furthermore, since the majority finds it logical to carry on doing what has always been done (subsidising water regardless of its use) the politician does not find enough reasons to implement far-reaching changes in the traditional policies. It is especially relevant to insist on this idea for its importance, because when the term of office allows politicians to execute the promised works, to show off in the short term, the achievement of their main aim is guaranteed.

If we want to change the current dynamics, it becomes essential to educate citizens environmentally. They must be taught why it is not advisable to look toward the future from the past, no matter how proud citizens can be of the history of their nation – because they can certainly be proud of that. And they also need to understand very clearly the whys and wherefores for that which most annoys them: having to dip into their pockets. That becomes essential to implement rates that permit to recover all the costs. And if there are reasons of any kind that prevent it, we should establish subsidies which favour efficiency and are not detrimental to the environment. Water is, without a doubt, the only manna which falls from heaven. And being free at its origin, nobody can or must put a price on it. But the convenience of having water available in one's own dwelling and managing it sustainably so as not to compromise the future of the next generations has a cost that the user has to assume. Environmental education is very important because, within a democracy, politicians are not going to promote actions which are not supported by the majority. That education has to eradicate the idea which has landed on nations located on the shores of the Mediterranean with technological development: that economic growth demands mobilising more water and that the latter is in unlimited supply. There is as much water as we may need. And the State has the responsibility to bring it from where there is abundant water and – should there not be any available – to desalinate sea water. That is essentially the mentality which guides our action nowadays.

All the existing economic, control and management mechanisms must be implemented beforehand in order to conclude that a certain territory needs more water. For instance, it seems paradoxical to declare that the water is a scarce, precious good while we ignore economic policies which favour efficiency and carry on contaminating and subsidising water without monitoring its use. Only when these measures have been implemented, when both the resources available and the consumption are accurately known, and when the possibilities for saving have been explored, can we conclude that a specific region has not enough water. Moreover, if we really want to be sustainable, we cannot authorise new uses which are not duly justified.

One of the greatest historians ever, Edward Gibbon, did his best to explain and understand how a unique culture like that of the Roman Empire could collapse as it did. Six long volumes, to which he dedicated nearly twenty years of his life, shape his work "The History of the Decline and Fall of the Roman Empire". And the executive summary of his diagnosis is: "what does not evolve, is decadent." Society, above all future society, neither can nor must permit a decadent water policy, no matter how 'brilliant' it might have been.

6 CONCLUSION

Water policy has some formidable challenges to face. Among them stand out betting no longer on the short term and thinking more about the coming generations. After all, in most cases throughout the 20th century, the lack of perspective has prevented the adoption of the measures which are really convenient for the future. Once again, we cannot see the wood for the trees.

This chapter in particular – and this book in general – seeks to provide the reader with the perspective needed to help identify the adequate strategies for us to start walking on the path that leads to sustainability. Insofar as we can contribute to that, even if it is very modestly, we will value as useful the remarkable effort made to allow this book to see the light at last.

REFERENCES

Allan, J.A. (2003) Virtual water – the water, food, and trade nexus. Useful concept or misleading metaphor? *IWRA, Water International*, 28(1), 4–11.

Armengol, J., Palau, A. & Dolz, J. (2008) Gestión Sostenible de Embalses. Numero Monográfico. *Revista Ingeniería del Agua*, 15, 221–267.

Arregui, F., Cabrera Jr, E. & Cobacho, R. (2007) *Integrated Water Meter Management*. London, IWA Publishing.

ASCE Task Committee on Hydraulic Engineering Research Advocacy (1996) Environmental hydraulics: new research directions for the 21st century. *Journal of Hydraulic Engineering. ASCE*, April, 180–183.

Asmal, K. (2000) Water is a catalyst for peace Remarks by Professor Kader Asmal at the opening session. *Stockholm Water Symposium Laureate Lecture*, August 14, 2000.

Barbera, G. (1983) Leonardo e le vie d'aqua. Leonardo a Milano 1482–1982. Comune de Milano.

Beach, L., Hamner, J., Hewitt, J., Kaufman, E., Kurki, A., Oppenheimer, J. & Wolf, A. (2000) *Transboundary Freshwater Dispute Resolution: Theory, Practice and Annotated References*. Tokyo and New York, United Nations University Press.

Bonnnin, J. (1984) L'eau dans l'antiqueté. *L'hydraulique avant notre ère. Collection de la Direction des Études et Recherches d'Electricité de France*. Paris, Editions Eyrolles.

Cabezas, F., Cabrera, E. & Morell, I. (2010) *Agua y estatutos de autonomía. El caso castellano-manchego*. Valencia, Spain, Asociación Valenciana de Empresarios.

Cabrera, E. & García-Serra, J. (1997) *Problemática de los abastecimientos urbanos. Necesidad de su modernización*. Spain, Universidad Politécnica de Valencia.

Cabrera, E., Cobacho, R. & Dubois, M. (2004) *La problemática de los abastecimientos urbanos Revista de la Real Academia de Ciencias*, 98(2), pp. 271–285.

CEC (California Energy Commission) (2005) California's water–energy relationship. *Energy Policy Report Proceeding (04-IEPR-01E)*, November 2005, CEC-700-2005-011-SF.

CEC (Commission of the European Communities) (2007) Addressing the challenge of water scarcity and droughts in the European Union. COM(2007) 414 final. Brussels.

Chapagain, A.K. & Hoekstra, A.Y. (2004) Water footprints of nations. *Volume 1: Main Report*. The Netherlands, UNESCO-IHE, Delft.

CMMAD (Comisión Mundial del Medio Ambiente y del Desarrollo) (1988) *Nuestro Futuro Común*. Madrid, Editorial Alianza.

Costa J. (1911) *Política Hidráulica. Misión social de los riegos en España. Colegio de Ingenieros de Caminos Canales y Puertos*. Madrid.

Delgado, J. (2007) Reseña de? 'Guerra por el agua en el Valle de México? Estudio sobre las relaciones hidráulicas entre el Distrito Federal y El estado de México de Perló Cohen y González Reynoso. Investigaciones Geográficas, April, 062. UNAM, México DF. pp. 158–163.

Dembskey, E.J. (2009) The Aqueducts of Ancient Rome. PhD University of South Africa.

EC (European Commission) (2010) Options for an EU vision and target for biodiversity beyond 2010. COM (2010) 4 final. Brussels.

EEA (European Environmental Agency) (2009) Water resources across Europe – confronting water scarcity and drought. Copenhagen, Denmark, EEA.

Embid, A. & Hölling, M. (2009) Gestión del Agua y Descentralización Política. *Conferencia Internacional de gestión del agua en países federales y semejantes a los federales*. Zaragoza, July 2008. Editorial Aranzadi. Cizur Menor, Navarre.

FMA (Foro Mundial del Agua) (2006) *El agua: Vital liquido para la justicia y la paz*. México, Declaración conjunta de los grupos de fe dirigida al IV Foro Mundial del Agua.
García, E. (1997) *La sentencia del conde de Ribagorza para el reparto de las aguas del río Mijares de 20 de marzo de 1347*, Castellón, Junta de Aguas de la Plana.
Gil, A. & Rico, A. (2007) *El problema del agua en la Comunidad Valenciana*. Spain, Generalitat Valenciana.
Giner Boira V. (1997) *El Tribunal de las Aguas de Valencia*. Valencia, Spain, Fundación Valencia III Milenio.
Gleick, P.H. (1998) *Water and conflict. The World's Water 1998–1999*. Washington, DC, USA, Island Press, pp. 105–135.
Gleick, P.H. (2009) *China and Water. The World's Water. 2008–2009*. Washington, Covelo, London, Island Press, pp. 79–100.
Hermosilla J. (2006) *Las galerías drenantes del sureste de la península Ibérica. Uso tradicional del agua y sostenibilidad en el Mediterráneo español*. Madrid, Ministerio de Medio Ambiente.
Hoekstra A.Y. & Hung, P.Q. (2002) *Virtual water trade. A quantification of virtual water flows between nations in relation to international crop trade*. Research Report Series No. 11, IHE Delft, The Netherlands.
IAHR (International Association of Hydraulic and Engineering Research) (2003) *Newsletter 4*, Vol. 20/2003 (Supplement to JHR – Vol 41 – n 4). Madrid, IAHR Secretariat.
IEEP (Institute for European Environmental Policy) (2009) *Environmentally harmful subsidies: Identification and assessment*. Final report for the European Commission's Environmental DG, November 2009. IEEP London-Brussels.
Kerisel J. (1999) *Le Nil. L'espoir et la colère. De la sagesse à la démesure*. Paris, Presses de l'École Nationale des Ponts et Chaussées.
Martínez, F.J. (1997) *La nueva cultura del agua en España*. Bilbao, Ed. Bakeaz.
Milly, P.C.D., Betancourt, J., Falkenmark, M., Hirsch, R.M., Kundzewicz, Z.W., Lettenmaier, D.P. & Stouffer, R.J. (2008) Stationarity is dead: Whither water management? *Science*, 319(5863), 573–574.
MIMAM, Ministerio de Medio Ambiente (2000) El Libro Blanco del Agua en España. Madrid.
Molle, F. & Berkoff, J. (2006) *Cities versus agriculture: Revisiting intersectoral water transfers, potential gains and conflicts*. Comprehensive Assessment Research Report 10. Colombo, Sri Lanka, Comprehensive Assessment Secretariat.
NAAA (Netherlands Environmental Assessment Agency) (2009) *Growing within limits*. A Report to the Global Assembly 2009 of the Club of Rome. Netherlands Environmental Assessment Agency (PBL), Bilthoven, The Netherlands, October 2009.
Ostfeld A. (2006) Enhancing water-distribution system security through modeling. Editorial. *Journal of Water Resources Planning and Management*, ASCE, July, August, 209–210.
Palaniappan M., Gleick P.H. (2009) *Peak Water. The World's Water 1998–1999*. Washington, DC, USA, Island Press, pp. 1–16.
Pascual, B. (2009) *Agricultura y Contaminación de suelos y aguas. XVII Foro Universitario Juan Luis Vives. Curso 8. Agua Perspectivas de un recurso esencial*. España, Ayuntamiento de Valencia.
Perló, M. & González, E. (2005) *¿Guerra por el agua en el valle e México? Estudio sobre las relaciones hidráulicas entre el Distrito Federal y El estado de México Ed.* Mexico DF, UNAM, Coordinación de humanidades.
Phelps D. (2007) Water and conflict: Historical perspective. *Journal of Water Resources Planning and Management*, ASCE, September/October, 382–385.
Plate, E. (1987) Opening address. *Hydraulics and Hydraulic Research. IAHR. A Historical Review*, In: Garbrecht, G. (ed.) Rotterdam, The Netherlands, Balkema, p. IX.
Pryor, F.L. (2006) *Water stress and water wars*. Swarthmore College Report. PA, USA.
Rouse, H. (1987) *Hydraulics' latest golden age. Hydraulics and Hydraulic Research. A Historical Review*, In: Garbrecht, G. (ed.) Rotterdam, The Netherlands, Balkema, pp. 307–314.
Rouse, H. & Ince, S. (1963) *History of Hydraulics*. New York, Dover Publications.
Sahuquillo, A., Capilla, J., Martínez-Cortina, L. & Sánchez-Vila, X. (2005) *Valencia Declaration. Intensive Use of Groundwater*, The Netherlands, Balkema Publishers, pp. 385–386.
Sheer, D.P. (2010) Dysfunctional water management: causes and solutions. *Journal of Water Resources Planning and Management*, ASCE, January, February 1–4.
Sieker, H. (2008) *SUDS, Green roofs, Rainwater Harvesting, Experiences and Recent Development in Germany. Sustainable Urban Drainage Systems Training Event*. UK, University of Birmingham.
UE (Unión Europea) (2000) Directiva 2000/60/CE del Parlamento Europeo y del Consejo de 23 de Octubre de 2000. Diario Oficial de las Comunidades Europeas, de 22.12.2000.
UE (Unión Europea) (2009) Directiva 2000/60/CE del Parlamento Europeo y del Consejo de 23 de Octubre de 2000. Diario Oficial de las Comunidades Europeas, de 22.12.2000.

Viollet, P.L. (2000) Hydraulique dans les civilisations anciennes. Paris, Presses de l'Ecole National des Ponts et Chaussés.

Viollet, P.L. (2005) *Histoire de l'energie hydraulique*. Paris, Presses de l'Ecole National des Ponts et Chaussés.

WHO (World Health Organization) (2010) *Progress on Sanitation and Drinking Water. 2010 Update*. Geneva, Switzerland, WHO Press.

Wildman, L., Klumpp, C., Greimann, B. & MacBroom, J. (2008) Sediment dynamics post dam removal: state of the science and practice. *Proceedings of the World Environmental & Water Resources Congress*, 12–16, May 2008, Honolulu, Hawaii, USA.

Wolf A.T., Kramer A., Carius A. & Dabelko G.D. (2005) *Managing Water Conflict and Cooperation. State of the World 2005: Redefining Global Security*. Washington, DC, USA, World Watch Institute, pp. 80–99.

Worm, B., Barbier E.B., Beaumont, N., Duffy, J.E., Folke, C., Halpern, B.S., Jackson, J.B.C., Lotze, H.K., Micheli, F., Palumbi S.R., Sala, E., Selkoe, K.A., Stachowicz, J.J. & Watson, R. (2006) Impacts of biodiversity loss on ocean ecosystem services. *Science*, 314, 787–790.

Part B
Water engineering and management through time

CHAPTER 2

Water engineering and management in the early Bronze Age civilizations

Pierre-Louis Viollet
EDF R&D
Ecole Nationale des Ponts et Chaussées, Clamart, France

1 INTRODUCTION

The paper addresses the question of water management in the Bronze Age civilizations of the Middle East, Central Asia and the Aegean world, for the period lying between approximately between 4000 BC, and 1100 BC (Egypt is addressed in another paper of the present conference).

Irrigation probably began to develop at a small scale as early as in the Neolithic period, in the so-called "fertile crescent" – an arc constituted of the hills of Syria-Palestine and the foots of the Taurus and Zagros mountains –, as well as south-east of the Caspian sea, when agriculture spread out from its initial Levantine birthland. But the conditions leading to large-scale hydraulic engineering and water management really appeared when early cultivators settled in the low plain where the Tigris and Euphrates join. There, towns and cities appeared during the IVth millennium BC, with the Sumerian and Akkadian civilizations – when writing who also invented. Civilization development followed in Egypt, in Central Asia, where are the earliest known traces of irrigation after Mesopotamia, and also in the Indus valley with the Harappean civilization which developed during approximately 10 centuries from the beginning of the IIIrd millennium. In Crete the Minoan civilization started from the end of the IIIrd millennium, and ended by 1500 BC, and in continental Greece there was the Mycenaean civilization during the IInd millennium, until 1100 BC, where it collapsed with the beginning of the dark ages of Greece. The end of the Bronze Age, by 1200–1100 BC, was marked by a lot of destructions, invasions and troubles, and marked also in a sense the end of many of the early civilizations[1].

Further civilizations (esp. those of the "classical" period) where their successors and used basically techniques derived from those of the early civilizations. All these civilizations between the Aegean sea and central Asia were connected to each other by trade, migration and invasion routes, and show many common features in water engineering.

The Bronze Age civilizations were highly centralized, at the scale of a city-state, a country, or an empire, the "palace" controlled the collection and distribution of the main products, and managed large public works. This came together with their capacity to develop large-scale water-management technique and works. Among those civilizations, the Sumerian and Akkadian civilizations of Mesopotamia are particularly outstanding, not only because they can be regarded as precursors in many aspects – including use and management of water of the two large rivers Tigris and Euphrates – but also considering the incredible number of texts of all sorts written in cuneiform symbols, in the Sumerian or Akkadian languages, that they left to us, buried under the mud and the sand covering the lost cities of Mesopotamia.

---

[1] The cities and palaces all around the Aegean sea were either destructed by fire (Troy, Mycenae, Pylos, Ugarit...), or abandoned, by 1225–1175 BC, after raids of mysterious "people of the sea", to whom only Egypt could resist; for a historical discussion of this catastrophe, see Drews (1993).

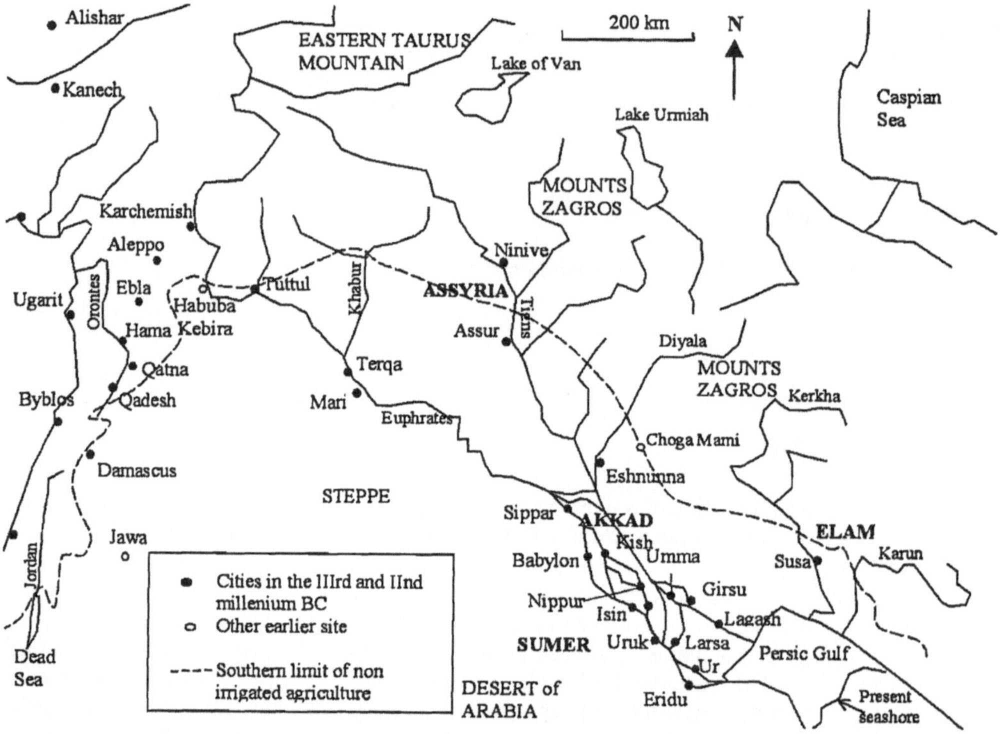

Figure 1. The system of the two Rivers Tigris and Euphrates in early Antiquity, with major cities of the same period.

2 THE EARLY CIVILIZATIONS IN THE TIGRIS AND EUPHRATES RIVER SYSTEM

2.1 *The Euphrates and Tigris river system and the early cities in the IVth and IIIrd millennium BC*

The Tigris and Euphrates rivers, issuing from the Taurus mountains in Anatolia and Armenia, flow through high grounds before reaching the low plain of lower Mesopotamia. This low country used to be 400 km long in the bronze age – it is now 575 km long, because of the constant flow of sediments issuing from the two rivers which, century after century, have deposited into the Persic Gulf. This long and relatively narrow plain has a very low downstream slope (0,1 to 0,2 m/km), which makes the river courses meandering and unstable. When entering the low country, in the area of the ancient Sippar (see Figure 1), the courses of the two rivers are very close to each other, and there is evidence that in early Antiquity there was a convergence of the two rivers according to two channels[2], as can be seen on Figs. 1 and 6.

South of the line which roughly corresponds to an annual rainfall of 200 mm/year (the doted line on Figure 1), agriculture is not possible without irrigation. By the beginning of the so-called Ubaid period, by 6400 BC, human settlements began to appear close to what used to be the shoreline of the Persic Gulf, in Eridu, a place which has always been considered by the Sumerians as one of the historical roots of their civilization. By that time, the river system in lower Mesopotamia

[2] This convergence is attested in the IInd millennium BC, but may have existed since the IIIrd millennium BC, or even earlier; the Sumerians of this time used to call the main southern channel the "Sippar River", and the northern one the Irnina. (Gasche, Tanret, Cole, Verhoeven, 2002).

was flowing according to multiple channels, along which many new settlements appeared. The maximum dispersed development of this habitat looks to have taken place in the beginning of the IVth millennium BC, when the so-called Uruk period begins, with the development of the earliest cities. During the IVth millennium BC, irrigation developed at the scale of the country surrounding each city. The initial cities, from the ancient Sumerian lists, were Uruk, Shuruppak, Lagash, Girsu, Ur, Eridu, Nippur. In these cities, writing (using cuneiform symbols), mathematics, soon became invented. A little bit on the north of the plain, in the country of Akkad, there were Kish, Sippar. The famous city of Babylon lies also in this area, but this city will only become important in the second millennium BC, with the reign of king Hammurabi, as we will see further. Some of these oldest cities covered a large area: the walled area was 50 hectares for Ur, 400 hectares for Uruk, 500 hectares for Lagash. It is in this area that the legend of the Biblic Flood came out. Actually there are acheological proofs that not all, but some of theses cities were flooded, at different times: in Ur at a date between 4500 and 4000 BC, then at another time between 2800 and 2600 BC; in Shuruppak by 2900; in Kish three times between 2800 and 2600 BC[3].

The large watercourses used to play a very important role as a source of water for the cities, for irrigation, and also for navigation. The annual floods of the Tigris and Euphrates rivers come at their peak in April and May, when the grain may still be in the fields (harvesting is between February and May), and can be taken over by the flow if the flood is not kept under control. And when irrigation is needed, in the summertime, then water in the two rivers is at its lowest level[4]. Irrigation used mostly gravity flow with several classes of secondary canals, with techniques which will be described further in the present paper (section 4, item *canals*). The importance of the watercourses for navigation is enhanced by many ancient texts: an inscription dated as early as 2490 BC, for instance, relates that wood had been brought by ship from Bahrein (called *Dilmun* in Sumerian) for the construction works of Ur Nanshe, prince of Lagash[5].

Comparing the two maps on Figure 2, with the area surrounding Uruk reconstructed at the beginning and in the second half of the IIIrd millennium BC, an eastern displacement of the main channels in activity is apparent. It may be assumed that the convergence of the Tigris and Euphrates close to Sippar (with the Irnina and the "Sippar River") either appeared at that time, or became more active than it used to be before. The cities of Adab, Zabalam, Umma, which lye along this eastern branch showed as a consequence a strong development in the 2nd half of the III millennium. And the oldest cities of Shuruppak, Uruk, and even Ur remained fed only by a single channel, the Puratum of the Sumerians. This channel has been since that time very steady and regular, without meanders, which strongly suggests that it has been artificially regulated and controlled by man already. Later evidence of this control, from texts dated 2200–1500 BC, will be given further.

By 2900 BC, a new city was founded in the middle Euphrates valley, Mari. The reasons of this foundation may have been to control the communication route between Mesopotamia on one side, the Syrian cost and the fertile valley of the upper course of the Khabur on the other side. There may also have been the purpose to create there a centre for metallurgy, with wood and minerals carried by boat to Mari. Together with the foundation of Mari, very large scale hydraulic structures were created, with a 30 km long irrigation canal, as well as a large navigation canal parallel to the Euphrates and issuing from the Khabur, 120 km long. Another large-scale canal was also created upstream of the Mari irrigation system, in order to provide irrigation water to the area of the ancient city of Terqa, as known from the texts of the IInd millennium BC. Figure 3 shows the map of the area surrounding Mari, with the bronze age hydraulic structures.

Managing long courses of rivers inside a coherent water-management policy is possible only if there is cooperation between the riverside cities, and if the upstream source of this course can

---

[3] Roux, 1964.

[4] This is an important difference with the irrigation calendar in Egypt: the flood of the Nile lasts from July to October, and allowed the Egyptians just to use the flood to water the fields prior to ploughing and seeding.

[5] Sollberger & Kupper, *Royal Inscriptions*, IC3c &IC3d.

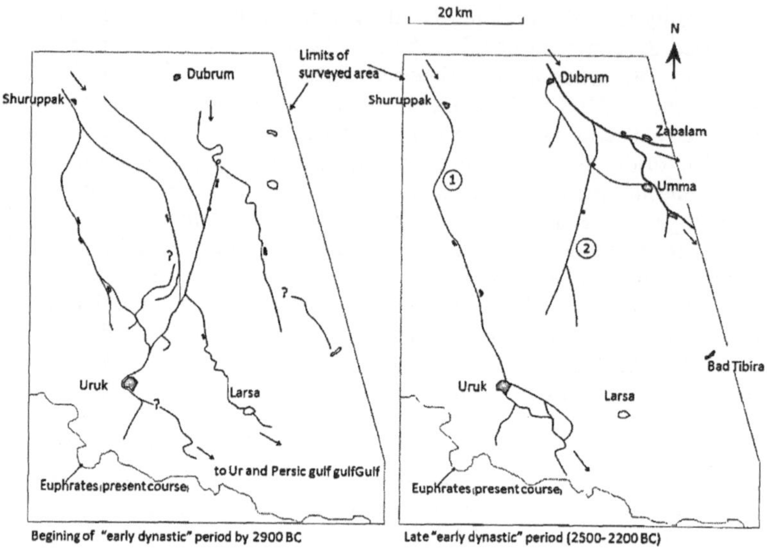

Figure 2. The human settlement and the probable river system in the area of Uruk, by 2900–2200 BC – drawn by the author using data from field survey by Mac Adams and Nissen (1972). These charts show how the river system in the south was progressively organized according to straight channels stabilized under human control: canals n° 1 and 2 may be followed during many centuries (see Figure 4 for the following period).

be kept under control. Thus, water management issues have always been dependent upon political issues. By 2300 BC there came a leader from the area of Kish, Sargon of Akkad, who unified under his leadership the whole country of Sumer and Akkad, as well as farther countries like Mari, in the middle Euphrates valley, and even Ebla, in Syria. The importance of the water system for navigation is again underlined by this citation: "Sargon made the ships from Meluhha (Indus valley), Magan (Oman), and Dilmun (Bahrein) anchor at the quay of Akkad"[6]. This first empire of Mesopotamia lasted until 2193.

2.2 *The control of the lower Mesopotamia river system from texts of the Ur, Lagash and Babylon dynasties (2200–1700 BC)*

It is at the transition between the IIIrd and the IInd millenniums that we have the best textual evidence of large-scale fluvial works, both from political and mathematical texts. Man control upon the fluvial system of lower Mesopotamia was particularly apparent during the IIIrd dynasty of Ur (a Sumerian dynasty which may have originated in Uruk some time after the fall of the Akkad empire, then took Ur as a capital, and reigned over a large part of Mesopotamia in the period 2112–2004 BC), and during the further dynasties of Larsa (1932–1763), and Babylon (1782–1700)[7].

In this period covering 4 or 5 centuries, two objectives seem to be underlying the largest hydraulic works. The first is to keep the access to the sea free for the harbour cities of Ur and Lagash, who keep active sea-trade with the countries neighbouring the Persic Gulf (especially Bahrein). The second objective is to secure the water supply of the cities of the south, Uruk, Ur, Larsa, as the main flow of the river system has been now in the ancient Tigris, close to Umma, as we have reported above. The Euphrates course in this area (*Puratum*), is flowing from Nippur, through Shuruppak (which had declined after what seems to have been a great fire in 2300), Uruk, and farther Ur. It is known that this course has been restored by Amarsuena, king of Ur (2046–2038). Babylon is

[6] Ancient text reported by Nissen (1984), p 168.
[7] See Renger (1990), Charpin (2002).

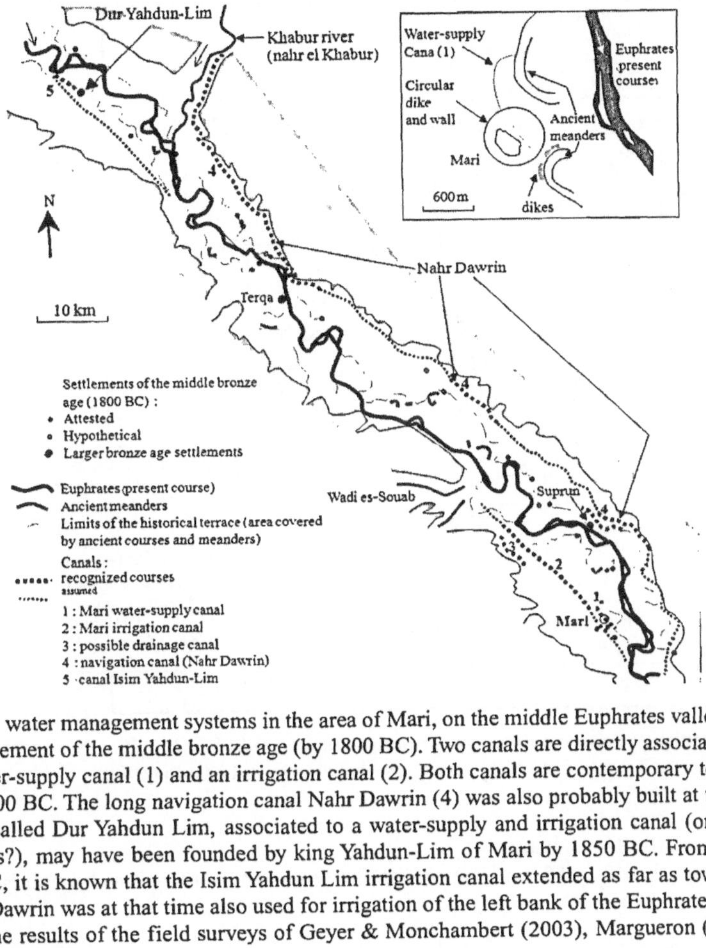

Figure 3. The water management systems in the area of Mari, on the middle Euphrates valley, together with the known settlement of the middle bronze age (by 1800 BC). Two canals are directly associated with the city of Mari: a water-supply canal (1) and an irrigation canal (2). Both canals are contemporary to the foundation of Mari, by 2900 BC. The long navigation canal Nahr Dawrin (4) was also probably built at the same period. North, a city called Dur Yahdun Lim, associated to a water-supply and irrigation canal (one single or two different canals?), may have been founded by king Yahdun-Lim of Mari by 1850 BC. From the texts dated 1800–1760 BC, it is known that the Isim Yahdun Lim irrigation canal extended as far as towards Terqa, and that the Nahr Dawrin was at that time also used for irrigation of the left bank of the Euphrates – drawn by the author using the results of the field surveys of Geyer & Monchambert (2003), Margueron (2004) and other sources (see Viollet, 2000).

fed by another course of the Euphrates, called *Ahratum*, which joins the *Puratum* either north of Uruk (which we have assumed when drawing the course of the rivers on Figure 1), or farther south. Another important river course looks as a truly artificial one, it is the large course derived from the main River close to Nagsu, flowing to the south-west direction, and joining downstream the *Puratum* coming from Uruk, and then flowing south-east towards Ur (n° 4 on Figure 4). Along this channel lye the cities of Bad-Tibira and Larsa. This course may have been dug by the kings of Ur (2112–2004). The first king of this dynasty of Ur, Urnammu, is known from his royal inscriptions to have done important works upon nine major canals of the system.[8]

After the fall of Ur in 2004, Larsa took the leadership upon lower Mesopotamia (1932–1763). Its kings' inscriptions[9] relate a large number of works on the Rivers. King Sin-Iddinam (1849–1843) "restored the Tigris". His successor, Rim Sim (1822–1763), along the 59 years of his reign, is known from his inscriptions to have undertaken to reshape the system, starting from east to west:

– in his 9th year of reign, the "Lagash canal",
– in his 16th year, the "Steppe canal" (in the area between Umma and Girsu)

[8] Renger (1990).
[9] Powell, 1990.

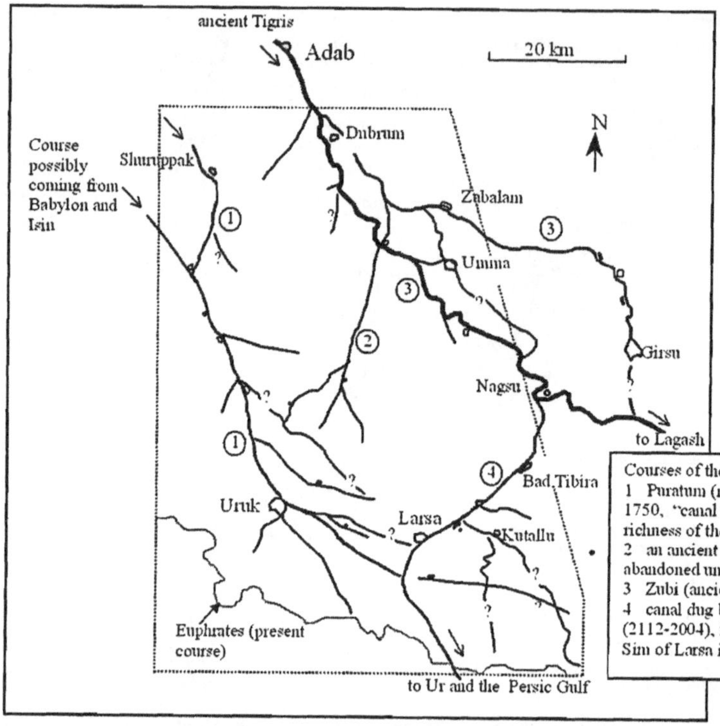

Figure 4. The settlement and the river system in the area of Uruk, Larsa and Umma by 2100–1700 BC – drawn by the author using data from field survey by Mac Adams and Nissen (1972).

– in the 19th year, the Tigris itself,
– in the 23rd year the 40 or 50 km long course flowing to the south-west through Bad Tibira and Larsa to Uruk and Ur.

All those four waterworks were said to have been established or restored "as far as to the sea". In Larsa, by 1900 BC, there was a corporation of "Dilmun travellers", merchants who where managing sea-trade of copper with Bahrein (*Dilmun*)[10]. On the total, nine of the years of Rim Sim's reign are celebrated for extraordinary waterworks: there are the years 7, 9, 16, 19, 22, 23, 24, 26, and 27.

At this time Babylon enters into the front scene. Hammurabi, king of Babylon, took over Larsa in 1763, and destroyed Mari in 1761. He established then an empire over the whole Mesopotamia. In 1750, which was actually the 33rd year of his reign in Babylon, he restored the Puratum, as reported by the following inscription: "Hammurabi has dug the canal "Hammurabi is the prosperity of the people" – the canal for which the gods An and Enlil take care – and thus provided the cities Nippur, Eridu, Ur, Larsa, Uruk and Isin with a steady supply of water for their prosperity and made it hence possible for the inhabitants of Sumer and Akkad, who had been scattered (by war) to return to their settlements". The list of the cities mentioned here suggests that the waterworks on this course of the Euphrates may have concerned a part of the course as long as 150 km[11]. Hammurabi's successor, Samsu-iluna (1749–1712) had to face repeated floods in the area of Babylon, his capital city. In the 3rd year of his reign he created (or possibly re-created) a diversion of the Euphrates allowing to store excess water during floods into a natural depression, now lake Habbaniyah north of Babylon; and later, in the 26th year he made a communication through a rocky barrier connecting

[10] Oates (1986).
[11] Renger (1990).

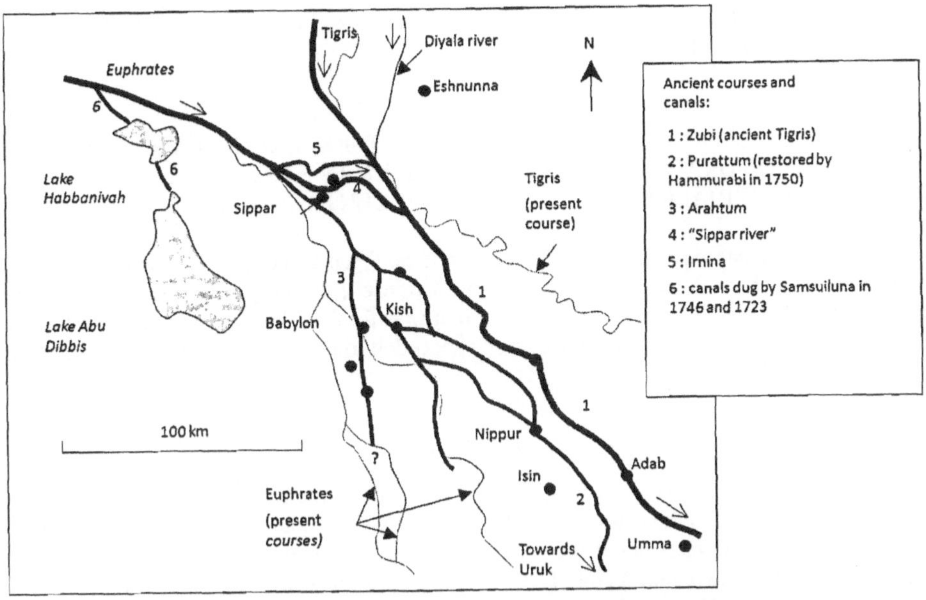

Figure 5. The river system in the area of Babylon by 2500–1500 BC, showing the works probably done by Hammurabi's successor, Samsu-iluna in the 3rd and 26th years of his reign, in order to protect Babylon against the floods of the Euphrates (adapted from Gashe, Tanret, Cole & Verhoeven, 2002).

lake Habbaniyah to lake Abû Dibbis (these are modern names), in order to increase the storage capacity (Fig. 5). Herodot, the Greek traveller and historian, related this story and attributed these waterworks to the legendary queen Semiramis[12].

This Babylonian empire soon weakened and fell after a raid of the Hittites from Anatolia, by 1595 BC, followed by invasions of barbarian invaders from the Zagros mountains, the Kassites, who settled in lower Mesopotamia, and established a stable dynasty untill 1235–1160. Under the Kassites, irrigation in the country of Sumer and Akkad seemed to be even more dependant of very long canals running from north-west to south-east. A ruler in the area of Babylon reported in despair to his Kassite master: "The town which my lord granted me is abandoned for lack of water, where should I go next year?"[13]

Then came the time for new powers upon Mesopotamia, from the northern and eastern higher countries of Assyria and Elam. New forms of water management came in these countries, using derivation structures from small rivers in the mountain with canals and aqueducts towards the cities lying at the foot of these mountains. The oldest work of that kind may be the water-supply of Dur-Untash, a new city founded in Elam by 1275–1240 BC, 40 km south-east of Susa: the king of Elam, Untash-Gal, created a 50 km long canal derived from the Kherka river. Later, in the iron age, similar water engineering works will be done by the Assyrians, and by the Urartians of Armenia. In the lower country of Sumer and Akkad, the recession did not mean, however, any real discontinuity in agricultural works, neither in the use – and obviously of the maintenance – of the irrigation structures in Mesopotamia. The task of maintenance of the irrigation structures and of controlling the Rivers in fact never ended, as it is reported that 2000 years later, in the VIIIth century AC, the Sassanid Persic dynasty who was ruling Mesopotamia by that time failed in keeping the Tigris and Euphrates controlled inside their dykes, in the lowest part of Mesopotamia; then a large area of

12 Herodot, book I, 184–187.
13 Van Soldt (1988).

fertile land got submerged, and this is the time when the Arabs invaded Mesopotamia and put an end to the Sassanid dynasty.

3 LAND RECLAMATION FOR AGRICULTURE OUTSIDE LARGE RIVER BASINS: THE QUESTIONS OF DRAINAGE AND WATER-SUPPLY IN EAST-MEDITERRANEAN COUNTRIES (SYRIA, JORDAN, GREECE) AND CENTRAL ASIA

The previous section has explained how the courses of the Tigris and Euphrates used to be controlled in Mesopotamia, where, because of the scarcity of rainfall, the large rivers used to be the only possible source of water. In many other areas, sometimes far from large rivers, but with temporary seasonal sources of water, the early civilization succeeded in building efficient water – and land – management techniques.

3.1 *Early water management systems in arid countries*

Areas located in modern Syria or Jordan, not very far from the southern limit of non-irrigated agriculture, are very arid in the dry season. The temporary watercourses (wadis) provide important but unsteady sources of water under the form of flash-floods. These wadis are completely dry in normal time, but may have a high peak flow-rate: 200 m^3/s for instance for the wadi Musa which passes through the site of Petra in Jordan. Descending from mountains whose rainfall may be strong and sudden, the slope of their bed is high; then the velocity of water during the flood is very high as well.

Bronze aged settlements with sophisticated water management techniques are found in the so-called "black desert", a basaltic area lying at the foot of Djebel el Arab, a volcanic mountain culminating at 1850 m at the boundary between modern Syria and Jordan (Figs. 6ab&c). These techniques include water diversion from the flash-floods of a wadi, water runoff catchment, and water conservation in reservoirs either made in the bed of the wadi or in natural depressions, with dams and dykes. Jawa in the Jordan side of this mountain, is the oldest of these sites[14], dated around the middle or the end of the IVth millennium. As can be seen from Figure 6b, a walled town was built there, with three independent systems of catchment from the wadi Rajil to canals deriving water into reservoirs (R1 to R10 on Figure 6b). One of these reservoirs (R4, the closest to the city), is constituted with a small depression closed by a dam. This dam which is 4,5 m high and 80 m long, is made of stone walls with earth and ash inside (Fig. 15), and is presently the oldest known dam. The runoff collection system used deflection walls and small furrows directing water to the canals of the main catchment system. 3000 to 6000 people, maybe refugees from a more civilized area, may have built the town and lived here some time. The total volume of the ten reservoirs is estimated as 42000 m^3, which, assuming an individual consumption if 1,2 m^3/month, and taking into account the seasonal cycles of the wadi Rajil (floods in may and november), as well as the winter/spring rain cycle, seemed able to fullfill the needs of the habitants and their cattle. Two other sites, Hebariyeh and Khirbet el Umbashi[15], at the foot of the same mountain but on the Syrian side, show similar water management technique, dated around 3000 BC: diversion canals, reservoirs made with earth dikes or earth-made dams. In Khirbet-el Umbashi (Fig. 6a) the settlement began by 3300–2900 BC, with the construction of a walled area (which curiously remained empty) together with a reservoir made in the bed of the wadi, closed with a dam (actually the oldest known earth dam); later, a dispersed settlement area developed outside the walled area, north of it, and at an unknown date in the bronze age a large reservoir (30 000 m^3) was made in a natural depression closed with massive earth dikes, and fed by a derivation from the wadi Umbashi.

[14] Helms (1987).

[15] Braemer, Echallier, Taraqji (1996).

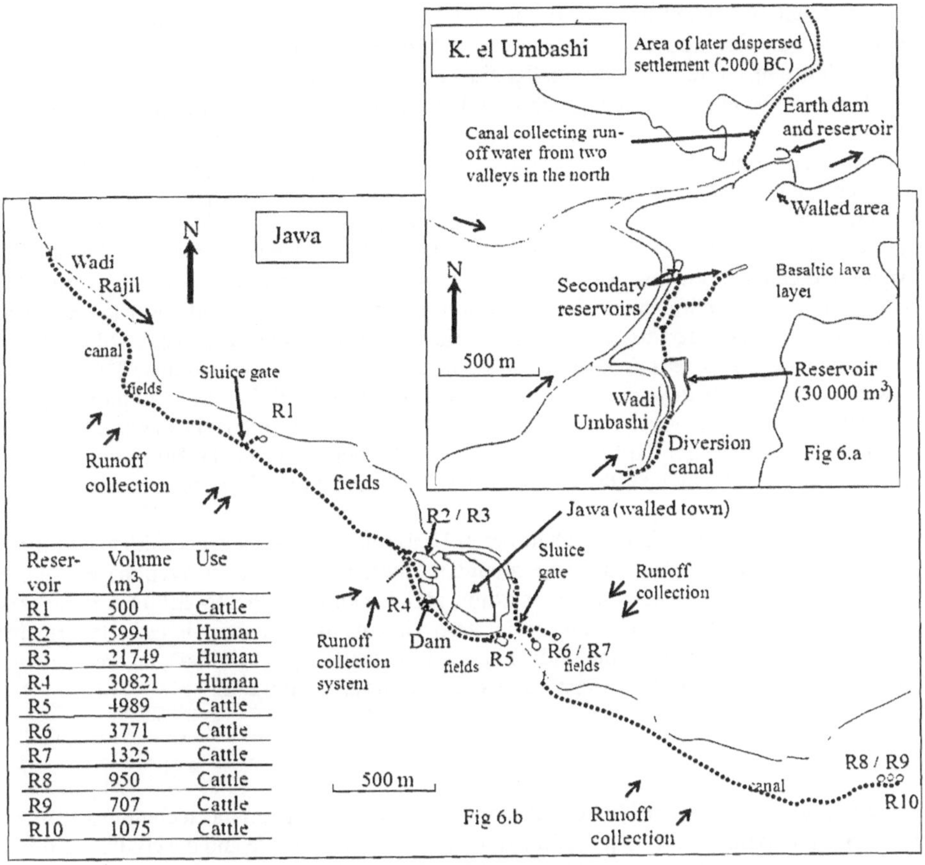

Reservoir	Volume (m³)	Use
R1	500	Cattle
R2	5994	Human
R3	21749	Human
R4	30821	Human
R5	4989	Cattle
R6	3771	Cattle
R7	1325	Cattle
R8	950	Cattle
R9	707	Cattle
R10	1075	Cattle

Figure 6a & 6b. The water management systems of Kirbet el Umbashi and Jawa (3500–3000 BC).

Figure 6c. The reservoir R5 at Jawa, photo taken from the upper walled town (photo P.-L. Viollet).

The permanent settlement of Jawa in fact did not last long, maybe only half a century, and this small city was soon abandoned by those who had settled there. On the contrary, Hebariyeh, Khirbet-el-Umbashi, as well as other smaller sites east of the Dead Sea, were associated only with dispersed and maybe temporary settlements, and lasted much longer (untill 1500 BC for Khirbet el Umbashi).

South Yemen, an area known in classical Antiquity as *Arabia Felix* (happy Arabia) shows a similar geographical and hydrological pattern: in the edge of the desert, there are wadis which come from high mountains and get lost in the desert, and which occasionally carry high flow rate during the scarce, but sudden floods. There, people used to occupy the valleys of these wadis, with weirs and canals allowing to distribute water loaded with sediments directly to the fields. The distribution of water was managed so that a thin sheet of water would cover a large surface of land. The sediments used to deposit on the fields, creating year after year a thick layer of fertile land – whose thickness growing at a rate evaluated as 0,7 m per century. This technique, which will have its larger development from the VIIIth century BC (with the kingdoms of Saba, Qataban, Hadramawt from where incense will be exported to the Mediterranean world), has begun to develop as early as in the IIIrd millennium BC or the beginning of the IInd millennium BC: Shabwa, which will become later the capital of the Hadramawt kingdom, as been occupied since 1800–1900 BC[16].

Less than 200 km north of Jawa and Khirbet el Umbashi, Damascus is also situated at the foot of a mountain, but here, coming out of the mountain, there is a permanent river, the Barada. This favourable place has been occupied for a long time, and it is known[17] that by 1500 BC there existed two canals derived from the Barada at the point where it exits the mountain, in order to irrigate the area around Damascus (which, in the middle age will be the famous "ghouta", an area known for its fruits and flowers, thanks to the many canals derived from the Barada).

3.2 *Central Asia*

Central Asia was the second historical area of water management techniques following Mesopotamia and Syria. Unfortunately, no written sources are available and the civilizations which early developed east of the southern shores of the Caspian Sea are poorly known. In the Oasis of Geoksyur, the Tedzhen river ends in the Kara Kum desert, and artificial irrigation is known as early as in the IVth millennium BC, with small canals (3 to 5 m wide, a few km long) derived from the Tedzhen and extending the area where its water could be used. Later in the bronze age, the same water management techniques extended towards the Murgab river in the Merw oasis, east of Geoksyur, as well as towards the riverine areas of the Amu Darya, and towards the Zeravchan river (in the area of the present Samarkand), with extended networks of canals. Figure 14, further in the present paper, gives another example related to eastern Bactria (north of modern Afghanistan).

3.3 *Land reclamation and drainage in continental Greece*

In continental Greece, there exist a number of marshy depressions where runoff water accumulates during the rain season. Those depressions used to be too wet for agriculture, while the surrounding higher lands were on the contrary too arid in the dry season. The typical water management method used by the Mycenaeans was to build long low dams allowing to separate such depressions into two parts, diverting all run-off waters to one side of the depression, and draining the other side: the depression is then separated into a polder on the one half and a lake on the other one, whose water can be used for controlled irrigation in the polder during the dry season. The largest-scale example of such land and water management is lake Copaïs, between the city-palaces of Orchomenos and

[16] Breton (1998).
[17] Kamel (1990).

Table 1. Mycenaean dams (from Knauss, 1991).

Name of the dam	Height (m)	Length (m)
Beotia (north to south):		
Boedria	2	1250
Thisbe I	2,5	1200
Thisbe II	4	200
Peloponnese (north to south):		
Pheneos	2,5	2500
Stymphalos	2,5	1900
Orchomenos	2	2100
Mantinea	3	300
Taka	2	900

Figure 7. Palaces, dams and reservoirs in Mycenaean Greece.

Gla in Beotia[18]. The dams are long low dikes, as can be seen from table 1, made of earth supported by cyclopean stone walls, and their locations are reported on Figure 7.

One other particular dam is encountered close to Tiryns. There, a small river with high slope (about 15 m/km) is descending from high grounds and passing close to the palace of Tyrins and the city surrounding the palace. The palace itself is on a hill, but the city was close to the river level, and was probably protected by dikes. By 1200 BC, it happened that the river came over the city and buried its major part under a thick layer of sediments, up to 4 m height at some places[19]. The city was further rebuilt, above the new sediment layer, and in order to avoid any risks of similar events in the future, the river was dammed and diverted 3,5 km upstream, with a 1,5 km long canal leading to another river flowing south, far enough from the city, as shown on Figure 8. The dam was made with an earth core and two walls of the cyclopean masonry usually used by the Mycenaeans. It was about 10 m high, 100 m long, and thicker on the right bank (103 m) than on the left one (57 m).

[18] These dams have been studied by Knauss (1991), and described later in review books like Schnitter (1994), Viollet (2000).

[19] Zangger (1994).

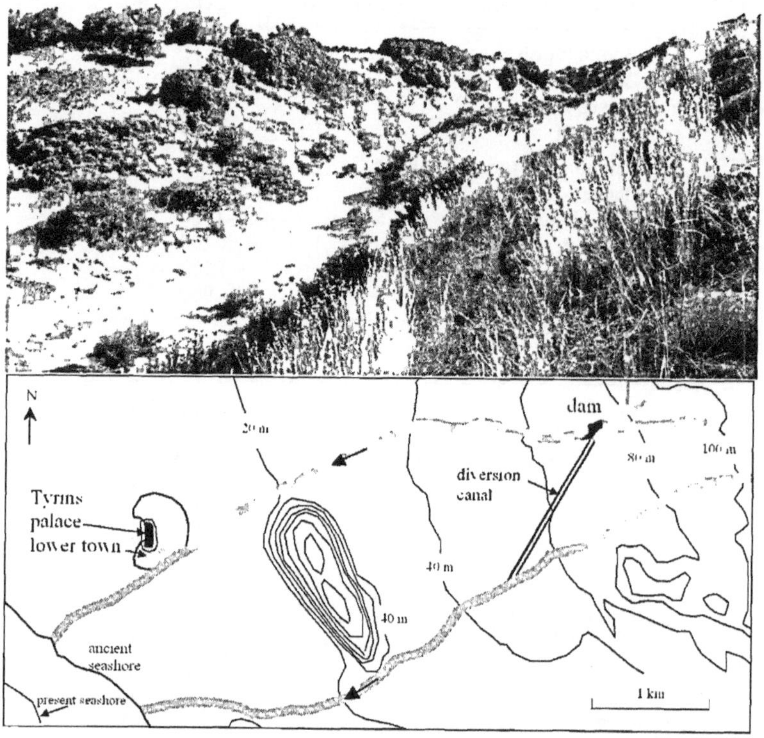

Figure 8. The diversion dam and canal in Tyrins (1200 BC): map adapted from Zangger (1994), and photo of the diversion canal taken from the dam (photo P.L. Viollet).

4 WATER MANAGEMENT IN THE EARLY CITIES

4.1 *Urban drainage*

Architecture in the bronze age civilizations developed flat roofs with terraces, and large courtyards paved with stone (Crete) or bricks (Mesopotamia, Indus valley). The smallest courtyards were used as light wells, bringing light to the lower floors of the palaces. In those countries, rain is rare, but falls very strong when it occurs. From the flat surfaces of the roofs and courts, rainwater had to be evacuated and sometimes diverted into reservoirs for a further use.

As early as in the neolithic period, in villages of Syria or on the middle course of the Euphrates, there have been conduits or furrows designed as to evacuate water from the houses. Following this early tradition, civilizations of the bronze age developed urban drainage systems for excess rainwater, and also for the evacuation of water from latrines and bathrooms: bathrooms with terracotta tubs have been found in Minoan Crete, in the ancient city of Akrotiri in Santorini island, in Ur and in Mari, in Mycenaean Greece (Pylos).

By 3500 BC, during the *Uruk period*, the Sumerians had founded a small town in the middle Euphrates valley, on the site of Habuba Kebira, in modern Turkey. This town, which has been occupied during approximately 150 years, may appear of a first prototype of the cities founded later in order to control the trade between Mesopotamia and Syria – in a way similar to the later foundation of Mari (2900 BC). This city was situated on a terrace 10 m above the river bed, and may have been occupied by about 1500 persons at its maximum state of development, 150 years after its

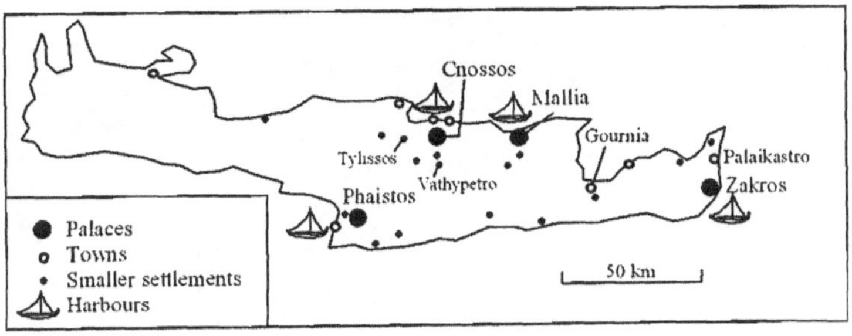

Figure 9. Minoan Crete.

foundation. There was a Main Street, whose width was 5,8 m at the latest state of development of the city, covered with a conglomerate of river gravels. The streets were equipped with a gutter forming a drainage network which allowed to evacuate water out of the city walls, using U-shaped terracotta conduits (made of 0,64 m long elements), using also pipes made of conical clay elements fitted to each other (Fig. 16), or channels with a bottom of clay and limestone walls. This network system evacuated both rainfall water and water from the houses[20]. Other sites of Mesopotamia show urban drainage systems. In Ur, archeological findings include different sorts of soakaway drains at the scale of each house: pottery jars with holes in the bottom, shafts coated with bricks[21].... In Mari, it is also known from archeology that shafts coated with bricks acting as soakaways allowed to evacuate water from bathrooms and latrines, and also excess rainfall. In the last state of the palace of Mari there were pipes descending from the upper floors, where there may have been latrines and bathrooms[22]. In the temples and palaces areas of Mari there existed by 2500–2250 BC networks of channels acting as drain, but probably more for rainwater evacuation (and possibly conservation, as it will be the case in the last state of the palace of Mari) than as a sewage. Ugarit was a city built upon a hill close to the Syrian shore, which was destroyed by 1190 BC by the Sea People; the drainage systems were designed as to evacuate both rainwater and used water from houses: in the palace; there was a large drain under the pavement, leading water towards a small river north of the city, and in private houses there were terracotta gutters leading to soakaway pits. Both in Mari and Ugarit the streets used also to evacuate directly rainwater outside the city. In Mohenjo Daro, a city of the Indus civilization, water was evacuated from the houses using terracotta conduits into covered gutters passing under the soil of the streets; houses which were located too far from this drainage network used to have soakaway pits made of terracotta bottomless vessels[23].

The first half of the IInd millennium was the golden age of the Minoan civilization in Crete. This civilization is known for its sophisticated urban architecture, together with urban water management systems. The major cities in Crete by that time were Cnossos, Mallia, Zakros, Phaistos, all four places with a large palace, but there were also smaller cities and settlements (Fig. 9). In almost all Minoan sites (Cnossos, Phaestos, Mallia, Gournia, Zakros), and even in Akrotiri on the isle of Thera (Santorini), there were gutters covered with stone which passed in the streets, and under the pavement of corridors and courts (Fig. 10).

An extended drainage system has been discovered in the eastern side of the palace of Cnossos, dating of the "middle minoan" period (by 1800 BC). A bathroom (called "the Queen's toilet") is known close to the SE corner of the central court. Close to it, a water-closet has been found, in a rooom called "room of the plaster couch": a seat is at the end of the room, and a flushing channel comes from the entrance of the room (so that it was possible to drop water into it from the outside

[20] Vallet (1997).
[21] Wilson (2000).
[22] Margueron (2004).
[23] Janssen (1988).

Figure 10. Drainage gutters in Minoan palaces: (a) left, under the pavement, in a court in the workshop area of the palace of Mallia; (b) right, in an open-air corridor connected to the central court of the palace of Phaestos (photos P.L. Viollet).

of the room), and under the seat this flushing channel communicates to the large drain of the eastern area of the palace. Other latrines probably existed on the second floor[24]. Figure 12 shows how this sewage drain served different rooms and light-wells (with terracotta pipes descending from the roofs) of the east side of the palace, including two other water-closets from the second floor. This drain was also used to evacuate excess rainfall water from the roofs. At the outlet of the eastern side of the palace, towards the Keratos river, this drain is 79 cm high and 60 cm wide. Drainage systems, probably inherited from the Minoan model, also existed in Mycenaean palaces of continental Greece (in Tyrins for instance).

4.2 *Water resources for the cities*

Water can be available from a permanent or non-permanent river, from underground water, from natural sources, and from rainwater collection and conservation. All those sources of water have been used in the cities of the early civilizations, depending on their particular situations. In the bronze age, and in fact as long as untill the IIIrd century BC, there had been no large-scale water lifting technique existing. Thus, water distribution in the cities needed either people to carry water from the place where it was available, or a canal or an aqueduct coming from a source at a higher altitude than the city. But in lower Mesopotamia – as well as in Egypt, actually – no source is available in altitude. Even in countries were such natural sources may be found, safety reasons often led to establish cities on a hill, which, in the absence of pressure pipes, was making almost impossible to deliver running water.

Every city of the country of Sumer and Akkad, in ancient Mesopotamia, had a canal and sometimes several ones connected to the River, or to a major stream, both for navigation and for the purpose of water-supply. From these canals, water was taken for the daily uses. In Mari, there was a canal connected to the Euphrates from both ends, and passing through the city (Fig. 3); it is known that servant women were assigned to the task of filling the 25 m^3 cistern of the palace with water

[24] Castleden (1990).

Figure 11. The exit of the main drain at the southern end of the palace of Phaestos (photo P.L. Viollet).

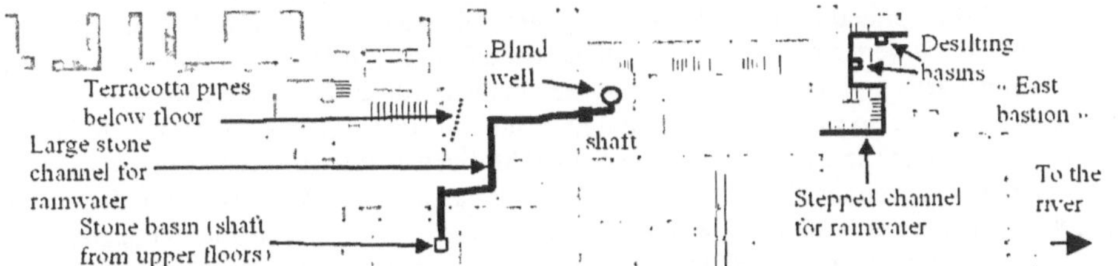

Figure 12. The known hydraulic systems in the eastern side of the palace of Cnossos (L = light-wells).

taken from this canal. There were also in the last state of the palace of Mari other cisterns which were connected to an extended rainfall collection system (Fig. 17).

In cities which were situated away from permanent rivers, there may have been dams and reservoirs for seasonal water storage: we have described earlier the water collection and storage system of Jawa, in the IVth millennium BC (Jawa is approximately contemporary with the end of the Sumerian city of Habuba Kebira, on the middle Euphrates valley).There were many wells in the

Table 2. Summary of sources of water for cities of the early civilizations (4000–1100 BC).

Sources of water	Cities
A short canal connected to a permanent river	Uruk, Ur, Mari, Babylon...(all cities in the Tigris and Euphrates valleys)
Canals and reservoirs storing floodwater of a non-permanent river as well as runoff from rain	Jawa, Khirbet el Umbashi
Gutters and cisterns collecting and storing rainwater	Mari, Cnossos
Underground cistern fed with from source captation or infiltration from the watertable, with stepped stairs to direct access to the water	Mycenae, Athens, Tyrins, Zakros
Wells	Ugarit (Syria), Mohenjo Daro (Indus valley),
Aqueducts or canal(s) from a source in altitude	Cnossos*, Mallia*, Pylos, Thebes, DurUntash (Elam)

* = probable

cities of the Indus civilization (Harappa, Mohenjo Daro), with an astounding density: in Mohenjo Daro, it is assumed that there were about 700 wells inside the city, many of these wells were inside the private houses[25]. In Ugarit, there were many wells, as well, most of them also inside private houses: in a 5000 m^2 investigated area in the southern part of Ugarit, 18 wells have been found; the bottom of some of these wells were dug lower than the phreatic layer, inside the impervious substrate, thus creating a kind of underground cistern with an extra storage, in case where the water table would be low, at the end of the dry season. Occasionally, in Ugarit, a gutter could be used in order to divert rainfall water into those wells[26]. In the Minoan palace of Zakros, there was a large cistern fed by underground water whose watertable was at low depth. In Phaestos, also in Minoan Crete, there were cisterns, but no well has been found[27]. In Mallia, a Minoan site lying in a narrow plain between high mountains and the sea, on the north shore of Crete, the source of water is presently unknown: no well has been found here, and it is questionable whether there have been or not cisterns. But it is known that the plain around Mallia was fed with small canals issuing from cisterns on the high grounds above Mallia[28] (the Lasithi area above Mallia is rich with sources). A similar water supply may well have been used to bring water to the palace and to the city.

In Cnossos, there were wells, and there was also an advanced system of rainwater collection (see Figs. 12 and 18). But a water distribution network probably existed inside the palace of Cnossos, which makes probable the existence of an aqueduct. On the other side of a small brook, south of the palace, there existed a building, the so called "caravanserai", which Arthur Evans (the discoverer of Cnossos) supposed to have been used for lodging travelers and palace visitors; and a strong viaduct upon the brook existed between this caravanserai and the south entrance of the palace. As this caravanserai was obviously provided with running water, for a basin where travellers could wash their feet, and for a drinking fountain, it is possible that the aqueduct, probably made with terracotta elements, which carried water from a source on the Gypsades hill to the caravanserai, could have crossed the bridge in order to feed also the palace with water[29]. It is also possible that the aqueduct distributing water to the palace, from higher grounds, for instance from the Iouktas mountain, a few kilometres SW of the palace, was distinct to the one feeding the caravanserai. According to Graham (1987), Angelakis and Koutsoyiannis (2003), there also existed aqueducts

[25] Janssen (1988).
[26] Calvet & Geyer, 1995; Callot & Yon (1995).
[27] Graham (1987).
[28] Muller (1997).
[29] Graham (1987).

in other Minoan settlements, for instance at a private villa in Tylissos, some kilometres south of Cnossos.

In Mycenae, in continental Greece (the site which gave its name to the Mycenaean civilization), an underground cistern, built by the end of the XIIIth century BC, was fed with water through a 200 m long subterranean conduit dug into the rock, deriving water from the Perseia spring – a terracotta pipe is said to be still visible at the roof of the cistern[30]. In the ancient Mycenaean palace on the acropolis of Athens, there was something similar, as well as in the palace of Tyrins. A 2 km long aqueduct used to bring water to the palace of Pylos (SW of the Peloponnese), and this aqueduct was made partly with U-shaped terracotta elements, partly with wood, and partly as a channel dug in the rock[31]. An aqueduct is also reported in Mycenaean Thebes[32].

5 THE BRONZE AGE TECHNOLOGIES FOR WATER MANAGEMENT AND CONTROL

The technologies for the control and use of water show a lot of common features in all early civilizations of the bronze age.

5.1 *A lifting device: the shaduf*

During the period of the early civilizations, devices for lifting water from a source, a well, or a river, made use only of the human strenght, with simple devices and did not allow any large-scale irrigation[33]. Nevertheless, an important progress came from the invention of the *shaduf*. The *shaduf* is made of a bag suspended at one end of a beam rotating around an axis. At the other end of this beam there is a counterweight. The man operating the *shaduf* pulls down the bag into the water, then lifts the bag full with water, and drops the water from the bag into a small channel or a furrow. The man operating the *shaduf* will use his own weight to compensate the counterweight, and lower the bag down into the water; then, thanks to the counterweight, only a small effort is necessary to lift the bag containing the water. This technique is known in Mesopotamia as early as by the times of Sargon of Akkad (2300 BC), according to a legend regarding the origins of Sargon: as a baby, he is supposed to have been abandoned by his mother in a canal (*naru*), and to have been saved by a man taking up a bag of water from this canal. There is also a *shaduf* represented on a cylindrical seal from Mesopotamia dated by 2200 BC. The *shaduf* probably appeared later in Egypt, as pictures representing the *shadouf* are known inside a tomb of the Ramessid period (1300–1100 BC). The *shaduf* was probably used for watering gardens close to rivers and canals.

5.2 *Dikes*

Dikes are necessary both to gain protection against floods, to protect polders which have been drained from excess of water and to create a canal. There were dikes on the Tigris and Euphrates rivers, and also in Mycenaean Greece. Dikes are called *kalu* or *kisirtum*[34] in Akkadian, which was the dominant language in Mesopotamia after the fall of Ur in 2004 BC. According to the texts, these dikes are made of earth, sometimes reinforced with reeds.

---

[30] Taylour (1983).

[31] Taylour (1983), Platon (1988).

[32] Dickinson (1994).

[33] The chain-pot wheel or *saqqya*, the Archimede's screw, the piston pump, the *noria*, were only invented in the IInd or IIIrd century BC.

[34] Van Soldt (1988). The term *kisirtum* was in use in Mari (see Durand, II, 804).

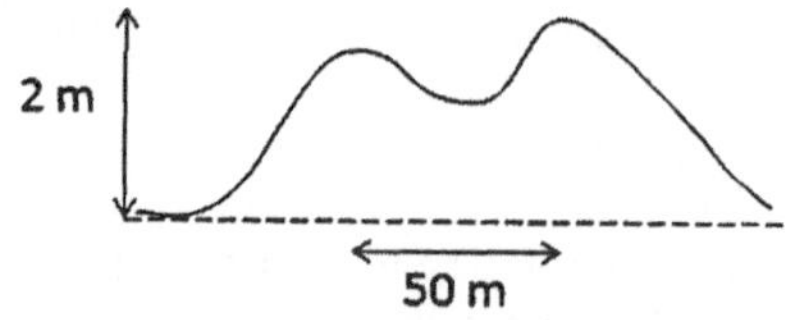

Figure 13. A typical cross-section of the Mari irrigation canal (rakibum) – the canal n° 2 on Figure 3; present state of conservation (after Geyer and Monchambert, 2003). Horizontal and vertical scales are different.

5.3 *Different types of canals*

Canals used to be built for navigation, irrigation, drainage, or in order to divert rivers[35]. They can be derived from either permanent or non-permanent rivers. The lands where canals were built to the highest development in the bronze age are Mesopotamia and Central Asia.

Many texts from Mesopotamia give information on the kind of canals which can be encountered in lower Mesopotamia. The major watercourses, either natural rivers or large canals, are called *id* in Sumerian, or *naru* in Akkadian. Navigation is possible on these *id-naru.* Upon these major watercourses, diversion structures allowed to feed secondary canals. These derived canals are called *pa* in Sumerian, or *namkaru* (and other terms like *atappu, palgu, pattu*) in Akkadian[36]. On these derived canals, one could find sluice gates (*hishtu* in Akkadian) for the regulation of the hydraulic system.

5.4 *Canals for large-scale gravity irrigation*

Lifting devices as the *shaduf* could not allow the large-scale irrigation which was needed for cereals cultivation. Thus, an irrigation canal which is supposed to carry water for a field must be build over the land where this field is. Let us consider a major river or canal (*naru*) flowing with some downward gentle slope, and with land situated above, on one bank of the canal, to be irrigated, it is then necessary to derive a secondary canal (*namkaru*) from the larger river, (*naru*), and then to give to this *namkaru* a downstream slope smaller than the slope of the *naru*; after some distance downstream the valley, the secondary canal is no longer dug in the surrounding ground, but rather built over the ground, with dikes. In the area of Mari, in the middle Euphrates valley, there was on the right bank two successive and independent irrigation systems, each one depending upon such a canal as long as 20 to 30 km, as can be seen from Figure 3. In the ancient texts from Mari, such canals are called *rakibum*, a name which means "the one which rides" (over the land): Figure 13 shows a typical cross-section, showing how massive the dikes used to be. These *rakibum* were not in use all the year long, but only during the irrigation season, as will be explain farther: this explains why there were no villages along their course[37].

Similar irrigation canals derived from large rivers have been also developped at large scale in Central Asia. In Shortughaï, in eastern Bactria (north of modern Afghanistan), a canal 35 km long was built by the middle of the IIIrd millennium in order to irrigate a large terrace which dominates the valley of the Amu Darya river (the ancient Oxus), where an ancient city pertaining to the Indus civilization had been founded. This canal was not derived from the Amu Darya itself, but rather from the Kokcha, which is another river joining the Amu Darya (Fig. 14).

[35] See ie Viollet (2000) for a review.

[36] Steinkeller (1988).

[37] This seems to be well attested for the bronze age period (Geyer & Monchambert, 2003). On the contrary, later in the early islamic period, in the same area, there will be an irrigation canal called the Nahr Saïd, along which there will be permanent human settlements; this has been interpreted as resulting from the differences in irrigation techniques, the irrigation in the islamic period using more lifting machines like chain-pot wheels (Berthier & d'Ont, 1994).

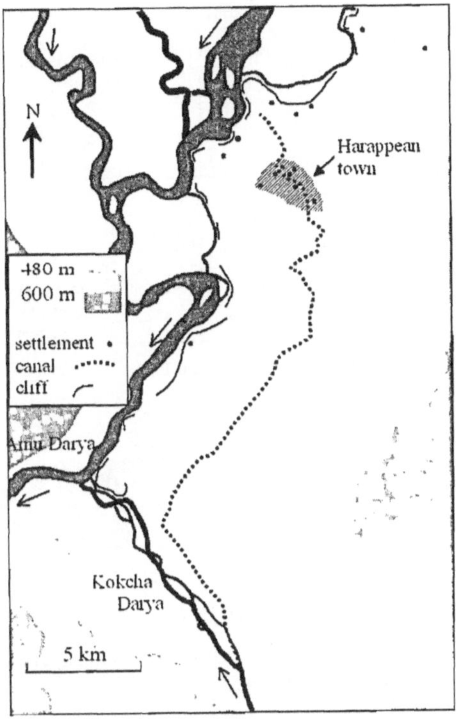

Figure 14. The irrigation canal of the ancient Harapean settlement of Shortughaï in eastern Bactria (north of modern Afghanistan), by 2500 BC. The canal is derived from the Kokcha Darya, an affluent of the Amu Darya, and allowed to irrigate a surface of 6000 ha. Note that the course of the Amu Darya is 20 m lower than the area to be cultivated, so that its water could not be used. This irrigation system will be extended during the further Iron Age and classical periods (under the Persian empire and the later Greek kingdom of Bactria). Adapted from Gardin (1998).

5.5 *Dams*

Dams used to be built in order to divert a river, or to constitute a reservoir. How were the ancient dams built? A very interesting mathematical text of the IInd millennium[38] describes a structure called *kasirum* (which means "blocker"), which is a obviously a dam: this *kasirum* is 120 m long, 30 m wide, and 9 m high. And it is made of two different materials: one third of the structure is made of earth, and two thirds of brick. There are no remains of ancient dams in lower Mesopotamia, because all possible remains would have been taken away for a long time by the powerfull floods of the rivers. But we do have some remains of ancient dams in other regions. The Sadd el Kafara dam of ancient Egypt (dating about 2600 BC) was made of two sides of stones, with an internal core of earth. And the smaller (but older) dam found in Jawa, in the desert area between Modern Jordan and Syria, was made of vertical stone walls with earth beneath (Fig. 15). So probably our *kasirum* was built in the same way, with vertical walls made of brick and an earth-filled core. In continental Greece, there are remains of a number of dams which were built by the Mycenaeans around 1200 BC: these dams are often made of earth supported by walls of big cyclopean stones (Fig. 15).

[38] Powell (1988), p 166.

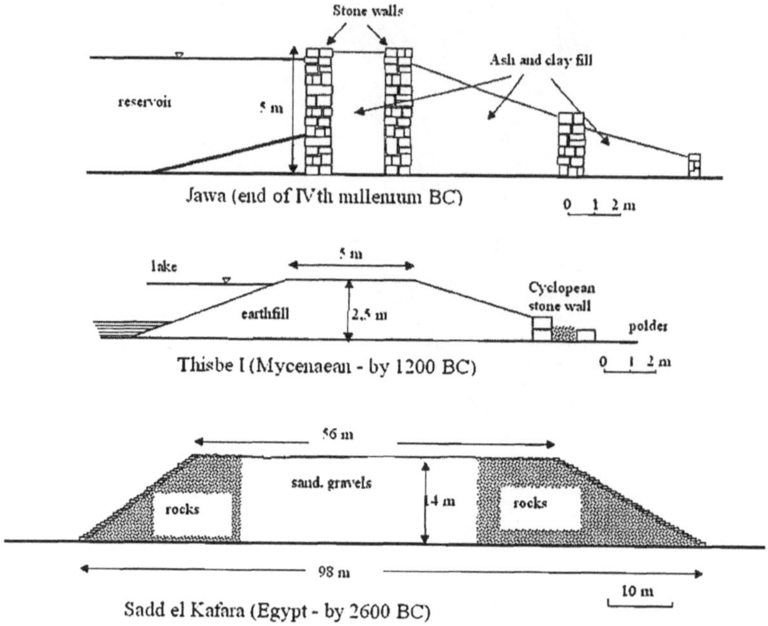

Figure 15. Cross section of some dams built by the early civilizations: Jawa, Thisbe, and Sadd el Kafara (after Helms, 1987, Knauss, 1991 and Garbrecht, 1985).

In Ugarit, on the Syrian Mediterranean shore, there was a dam made of wooden beams between two massive masonry piles, a seasonal storage device which was probably used more for agriculture than as a water supply for the city.

5.6 *Intake structures, structures allowing to feed a secondary canal*

The intake structure allowing to feed with water a secondary canal must ensure regular hydraulic conditions at the intake, raising the water level in the main watercourse. A dam or a weir built across the main watercourse can do this job, but has the disadvantage to be an obstacle for navigation. The techniques which were used in early Antiquity are not all clear, and there are very few archeological remains of these structures. But the existence of intake structures is proofed as early as in the IIIrd millennium BC, in the texts from Ur and Lagash[39]. In the area of Mari, the intake structures of the irrigation canals seemed to be located on the convex side of meanders of the Euphrates, especially in meanders stabilized by geological natural structures[40].

There is a structure which is called *appum* (which means "nose" in Akkadian), known from a mathematical text of the IInd millennium: it is 15 m long, 12 m wide "on its back", and 4,5 m high – its top is 1,5 m above the water surface. It is made for two- thirds with "reed bundle work", while the remaining 1/3 was made of earth. This structure looks to have a triangular horizontal section, and is probably a diversion structure, a reinforced part of a dike at the connection of a secondary canal to a main canal[41]. In a text from Mari, dated from the IInd millennium[42], it is reported how a

[39] Steinkeller (1988).
[40] Geyer & Monchambert, 2003.
[41] Powell (1988), p 167.
[42] Durand (1978), see II, 784.

big stone one day fell from the forepart of an *appum*, and that, consequently, the canal which was derived from this *appum* found itself closed.

5.7 *Gates*

Gates were called *erretum* in Mari. From the texts of the IInd millennium issuing from the palace of Mari, it is known that at that time there used to be many *erretum* on the long Nahr Daourin canal facing Mari, allowing to use this canal to irrigate the fields situated on the left bank of the Euphrates.

In the Akkadian texts of lower Mesopotamia, gates are known as *hishtu*. These *hishtu* may have large dimensions (10 m × 6 m according to a text dated from the IInd millennium!), and may include smaller mobile parts which can be opened and closed. Little is known about the way these gates were built. The only known remains of gates of this period are sluice gates made of two stones with slots in which wooden beams could be inserted to close the gate, or lifted in order to let the water go. Close to Ugarit, were was the dam described above which was exactly designed in this way[43]. There were 16 known sluice gates in the hydraulic system of Jawa (Syria), as early as in the IVth millennium.

5.8 *Irrigation structures for water distribution into the fields: repartitors and furrows*

Thanks to the large number of texts, Mesopotamia is the only area where details are known about the practical use of large-scale gravity-driven irrigation. These texts mention the existence of rectangular reservoirs, built over the plain with earth reinforced with reeds. These reservoirs were called *nag-kud* in Sumerian, in texts of the third millennium, and bear the name *natbatku* in later texts of the end of the IInd millennium in Babylonia. These *nag-kuds* are typically 12 to 72 m long, 1 to 12 m wide, and 1 to 5 m high. As can be seen from these dimensions, the storage capacity of the *nag-kuds* is small, which leads to the assumption that they were operated as repartitors, considering also that the name *nag-kud* means in Sumerian "the one which divides the water". These *nag-kuds* or *natbaktu* were built along the course of the canals, they were equipped with gates, and used to feed very small canals, furrows known as *eg* in Sumerian or *iku* in Akkadian[44]. These furrows (*egs* and *ikus)* used to be built at the top of small earth walls, then they were able to provide irrigation water directly into the fields.

The fields appear to have been rectangular, with access of irrigation water (through an *eg* or *iku*) from one of the smaller sides. These fields are surrounded by earth walls, so that the irrigation water devoted to one particular field could be contained inside the boundaries of the field.

In Mesopotamia, the period for irrigation lied from june to august, which is well after the peak flow of the Rivers (april-may for the Euphrates). Before irrigation, it is necessary to undertake the maintenance works on the canal system: to clear from reeds and mud deposits the secondary canals (*namkaru* and *rapikum*), as well as their intake structures, but also the repartitors *nag-kuds* and the small *eg* or *iku*. From many texts it is shown that the yearly preparation of the irrigation system was a hard task, sometimes requiring a large number of workers for the clearing of a *rapikum* as long as 30 km, as can be read from the texts of Mari. There is a text from Mari describing the works aiming at the opening of the irrigation canals, dating this work of the end of the 4th month of the ancient calendar (august), a period when the Euphrates is almost at its lowest flow. When the irrigation season comes, the intake gates of the secondary canals *rapikum* or *namkaru* are opened, and the gates of the repartitors *nag-kuds* are operated in such a way that every field can be, in its turn, covered with a thin layer of water. After some time, the water covering a particular field is evacuated, and ploughing and sowing is now possible. It is usefull to irrigate again the fields after the grain has germinated, and then two times again during the growing of the plants. The grain are

---

[43] Calvet et Geyer, 1992.

[44] See Steinkeller (1988), for the *nag-kuds*, and Van Soldt (1988) for the *natbaktu*.

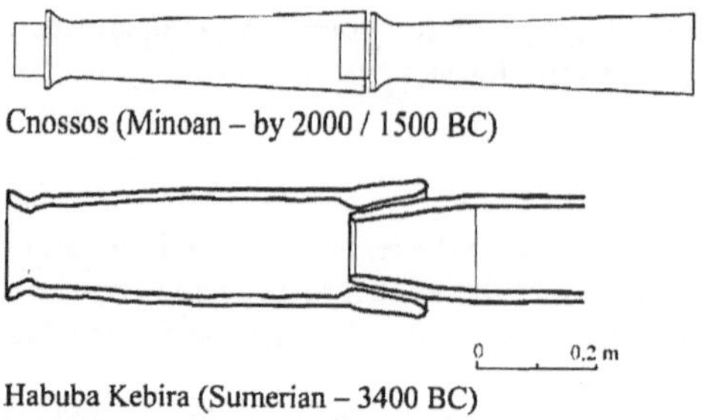

Figure 16. Elements of terracotta pipe from Cnossos(Graham, 1987) and Habuba Kebira (Ludwig, 1977).

collected between february and may[45], and then it is again the time for maintenance and clearing of the irrigation system.

5.9 *Pipes*

Terracotta pipes are known from archeology in Habuba Kebira, a Sumerian settlement on the middle Euphrates valley, by the middle of the IVth millennium BC, and also from the Minoan civilization of Crete by the middle of the IInd millennium, in the palaces of Cnossos, Zakros, and some other Minoan settlements. There were apparently used only for urban drainage and water distribution in cities. These pipes were made of conical elements fitted to each other. The length of these elements are between 50 cm (Habuba Kebira) and 75 cm (Cnossos) each.

5.10 *Aqueducts*

An aqueduct is an artificial channel designed as to bring water to a city. In fact, the concept of an aqueduct is possible only if a source of water is available in altitude, at a reasonable distance from the city: this situation is neither possible in the plain of Mesopotamia nor in Egypt, but only in areas situated at the foot of mountains having permanent sources. In Minoan Crete, aqueducts have been identified at different sites[46]; they were probably made with terracotta tube elements, of the kind shown on Figure 16. Aqueducts made of wood or terracotta elements (U-shaped or maybe pipes made of conical elements) are known from the Mycenaean civilization of continental Greece, in the IInd millennium BC, for instance in Pylos for supplying the palace with water[47].

6 RAINFALL COLLECTION AND CONSERVATION STRUCTURES

Rainfall collection from the flat roofs and terraces involves conduits descending from the roofs, often made of terracotta elements. These vertical conduits lead to gutters either on the ground level or under it. These gutters were either made of baked bricks or of carved stone elements, depending upon the available material on the area (Figs. 17 and 18)[48]. The network constituted by these gutters carry water to cisterns.

Three cisterns devoted to the conservation of rainwater have been found in Mari, they were called *iggum* in the texts of the second millennium BC, and were coated with bitumen; one of these cisterns appears as a pit 3 m high, with a diameter varying from 2,1 to 2,4 m[49].

[45] Van Soldt (1988) for Babylonia.
[46] Angelakis and Koutsoyiannis (2003).
[47] Taylour (1983).
[48] See Viollet (2003) for the description of another rainfall collection network at the "east bastion" in Cnossos.
[49] Margueron (2004).

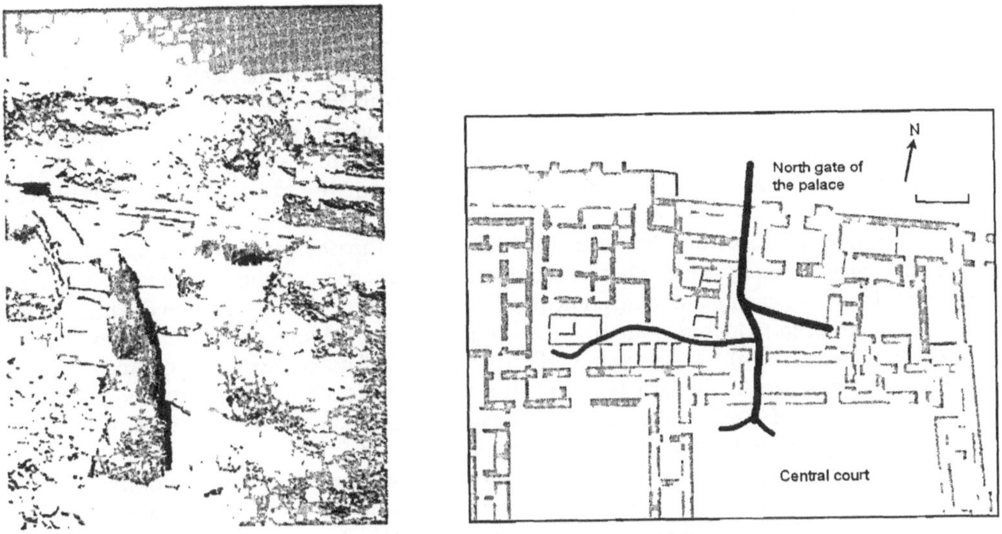

Figure 17. Gutters made of baked bricks for rainfall collection in Mari: photo (left) and sketch of the system which was discovered at the north gate of the palace (see Margueron, 2004, for detailed results).

Figure 18. A conduit in Cnossos made of carved stone elements; the square basin for collecting the water descending from the roofs is also visible (photo P.L. Viollet). See Figure 12 for sketch of the circuit.

7 THE PEOPLE IN CHARGE AND THE LAWS

7.1 *Specialists for hydraulic works and water management*

As we have seen earlier, large hydraulic works have often been decided and sometimes directly controlled by the political leaders themselves (kings, governors). They relied upon different kind of specialists. In Mesopotamia, the scribes were able to calculate the volume of earth to be removed or erected for digging a canal or building a dike. And, as a consequence, the quantity of manpower which should be necessary for one particular work, taking into account the depth from which earth is to be removed: there is a text showing that removing earth from two or three cubits depth (1 cubit = 0,5 m) is called "basket work" and evaluated at a rate of 3 m^3 per day, while digging at a depth smaller than 1 cubit is "through-out work", and evaluated as 6 m^3 per day[50]. Then the salary which would have to be paid to the workers (at the rate of 10 litres of barley per day) can be evaluated.

For the period lying between 2000 and 1600 BC, a lot of mathematical texts are known from Babylonia, many of them appearing as "exercises" for students, and many related to hydraulic structures. This is an example of a theoretical problem: "Cistern: it has 10 *nindan* (60 m) on each side, 10 *nindan* deep (which is quite non-realistic). Its water I drained and with this water to a depth of 1 finger (1,7 cm) how much land did I irrigate?"[51] Another text concerns the case of a village surrounded with a circular canal, and on the outer bank of the canal with a circular dike: "... beyond the ditch I made a dike, one cubit per cubit is the inclination of this dike. What is the base, the top and the height of it? And what is its circumference?"[52]

For the distribution of water through the irrigation network, there comes another different class of specialists. There are "canal inspectors" (*kug-gal*) in Sumerian texts, "irrigation supervisors" (*sekirum*) in Akkadian texts from Mari: their knowledge, which we are not able to qualify, was transmitted from father to son.

7.2 *Hydrological measurements*

The height of water in the major watercourses used to be measured. This is known in the area of Umma, by the time of Sargon of Akkad (end of IIIrd millennium), in a text showing data collected from water height measurements done twice a day, at midday and midnight, during 11 days, on one of the canals of the system. Measurements of the water level during flood is known also in the IInd millennium in the area of Mari. Similar measurements used to be done also in Egypt, with the nilometers on the Nile.

7.3 *Legislation regarding the use of water*

Early laws regarding water management are known from the Ur III and Babylon dynasties of Mesopotamia, with the Ur-Nammu (by 2100 BC) and Hammurabi (by 1800 BC) codes[53]. Those laws consider the cases where a man would have taken the water from another man's field, or when a man would have neglected to maintain the dike of his field, and that water would have caused damages in another man's field or to another man's cultures. In such cases, compensation in cereals would be due to the person who have suffered the damage. Other later texts of Arabia Felix (South Yemen)[54], dated in fact of the iron age, but probably inherited from a practice issuing from the bronze age, show that some areas where water should flow freely during the flood were classified according to a king's decision and that as a consequence neither construction nor cultivation was allowed in these areas.

It is generally considered that in the early civilizations the right to use water was associated with a particular area of land, that this right was indeed regulated because the operation of irrigation was complex and depending upon the use of a complex infrastructure, as we have seen above. Irrigation was under the control of the political power, and was managed through the "irrigation supervisors".

---

[50] Powell (1988), p 164.
[51] Powell (1988), p 162.
[52] Oates (1986), p 184.
[53] See ie Brunn (2000) for a review on water legislation. The integral code of Hammurabi has been published in English (Driver and Miles, 1955) and in French (Finet, 1996).
[54] See Pirenne (1982), and other citations collected in Viollet (2000).

Water in a sense was granted by the kings (or governors). In case the king would like to punish a city, access to water could be denied. But in general the ancient texts show the kings and governors always very anxious to be able to deliver the irrigation water in a proper way, so that there would not be any complaints. The ability to provide water was in a sense a proof of a king's ability.

8 CONCLUSION

The early civilizations of the bronze age developed a very high skill in water engineering using gravity-driven flow, for water-supply, irrigation, navigation, drainage, and protection against floods Those technologies include the control of large rivers, building canals, dikes, dams with composite structures, reservoirs, developing rainwater collection and conservation systems, as well as drains and soakaways in cities, wells and aqueducts. Some of those civilizations deserve the name of "hydraulic civilizations". They disappeared under violence of sank slowly under oblivion, but their hydraulic heritage is still present.

REFERENCES

Adams, Mc, R. & Nissen H. (1972) *The Uruk Countryside*, The University of Chicago Press.

Angelakis, A.N. & Koutsoyiannis, D. (2003) Urban water engineering and management in ancient Greek times, In: Stewrat, B.A. & Howell, T. (eds.), *The Encyclopedia of Water Sciences*. New York, Marcel Dekker. pp. 999–1008.

Angelakis, A.N., Koutsoyiannis, D. & Tchobanoglous, G. (2005) Urban wastewater and stormwater technologies in ancient Greece. *Water Research*, 39(1), 210–220.

Arnaud, D. (1982). La législation de l'eau en Mésopotamie du IIIème au Ier millénaire, *L'Homme et l'eau en Méditerranée et au Proche Orient – II – Aménagements hydrauliques, état et législation*. Maison de l'Orient/Presses Universitaires de Lyon.

Berthier, K. & d'Ont, O. (1994) Le peuplement rural de la moyenne vallée de l'Euphrate à l'époque islamique: Premiers résultats, *Archéologie Islamique*, 4, 153–175.

Bonneau, D. (1993). *Le Régime Administratif de L'eau du Nil dans l'Egypte Grecque, Romaine et Byzantine*. E.J. Brill.

Braemer, F., Echallier, J.C., Taraqji, A. (1996) Khirbet el Umbashi (Syrie): Rapport préliminaire sur les campagnes 1993 et 1994, *Syria*, 73, 117–127.

Brunn, C. (2004) Water legislation in the ancient world. In: Wikander (ed.) *Handbook of Ancient Water Technology*, pp. 537–604, Brill.

Callot, O. & Yon, M. (1995) Urbanisme et architecture, *Le pays d'Ugarit autour de 1200 av JC*. Edition Recherches sur les Civilisations. pp. 155–168.

Calvet, Y. & Geyer, B. (1992) *Barrages Antiques de Syrie*. Maison de l'Orient Méditerranéen.

Calvet, Y. & Geyer, B. (1995) Environnement et ressources en eau dans la région d'Ugarit, *Le pays d'Ugarit autour de 1200 av JC*. Edition Recherches sur les Civilisations. pp. 169–182.

Castleden R. (1990) *Life in Bronze Age Crete*. Routledge.

Charpin, D. (2002) La politique hydraulique des rois paleo-babyloniens, *Annales d'Histoire et Sciences Sociales, Politique et Contrôle de l'Eau dans le Moyen Orient Ancien*, 57(3), 545–559.

Delleur, J.W. (2003) The evolution of Urban Hydrology: Past, present and future, *Journal of Hydraulic Engineering*, August 2003, 563–575.

Dickinson, O. (1994) *The Aegean Bronze Age*. Cambridge University Press.

Durand, J.-M. (1997–98) Documents épistolaires du palais de Mari. Cerf.

Drews, R. (1993) *The End of the Bronze Age*. Princeton University Press.

Driver, G.R. & Miles, J.C. (1955) *The Babylonian Laws*. vol II. Oxford.

Finet, A. (1996) *Le Code de Hammurapi*, Paris, Cerf.

Garbrecht, G. (1985) Sadd el Kafara, the world's oldest large dam. *Water Power and Dam Construction*, July, 71–76.

Gardin, J.-C., (1998) Prospections Archéologiques en Bactriane orientale. vol III. De Boccard.

Gashe, H., Tanret M., Cole, S.W. & Verhoeven, K. (2002) Fleuves du temps et de la vie: permanence et instabilité du réseau fluviatile babylonien entre 2500 et 1500 avant notre ère. *Annales d'Histoire et Sciences Sociales, Politique et Contrôle de l'Eau dans le Moyen Orient Ancien*, 57(3), 531–544.

Geyer, B. & Monchambert, J.-Y. (2003) *La Basse Vallée de l'Euphrate Syrien*. Institut Français du Proche Orient, Bibliothèque archéologique et historique, T. p. 166.

Graham, J.W. (1987) *The Palaces of Crete*. Princeton University Press.

Helms, S. (1987) Paleo-Beduin and Transmigrant Urbanism. In: *Studies in the History and Archeology of Jordan, I*. Department of Antiquities, Amman, pp. 97–113.

Janssen, M. (1988) Mohenjo Daro: architecture et urbanisme. In: *Les Cités Oubliées de l'Indus*. Guimet Museum, Paris.

Kamel, M.W. (1990) L'importance structurelle du qanaye et de la noria en Syrie. In: *Techniques et pratiques traditionnelles en domaine irrigué. Proc. IFAPO Conf. in Damascus*. vol. II. Geuthner. pp. 383–393.

Knauss, J. (1991) Mykenische Talsperren in Arkadien und Beotien. In: *Historische Talsperren 2*. Wittner. pp. 19–41.

Ludwig, W. (1977) Mass, Sitte und Technik des Bauwens in Habuba Kabira Süd, In: Margueron, J. (ed.), *Le Moyen Orient, Zone de Contact et d'Echanges*, Actes du Colloque de Strasbourg. Brill Academic, Leiden. pp. 63–74.

Margueron, J.-C. (2004) *Mari, Métropole de l'Euphrate*. Picard/Editions Recherches sur les Civilisations.

Michailidiou, A. (2005) *Guide of the Palace of Knossos*. Ekdotike Athenon, Athens.

Müller, S. (1996) Prospection archéologique de la plaine de Mallia, *Bulletin de Correspondance Hellénique*, 120(2), 921–928.

Nissen, H.J. (1984) *The Early History of the Ancient Near East*. The University of Chicago Press.

Oates, J. (1986) *Babylon*. Thames & Hudson.

Pirenne, J. (1982) La juridiction de l'eau en Arabie du Sud d'après les inscriptions. In: *L'homme et l'Eau en Méditerranée et au Proche Orient*. Maison de l'Orient Méditerranéen. pp. 81–102.

Platon, N. (1988) *La Civilisation Egéenne*. Albin Michel.

Powell, M.A. (1988) Evidence for agriculture and waterworks in Babylonian mathematical texts. In: *Bulletin on Sumerian Agriculture vol. IV, Irrigation and Cultivation in Mesopotamia, Part I*. Cambridge. pp. 161–172.

Renger, J. (1990) Rivers, watercourses and irrigation ditches and other matters concerning irrigation based on old Babylonian sources (2000–1600 BC). *Bulletin on Sumerian Agriculture vol. V, Irrigation and Cultivation in Mesopotamia, Part II*. Cambridge. pp. 31–46.

Roux, G. (1964) *Ancient Irak*. Allen and Unwin, London.

Schnitter, N. (1994) *A History of Dams – The Useful Pyramids*. Balkema.

Sollberger, E. & Kupper, J.R. (1971) *Inscriptions Royales Sumériennes et Akkadiennes*. Cerf, Paris.

Steinkeller P. (1988) Notes on the irrigation system in third millennium southern Babylonia. In: *Bulletin on Sumerian Agriculture vol IV, Irrigation and Cultivation in Mesopotamia, Part I*. pp. 73–91.

Taylour, W. (1983) *The Mycenaeans*. Thomas & Hudson.

Vallet, R. (1997) Habuba Kebira, ou la naissance de l'urbanisme, *Paléorient*, 22(2), 45–76.

Van Soldt, W. (1988) Irrigation in Kassite Balylonia. In: *Bulletin on Sumerian Agriculture vol IV, Irrigation and Cultivation in Mesopotamia, Part I*. pp. 105–120.

Viollet, P.-L. (2005) *L'hydraulique dans les Civilisations Anciennes*. Presses de l'Ecole Nationale des Ponts et Chaussées, 2000.

Viollet, P.-L. (2006) *Water Engineering in Ancient Civilizations*. (translated into English by F. Holly). International Association for Hydraulic Research.

Viollet, P.L. (2006) The predecessors of European Hydraulic Engineers: Minoans of Crete and Mycenaeans of Greece (2100–1200 BC). *Proc. XXX IAHR Congress*, Aristotle University of Thessaloniki, theme E.

Wilson, A. (2000) Drainage and sanitation. In: Wikander (ed.), *Handbook of Ancient Water Technology*, Brill. pp. 150–182,

Zangger, E. (1994) Landscape changes around Tiryns, *American Journal of Archeology*, 98, 189–212.

CHAPTER 3

Water engineering and management in Ancient Egypt

Larry W. Mays
Arizona State University, Tempe, Arizona, USA

ABSTRACT: Throughout history humans have been fascinated with the Nile River, especially the Egyptian part of the Nile. The birth of this great civilization has been traced back to a time between 11,000 and 10,000 years ago. Around five thousand years ago this civilization started depending entirely on the Nile River and its annual inundation. The fascinating part is that the source of the flow in the Nile and the unknown cause of the annual inundation remained a mystery for the ancient civilizations, even with the level of literacy and highly organized society of these civilizations. Finally in the 19th century explorers resolved the mystery. This chapter traces the history of water engineering and management in ancient Egypt starting with the uses of water from the annual inundation of the Nile River for natural irrigation in the Predynastic period to the management of water through the Hellenistic period.

1 INTRODUCTION

For thousands of years the people of Egypt have owed their very existence to a river that flowed mysteriously and inexplicably out of the greatest and most forbidding desert in the world (Hillel, 1994). Herodotus said that "Egypt is an acquired country, the gift of the River" (Biswas, 1970). The ancient Egyptians depended upon the Nile not only for their livelihoods, but they also considered the Nile to be a deific force of the universe, to be respected and honored if they wanted it to treat them favorably. Its annual rise and fall were likened to the rise and fall of the sun, each cycle equally important to their lives, though both remaining a mystery. Since the Nile sources were unknown up until the 19th century, the Ancient Egyptians believed it to be a part of the great celestial ocean, or the sea that surrounds the whole world. Shown in Figure 1 is Hapi – the Nile God, first shown as one god and then as two gods, portrayed with breasts to show his capacity to nurture.

Margaret A. Murray(1949) in her book, The Splendour That Was Egypt, points out so eloquently that in Egypt are found the first beginnings of material culture (building, agriculture, horticulture,

Figure 1. Hapi – The Nile God First as one god then as two gods.

Table 1. Chronology of Ancient Egypt (after Butzer, 1976 and Strouhal, 1993).

Epi-Paleolithic Period	5000–4000 B.C.
Predynastic Period	5200–3050 B.C.
Early Dynastic Period	3050–2700 B.C.
Dynastic Period	2700–332 B.C.
Old Kingdom (Pyramid Age)	2700–2215 B.C.
First Intermediate, 7th–10th dynasties	2250–2040 B.C.
Middle Kingdom, 11th–12th dynasties	2040–1715 B.C.
Second Intermediate, 13th–17th dynasties	1715–1570 B.C.
New Kingdom, 18th–20th dynasties	1570–1070 B.C.
Late Period, 21st–31st dynasties	1070–332 B.C.
Graeco-Roman Period	332 B.C.–641 A.D.
Macedonian	332–323 B.C.
Ptolemaic	323–30 B.C.
Roman – Byzantine	30 B.C.–641 A.D.
Arab Conquest	A.D. 641

clothing), of sciences (physics, astronomy, medicine, engineering), of the imponderables (law, government, religion). To quote Murray: "The splendour of Egypt was not a mere mushroom growth lasting but a few hundred years. Where Greece and Rome can count their supremacy by the century Egypt counts hers by the millennium, and the remains of that splendour can even now eclipse the remains of any other country in the world."

Robert Payne (1959) in his book, The Canal Builders, states the following: "Plato says the Egyptians looked upon the Greeks as children, too young and innocent to be the creators of great things. The Greeks had no pyramids, no vast administrative buildings like the Labyrinth, no kings as splendid as the Pharaohs, no luxuriant Nile flowing at the foot of the Acropolis." Throughout the discussions in this chapter there will be references to various time periods associated with Egyptian history. A chronological framework for ancient Egypt is given in Table 1. Kitchen (1991) presents a more detailed chronology of ancient Egypt for those interested.

2 THE NILE RIVER

2.1 *Climate*

The configuration of the coastline of the southern edge of the Mediterranean Sea causes the main path of the rainstorms approaching from the west lie outside the deserts of northern Egypt, Sinai, the Negev, and southern Jordan (Issar, 2003). Rainfall scarcity a higher variability from year to year is greater the farther into the desert. Rainfalls are therefore scarce and random. Egypt is the northeastern edge of the Sahara belt with extreme aridity caused by the descent of tropical air masses, which cause them to be hot and dry. The annual average precipitation over most of Egypt is less than 50 mm. Only along the coast does it reach 100 mm.

2.2 *Geography*

The Nile Delta, named because its shape resembles the Greek letter delta, is comprised of an ancient gulf of the sea that has been filled in by river sediment (Hillel, 1994). The Nile Delta, which is also known as Little Egypt, is a giant triangle of land, as shown in Figure 2, and is approximately 200 kilometers wide along the Mediterranean Sea with the apex at Cairo, about 160 kilometers inland. To the south of the delta apex and to the west of the Nile is the Faiyum Depression, which is discussed later.

The Nile River, nearly 6,650 kilometers in total length, is the longest river in the world, draining an estimated 3,350,000 square kilometers. This area is about one-tenth of the African continent

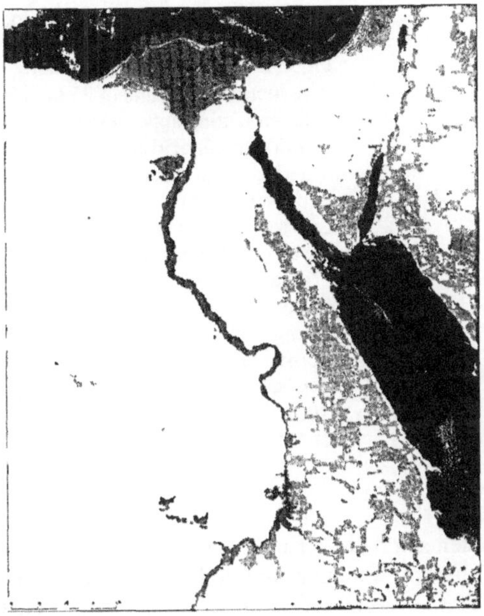

Figure 2. View of the Nile River in Egypt (NASA).

with catchments in nine different countries-Tanzania, Burundi, Rwanda, Zaire, Uganda, Kenya, Sudan, Ethiopia, and Egypt. In contrast the Amazon River has a length of 6,700 kilometers with an estimated drainage area of 7,050,000 kilometers. The annual discharge of the Nile is only 84×10^9 m^3 as compared to $5{,}518 \times 10^9$ m^3 for the Amazon River.

Main sources of the present-day Nile are the Sudan basin and the Ethiopian Highlands. The Atbara River, the Blue Nile River, and the Sobat River are the three main tributaries that emanate from the Ethiopian Highlands. These three rivers are highly seasonal, so that at their confluence with the Nile there is a peak flow to low flow ratio of about 40 to 1 and a peak sediment concentration of about 4000 mg/l in August as compared to only 100 mg/l in June. Tributary sources of the Atbara River, a very seasonal river, are not far from those of the Blue Nile in the Ethiopian high plateau east and west of Lake Tana. The Blue Nile basin covers most of Eithiopia between latitudes 9 and 12° north and west of longitude 40° east. The two major tributaries of the Sobat River are the Baro River and the Pibor River, which receive most of their discharges from the Ethiopian Highlands falling rapidly into the wide plains of the Machar marshes.

2.3 *Hydrology of the Nile River*

The Nile River valley is a seasonally inundated floodplain. The seasonality of flows in the Nile results from 90 percent of the annual rainfall in the Ethiopian Highlands falls between June and October with the peak during July and August. Average rainfall over the Ethiopian Highlands ranges from 1000 to 1400 milliliters/year and in the southwest, where the Baro of the Sobat has it source, the highest rainfall ranges from 1400 to 2200 millimeters/year (Said, 1993). Ideally the Nile would rise (under natural conditions) to bank-full stage by mid August in southern Egypt, then spread out through various overflow channels and/or through breaches of low levees to successful flood basins. The northern most basins of the Nile would be flooded four to six weeks later. Seasons with poor floods would result in many basins being dry or for smaller flood stages only portions of basins would be flooded.

Being one of the most predictable rivers in the world, the Nile flood is seldom sudden or abrupt and is timely, in contrast to the floods of the Tigris and the Euphrates which have more abrupt

Table 2. Nile River flows in Egypt and the lower Nubia during the upper Holocene period.

3000 to 2800 B.C.	Flood levels declined significantly, representing an overall reduction in volume of 25 to 30%. The concomitant down-cutting appears to have initiated the modern flood plain downstream of Wadi Halfa (Bell, 1971).
2250 to 1950 B.C.	A period of catastrophically slow floods.
1950 to 1840 B.C.	Improved floods.
1840 to 1770 B.C.	Excessive floods are documented, reoccurring every 2 to 5 years, with peak discharges three times that of the ten greatest floods of the nineteenth century AD (Bell, 1971)
1770 to 1180 B.C.	Average levels remained high.
1180 to 1130 B.C.	Strong decline in levels.
1130 to 600 A.D.	Normal levels.
600 to 1000 A.D.	Generally high levels.
1000 A.D. to present	Normal levels.

floods. At the Roda Nilometer, south of Cairo, there were 820 floods recorded between the 7th and 15th centuries, of which 73 percent were normal floods, 22 percent were low, and five percent were destructively high (Said, 1993). Normal refers to floods that inundated all basins and subsided at the proper time for sowing of seeds. Average duration of a flood was about 110 days. The beginning of the rise of the Nile begins in June, with the maximum rise of the river usually occurring in the later part of September and the early part of October.

The Nile flood levels for ancient times have been summarized by Butzer (1976). Nile flood levels were substantially higher during the Predynastic times with floods declining during the Old Kingdom and even lower during the First Intermediate Period having catastrophic consequences. During the Middle Kingdom the Nile flood levels rose and were exceptionally high for a while and declined precipitously during the later Ramessid times. During the Saites and Prolemies times the good floods (Nile River flows) seem to have predominated. A more detailed attempt to describe Nile fluctuations from ca. 9000 to 332 B.C. is presented by Said (1993).

Main features of the Nile River flows in Egypt and the lower Nubia during the upper Holocene period have been summarized by Butzer (1980) and is presented by Issar (2003) as shown in Table 2.

The Nile River valley receives alluvial deposits from the Nile in Nubia and Egypt. These deposits consist of sandy and/or gravelly channel beds with silt and clay from the flood waters that spill over the river banks. The Nile has a convex floodplain accumulating primarily from the bank overflow of suspended sediments (silt and clay). This type of floodplain in comparison to a flat floodplain results from shifts and lateral accretion of bedload sediments of sand and gravel within the channels. Convex floodplains have natural levees that constitute the lower channel banks.

On broader floodplains, secondary channels or branches existed. One such example is the Bahr Yusef, which at one time in history flowed into the Faiyum Depression as shown in Figure 3.

2.4 *Theories of the origin and rise of the Nile*

The source of the flow in the Nile and the unknown cause of the annual inundation remained a mystery for the ancient civilizations, even with the level of literacy and highly organized society. Finally in the latter part of the 19th century explorers resolved the mystery.

Theories about the origin of the Nile involved the river traveling underground to lakes. The King of Mauretania, Juba II (20 A.D.) stated that the source of the Nile is in western Mauretania – not far from the ocean. From there it travels underground for several days' to a similar lake in Mauretania Caesaiensis; then it flows underground again for another twenty days' to the source of the Nigris – at the borderline between Africa and Ethiopia. From there it continues under the name Astapus, through Ethiopia (Biswas, 1970).

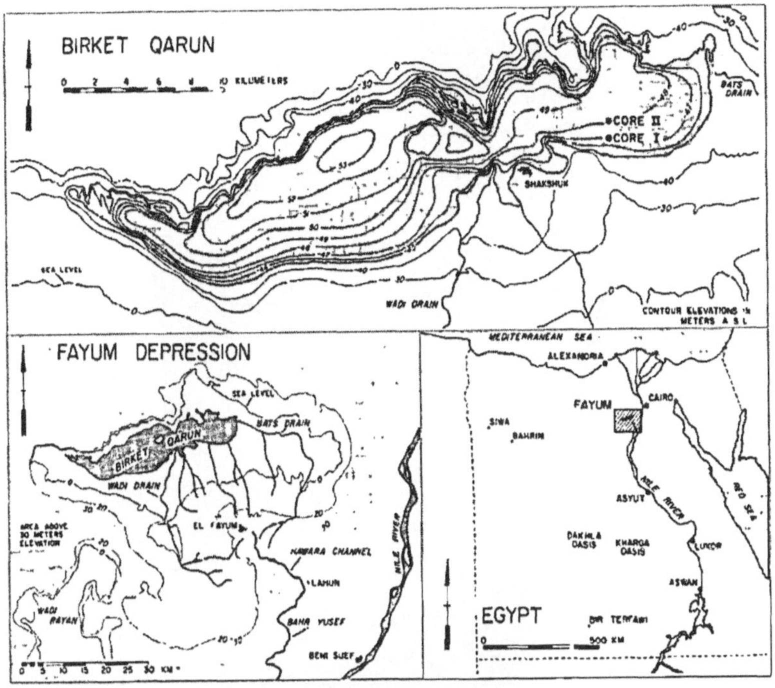

Figure 3. The Faiyum Depression Showing the Birket Qarun (Mehringer, et al., 1978).

The Roman Emperor Nero sent two centurions in 66 A.D. to explore Nubia (present day Sudan) to find the ultimate source of the Nile. As they traveled upstream the journey became more and more difficult. After they traversed the desert they encountered the morass (the Sudd swamps) and were forced to turn back.

Leonardo da Vinci, in the 15th century, remarked that the source of the Nile involves three very high lakes in Ethiopia:

"It issues forth from the Mountains of the Moon from diverse and unknown beginnings; and comes upon the said lakes high above the watery sphere at an altitude of about four thousand braccia, that is a mile and a third, in order to allow for the Nile to fall a braccia in every mile" (Maccurdy, 1956).

2.5 *Agricultural origins in the Nile valley/emergence of a floodplain civilization*

Agricultural origins of the Nile River valley are important to understanding the first water management. The oldest agricultural sites in Egypt (ca. 5200–4000 B.C.) are found in the Faiyum depression in scattered ancient lake settlements and on the western delta margin at Merimde during the Predynastic period. The importance to agriculture is the annual inundation by normal floods.

Radiocarbon dating shows that plant and animal domesticates that dominated ancient Egyptian agriculture first appeared in Egypt several millennia after they did in the Near East indicating that they were derived from Southeast Asia. It is also interesting to note that the spectrum of domesticates was modeled closely on that of Southeast Asia. To replace the hunting-and-gathering economy did early agriculture diffuse to Egypt or was it brought there by invaders from Asia? This is the general argument.

During early civilization in Egypt the population was the most dense in the south (between Aswan and Qift), and in the north (between the Faiyum entrance and the head of the Nile Delta) (Butzer, 1976). Later on the Faiyum became highly populated. The floodplain between the northern

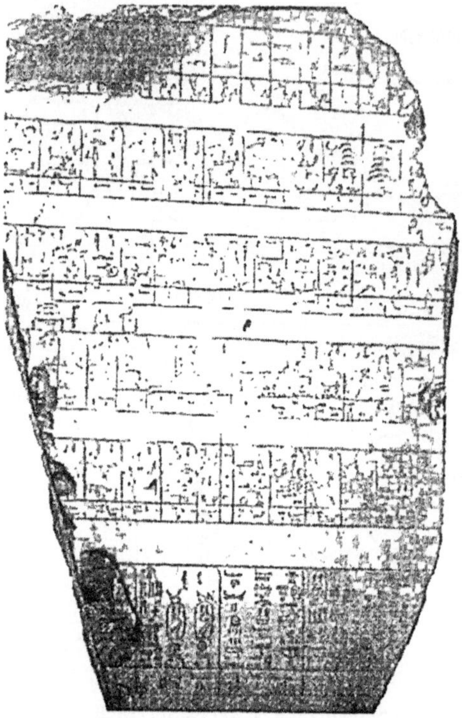

Figure 4. Palermo Stone.

and southern populated area was thinly populated throughout the Dynastic era. Butzer points out that "despite a great number of political and ecological variables, and some local problems of incomplete representation, density was an inverse function of floodplain width." Only in the Coptic times did the population density of the broader floodplain segments increase to relative proportion comparable to those of today (Butzer, 1976). Some three millennia elapsed between the political unification of Egypt and the filling out of the broad, immediate floodplain expanses and of the delta plain. One explanation is that it was much easier to implement artificial irrigation in the far south and on the eastern band of the Nile (Willcock, 1904, Butzer, 1976).

2.6 *Measurement and Long-term records of the Nile*

Because the well-being of Egyptian society relied on the annual flows in the Nile River, the ancient Egyptians developed methodologies to measure and record flood levels. Oldest records of Nile flood levels were carved on a large stone monument during the Dynasty V (2480 B.C.). The Palermo Stone (Fig. 4) is the most valuable surviving fragment of the monument, named after the capital of Sicily were it is located in a museum. The Palermo Stone also records a number of early gods, that copper smelting was taking place, records forty ships that brought wood from and unknown region outside of Egypt.

Nilometers were used to measure the levels of the Nile River. At Karnak, Nile levels were marked on the quay walls of the great temple, dating from about 800 B.C. Three types of nilometers were used: simply marking water levels on the cliff of the river banks; utilizing flights of steps that led down to the river; and using conduits to bring river water to a well or cistern.

The ancient lake sediments and shorelines in the Faiyum Depression in Middle Egypt also provide a record of Nile floods (Mehringer, et al., 1979; Hassan, 1986, 1998). Various lake levels are indicated by former shoreline features and deposits, as shown in Figure 6. These allow the inference of variations of the Nile flood discharge in prehistoric and Pharaonic times.

High-resolution palaeoclimatic data shows evidence of an abrupt global cooling event ca. 2200 B.C. which would have caused a drastic drop in the Nile flood levels (Hassan, 2002). Textual accounts, dating shortly after 2200 B.C., indicate that famines most likely were related to catastrophically low Nile floods.

The science of geometry arose from the need to perform new land measurements after every flood of the Nile. To quote Diadochus: "For the Egyptians had to perform such measurements because the overflow of the Nile would cause the boundary of each person's land to disappear. Furthermore, it should occasion no surprise that the discovery both of this science (geometry) and of other sciences proceed from utility . . . And so, just as accurate knowledge of numbers originated with the Phoenicians through their commerce and their business transactions, so geometry was discovered by the Egyptians for the reason we have indicated" (Cohen and Drabkin, 1948).

3 LAND USE IN ANCIENT EGYPT

There are many surviving sale documents of farm land in ancient Egypt, show that during all periods of ancient Egyptian history part of the land was owned by private individuals (Baer, 1962). The king was the sole owner of the land, however he had the power to bestow property as gifts to members of his family or to others whom he regarded as equals because of their place in society (Kees, 1961).

The main crops of basin irrigation in ancient Egypt included winter crops, cereals (barley, emmer and winter wheat) and flax. Other crops included beans, lentils, and onions. These crops had a growing season that matched the flood cycle of the Nile. Cultivable lands included all lands, which were inundated annually by the Nile River. Reasonably good floods inundated about 2.9 million hectares or 7 million feddans (1 feddan = 1.038 acres = 4,200 square meters) (Said, 1993).

Upper Egypt the area of cultivable floodplain remained constant at about 800,000 hectares (2 million feddans). Expansion of lands in upper Egypt was limited so that only after the introduction of lift irrigation allowed summer cropping on the levees increasing cultivable lands 10 to 15% (Said, 1993).

In the Nile delta cultivable land differed from one time to another depending on how much it was drained and reclaimed. The delta was settled from the earliest of times when the land was used for pasture. Cultivable and pasture land in the Nile delta increased from 800,000 hectares in Predynastic times to 1,000,000 hectares in 1800 B.C. to 1,300,000 hectares in the Ramesside period (1250 B.C.) to 1,600,000 hectares during the Ptolemaic Period ~150 B.C. (Said, 1993).

Total cultivable (including pasture) lands of Egypt increased from around 1,600,000 hectares (3.8 million feddans) in Predynastic time to around 1,700,000 hectares (4.08 million feddans) in 2500 B.C. to 1,800,000 hectares (4.3 million feddans) in 1800 B.C. to 2,200,000 hectares (5.28 million feddans) in 1250 B.C. to 2,700,000 hectares (6.55 million feddans) in 150 B.C. (Butzer, 1976).

The basic administration of the agricultural system remained local, even though the central government imposed a tax on the peasant farmers of about 10-20 percent of their harvest. As Hassan (1997) observed, "Egypt probably survived for so long because production did not depend on a centralized state. The collapse of government or the turnover of dynasties did little to undermine irrigation and agricultural production on the local level."

4 WATER ENGINEERING AND MANAGEMENT: PREDYNASTIC

First actual recorded evidence of water management was the mace head (Fig. 7) of King Scorpion, the last of the Predynastic kings, which has been interpreted as a ceremonial start to breaching the first dyke to allow water to inundate the fields or the ceremonial opening of a new canal (Strouhal,

(a) Photo of Mace-head of King Scorpian (photo courtesy of Henning Fahlbusch)

(b) The lower part of the mace head shows a waterway that bifurcates to two channels (irrigation canals). The king is holding a large hoe while attendants are ready with a fiber basket and broom. Two other workers, with hoes are working on the lower canal. This document has been interpreted as leaving little doubt that the transition from natural to a modified and ultimately artificial irrigation had been completed by the end of the Predynastic period.

Figure 5. Mace-head of the Scorpion King – The first recorded evidence of water management.

1992). Similarly others have interpreted the main part of the mace-head of the king as depicting irrigation work under his supervision. This mace-head indicates that the ancient Egyptians began practicing some form of water management for agriculture about 5,000 years ago.

One of the key unknowns in Egyptian history is the time when people began artificial irrigation, in particular canal systems, consciously as a means to improve the natural effect of the Nile (Strouhal, 1992). Canals allowed the flow of floodwater to locations that could not be reached otherwise, and

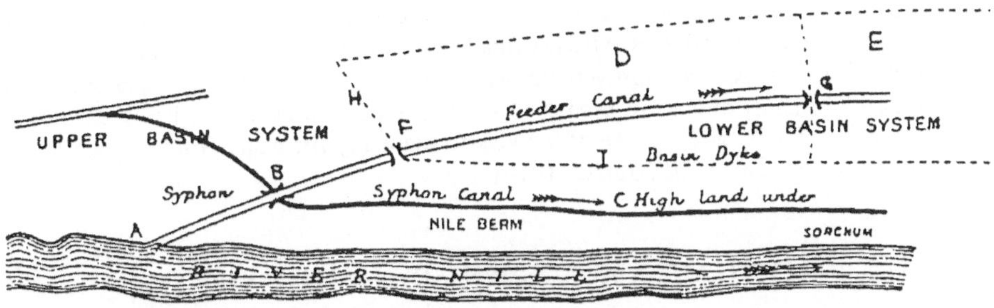

Figure 6. Plan of basin irrigation. (A) head of basin canal, (B) siphon of siphon canal, (C) High land under sorghum, (D, E) basins, (F, G) regulators, (H) transverse dike (modified after Willcocks and Craig, 1913, as presented in Said, 1993).

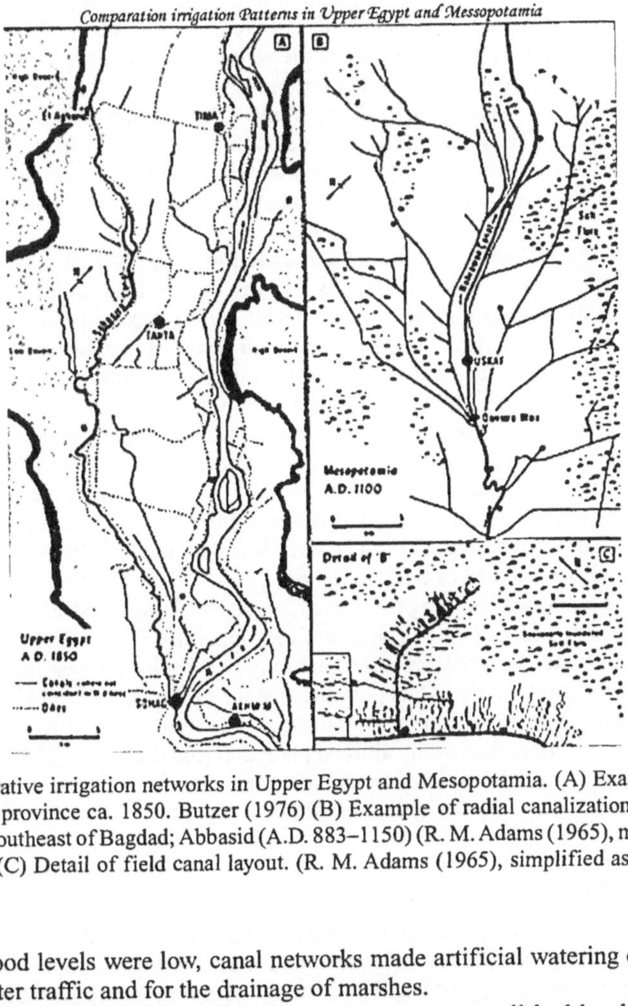

Figure 7. Comparative irrigation networks in Upper Egypt and Mesopotamia. (A) Example of linear, basin irrigation in Sohag province ca. 1850. Butzer (1976) (B) Example of radial canalization system in the lower Nahrawan region, southeast of Bagdad; Abbasid (A.D. 883–1150) (R. M. Adams (1965), modified as presented in Butzer (1976)). (C) Detail of field canal layout. (R. M. Adams (1965), simplified as presented in Butzer (1976)).

when the Nile flood levels were low, canal networks made artificial watering easier. Canals also were built for water traffic and for the drainage of marshes.

The shift from natural to artificial flood irrigation was accomplished by the late Predynastic times. As long as the annual Nile floods were persistently good, the Predynastic population density was not large enough to warrant artificial irrigation (Butzer, 1976). The average Nile flood would allow a single crop season over possible two-thirds of the alluvial surface (Butzer, 1976).

Artificial irrigation increased the area of annual cropland in relation to the floodstage and retained water in the basin after smaller floods. Also artificial irrigation allowed second and even third crops in some basins. The first level of improvement included annual dredging and deepening of natural channels, digging of short ditches to breach lower points of natural levees, using earthen dams to block gathering streams, and the use of buckets to manually lift water from residual ponds or natural channels to adjacent fields (Butzer, 1976).

5 WATER ENGINEERING AND MANAGEMENT: DYNASTIC

5.1 *Artificial basin irrigation*

Artificial irrigation was established by the 1st Dynasty (Butzer, 1976). This included deliberate flooding and draining using sluice gates and water contained by longitudinal and transverse dikes. This form of water management, called basin irrigation, consisted of a network of earthen banks, some parallel to the river and some perpendicular to the river that formed basins of various sizes. Floodwaters were diverted into the basins where the water was allowed to saturate the soil with the remaining water drained off to a down-gradient basin or to a canal. After the draining process was completed in a basin, crops were planted. King Menes, the founder of the first dynasty in 3100 B.C. traditionally has been known as the first to develop a major basin irrigation project. He also dammed the Nile in the vicinity of Memphis to protect the city from flooding and for purposes of defense. Basin irrigation was carried out on a local scale as opposed to being centrally managed on a national scale.

Artificial basin irrigation was based upon the inundation of the Nile floodplain starting in early August. The floodplain was divided into basins ranging in size from 2000 feddans in the upper part of Egypt to 20,000 feddans in the Nile delta (Said, 1993). Figure 6 illustrates the concept of basin irrigation in which the basins were supplied with water by feeder canals. The bed level of the feeder canal was midway between low Nile and ground level with a natural downstream slope less than the slope of the Nile. Each canal supplied water to an average of eight basins arranged in succession. Dikes (levees) separated the basins with controls (masonry regulators) on the earthen embankments to control the flow of water into the basin. Average depths of water in the basins varied according to the local flood volume and stayed in the basins between 40 to 60 days after which the basins were drained (Said, 1993). The basins were very level as a result of the water laden alluvium that deposited in the basins. During years of low flow in the Nile basins were drained into the next downstream basin instead of back to the river.

It is interesting to compare the irrigation used in ancient Egypt and Mesopotamia. Figure 7 compares irrigation networks in Upper Egypt and Mesopotamia. The hydrological characteristics of the Tigris and Euphrates Rivers were such that only the Euphrates was the significant source of irrigation water for most of the Mesopotamia history (Adams, 1981). The Euphrates flows were modest and had a wide variation in flow from month-to-month and from year-to-year. This resulted in much less sustainability of the alluvial plain between the rivers. The Tigris was not used extensively for irrigation until he Hellenistic or early medieval times. In comparison we know that a much different story took place along the Nile River in Egypt.

5.2 *Water management*

Mesopotamia experienced a rapid population growth that lead to greater competition for water, increased labor efficiency, intensified irrigation, and ultimately, state superstructures. This was not Egypt's experience. Instead Egypt had a multi-tiered economy as suggested by the architecture of the Early Dynastic Period with a complex urban social stratification as evidenced from written records of the Old Kingdom (Baer, 1960; Butzer, 1976). Only at the local level would competition for water be an issue. Irrigating in any one natural flood basin did not deprive the downstream basins of their direct access to the Nile River. In the Mesopotamian radial irrigation systems (see

Figure 7), water was artificially regulated at each distribution node, allowing for conflict. In the Egyptian system cooperation was only required within the natural flood basins.

Written regulations did not exist during the Dynastic period, suggesting that water legislation was not very complex and that it was administered locally. The Egyptians codified their civil and criminal laws repeatedly in response to economic and social changes. Because this was not done in the case of water laws argues that water legislation belonged within the oldest oral traditions of common law. This further implies that water legislation accumulated in prehistoric times, prior to the establishment of any centralized political superstructure and did not require any formal modification in later millennia (Butzer, 1976). The local rural population within a basin most likely managed both flood control and irrigation.

5.3 *Lift irrigation*

Improvement of irrigation technology continued through the Dynastic period. The shift to lift irrigation was well under way during the 18th Dynasty and was effective by Roman times (Butzer, 1976). Sometime after 1500 B.C. the ancient Egyptians began lift irrigation with the shaduf (shadouf), already in use in Mesopotamia, is shown in Figure 8 for irrigating small plots. This device allowed the irrigation of crops near river banks and canals during the summer. The shaduf had a bucket and rope attached to the one end of a wooded arm with a counter balance at the other end of the arm. This device typically lifted water up to 1.5 m. One shaduf could irrigate approximately 0.3 acres of land in 12 hours.

5.4 *Sadd-el-Kafara Dam*

The Sadd-el-Kafara dam (Dam of the Pagans) was constructed about 2600 to 2700 B.C. (Garbrecht, 1985). Henning Fahlbusch (personal communication) has confirmed that the dam was constructed around 2650 from his dating studies. This dam was the first attempt at storing water on a large scale (Murray, 1955; Garbrecht, 1985). Possibly older dams include the Jawa reservoir in Jordan and diversion dams on the Kasakh River in the southern part of the former Soviet Union. However these structures were much smaller than the Sadd-el-Kafara dam allowing us to refer to this dam as the world's oldest large-scale dam (Garbrecht, 1985, Schnitter, 1994). This dam was constructed in the Wadi Garawi, seven miles southeast of Helwan (also Heluan) and approximately 30 km south of Cairo on the eastern Nile bank for flood protection of installations in the lower wadi and in the Nile valley (see Figure 9).

This dam was probably still in the construction phase (about eight to ten years) when it failed as a result of a flood catastrophe. There was no channel or tunnel to divert the river around the construction site (Schnitter, 1994). It was another eight centuries before the Egyptians constructed another dam. Sadd el-Kafara dam was discovered in 1885 by the German archaeologist G. Schweinfurth (Smith, 1971).

Dimensions of the dam were 14 m in height and 113 m in crest length with a 0.5 million m^3 storage capacity (Schnitter, 1994). The dam consists of two rock fill sections with the space between them filled with a core of silty sand and gravel (Garbrecht, 1985). The volume of the fill was 87,000 m^3. Revetment (limestone) blocks (around 17,000 according to Schnitter, 1994), of about 50 lbs each arranged in the form of stairs 11 inches high, were used to cover the surface of the rock fill (Murray, 1955). The facing of the dam with the revetment blocks is shown in Figure 10. Smith (1971) estimates the catchment area above the dam to be 72 square miles.

A sag in the structure existed along the top of the dam that diminished the effective height of the dam. The maximum amount of the breach cannot be determined because of the extent of the breach, 44 yards wide, and the sag probably did not occur after the breach because there is no sign of slip in the abutments (Murray, 1955). Possibly the sag was caused by general settlement in the loosely compacted structure and being constructed, as no mortar was used in the dam. Ancient Egyptians did not use mortar as a cementing material but used it as a lubricant in moving heavy blocks and for purposes of leveling.

(a) Shaduf (http://journeyintohistory.com/time/world/AncientWorld/)

(b) Modern day use of lift irrigation (photo courtesy of Henning Fahlbusch)

Figure 8. Illustrations of lift irrigation.

No spillways were provided for the dam indicating that the reservoir was not built for the purposes of irrigation. This is even more evident in the fact that there is no land capable of being cultivated in the Wadi el-Garawi. Because there was no spillway, most likely the reservoir was not to be completely filled. Murray (1955) believed that they never intended to fill the reservoir but merely intended to contain the largest flood for the Wadi el-Garawi. However as Falbusch (2004) points out, it is assumed that the dam was constructed as a flood protection measure, but what was it protecting? In other words the purpose of the dam is still controversial.

Fahlbusch (2004) gives us something to think about, "It seems to be fact that the Sadd el Kafara dam was destroyed before its completion. But was this dam, with its extremely large dimensions really the first of its kind in Egypt?" He also states, "the courage and daring of the master-builders

Figure 9. Sadd el-Kafara – Dam of the Pagans (photo courtesy of Henning Fahlbusch).

Figure 10. Facing of the Sadd el-Kafara – Dam of the Pagans (photo courtesy of Henning Fahlbusch).

are still admirable even after more than 4000 years." Other excellent readings on the Sadd el Kafara dam include Falbusch (1996, 2004).

5.5 *Faiyum depression*

The Faiyum (or Fayum) Depression (Fig. 3), located about 60 kilometers southwest of Cairo, is a huge (1700 km^2), geological depression (below sea level) that begins 20 kilometers west of the Nile Valley, extending into the Western Libyan desert region. A vast saltwater lake (Lake Moeris) was in the heart of this region until the Paleolithic Period. Historically, a natural channel, the Bahr

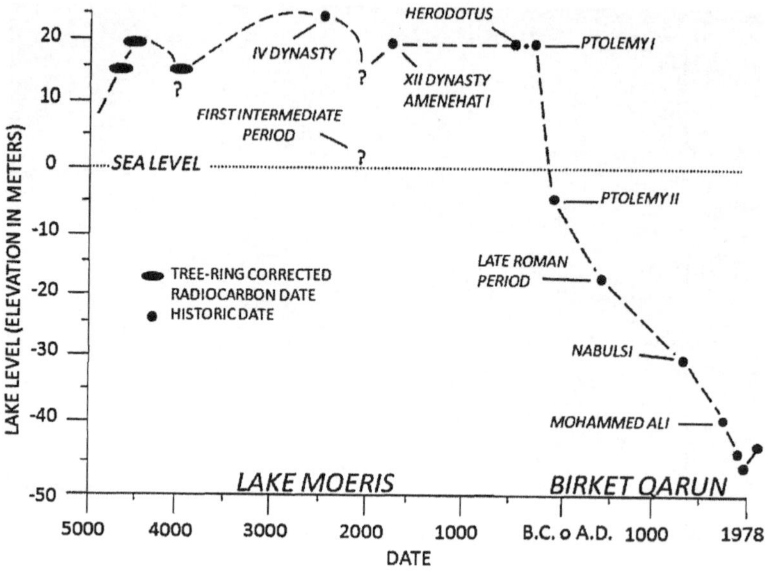

Figure 11. Trends in Lake Moeris levels from 5000 B.C. (Mehringer, et al., 1978).

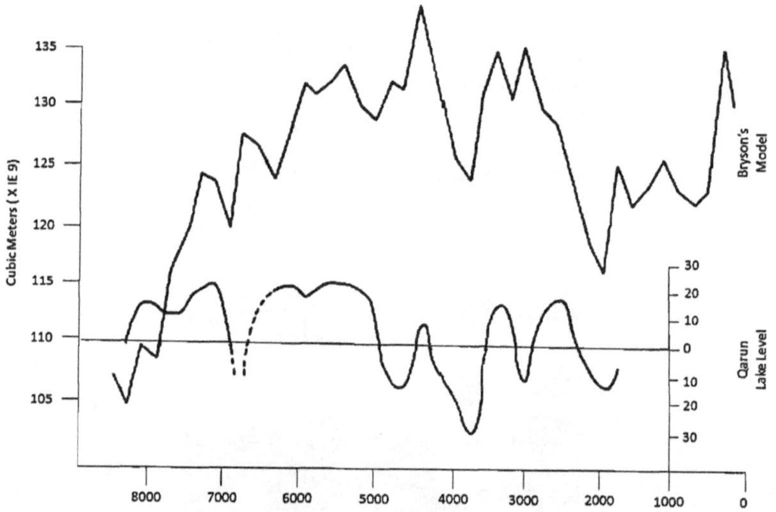

Figure 12. Fluctuations in Nile flow as indicated by variations in the lake level in the Faiyum Depression from palaeoclimatic model by R. Bryson. (Hassan, 1997).

Yusuf, branched off the Nile River about 334 km south of the Faiyum Depression and located along the valley's western escarpment and connected the Faiyum to the Nile River through the Hawara Channel. Figure 11 shows trends in Lake Moeris levels from 5000 B.C. to present. High water levels in the Nile resulted in the formation of a lake within the Faiyum. During the Old Kingdom a permanent lake existed in part of the depression. In the Middle Kingdom the kings directed that the Hawara Channel be cleared to permit excess flood waters from the Nile to enter the depression, sparing the Delta from flooding. After the flood the water drained from the Faiyum back to the Nile. Figure 12 shows fluctuations in Nile flow as indicated by variations in the lake levels of lake levels in the Faiyam Depression and from palaeo-climatic model by R. Bryson (Hassan, 1997).

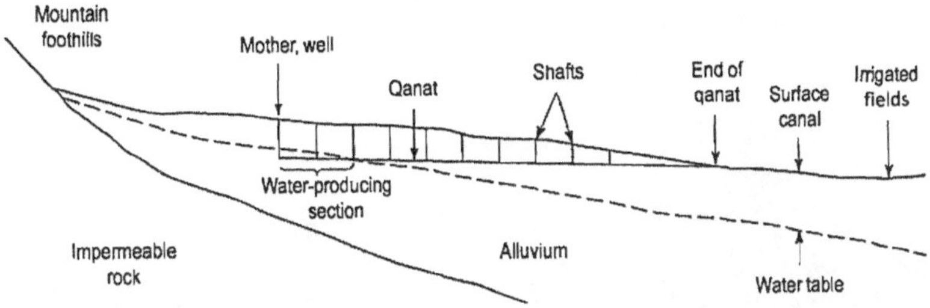

Figure 13. Qanats.

Flood control was no longer deemed necessary by the time of the Ptolemaic kings (Graeco-Roman period) and the Faiyum was exploited for agriculture. The Bahr Yusef was used to convey irrigation water into the depression and then was dispersed by canals across the fields. Drainage water was conveyed to the deepest part of the depression to collect in the Lake Qarun. Prior to the time of the lowering of the lake, the Faiyum Depression was a natural storage for a large portion of the floodwaters that protected lands of Lower Egypt. Reclamation of the Faiyum depression is discussed later in this chapter.

5.6 *Qanats*

A qanat is a subterranean system of tunnels, connecting shafts designed to collect and transport water for distribution as shown in Figure 13. The rate of flow in a qanat is controlled by the elevation of the water table. Qanats have a series of vertical shafts that were used for excavation of the tunnel and provided air circulation and lighting. The oldest qanats have been found in the northern part of Iran and date back to around 3,000 years ago. From 550–331 B.C. the Persian rule extended from the Indus to the Nile, during which time qanat technology spread. Qanats were constructed to the west of Persia from Mesopotamia to the Mediterranean and southward into parts of Egypt. Qanats were also constructed to the east of Persia in Afghanistan, in the Silk Route oases settlements of central Asia and to Chinese Turkistan. The Persians introduced qanats to Egypt around 500 B.C. As this technology transferred to other civilizations, it was known by different names: karez (Afghanistan and Pakistan), kanerjing (China), falaj (United Arab Emirates), and foggara and fughara (North Africa).

6 WATER MANAGEMENT: GRAECO-ROMAN PERIOD

Alexander the Great, King of Macedon, conquered Egypt in 332 B.C, with little resistance from the Persians and was welcomed by the Egyptians as a deliverer. After Alexander's death in 323 B.C. Ptolemy ruled Egypt and in 305 B.C. took the title of King. As Ptolemy I he founded the Ptolemaic dynasty that ruled Egypt for nearly 300 years. Egypt became part of the Roman Empire as the province Aegyptus in 30 B.C. The Roman's interest in Egypt was the reliable delivery of grain to Rome. As a result Rome made no change to the Ptolemaic system of government. The Romans replaced Greeks in the highest offices but used Greeks to staff most of the administrative offices. Greek remained the language of government except at the highest levels. Unlike the Greeks, the Romans did not settle in Egypt in large numbers. Culture, education and civic life largely remained Greek throughout the Roman period. The Persian occupation of Egypt began in 641 with the Arab Conquest, ending the Graeco-Roman Period.

Irrigation of larger plots became possible using the Archimedes screw or tanbur (Fig. 14) and the waterwheel or saqiya (Fig. 15), which were introduced in the Ptolemaic times. The Archimedes screw consists of a water tight cylinder enclosing a chamber walled off by spiral divisions running

Figure 14. Archimedes screw (photo courtesy of Henning Fahlbusch).

Figure 15. Water wheels (photos courtesy of Henning Fahlbusch).

from end to end. Water is lifted by turning the so that water is raised in successive divisions of the spiral chamber from the lowest portion first.

Introduction of the saqiya (or waterwheel) during the early Ptolemaic times revolutionized lift irrigation. This device (Fig. 15) consists of a row of pots attached to the rim of a revolving wheel. The pots when dipped into an irrigation canal fill with water and are then lifted on the wheel to a height the diameter of the wheel (usually 3 m to 6 m). The wheels were turned by oxens.

Lift irrigation coupled with the entrepreneurial system allowed Egyptian agriculture to expand and intensify during the Ptolemaic times. The waterwheel (saqiya) enabled the reclamation of the Fayum depression, which had formed the lake and allowed the storage of high floods since the Middle Kingdom. The spread of the saqiya during the Ptolemic and Roman times led to the introduction of summer crops as well as flood crops. This in turn led to increased wealth of Egypt. The expansion was to a degree unmatched until a century ago, after the introduction of perennial irrigation a century ago (Butzer, 1976).

6.1 *Reclamation of the Faiyum Depression*

The reclamation project of the Faiyum Depression was performed by the Ptolemaic engineers in which Lake Moeris was dried up from a previous level of 20 meters above sea level to about 2 m above sea level during the reign of Ptolemy II (323–285 B.C.). The dried up level of the lake has been surmised by Caton-Thompson and Gardner (1934) from the inferred level of a saqiya well northeast of Birket Qarun. The drying up of the lake during early Ptolemaic times could not have been caused from low flows in the Nile because the flows were adequately high during this time period (Said, 1993).

To lower the lake Ptolemy I constructed an embankment (near Lahun before the first century B.C.) in order to control the flow of water from the Nile into the Hawara channel that flows to the depression. This embankment (dike today measures some 5000 m in length and up to 4 m height, Schnitter, 1994) closed the gap between two hills with the exception a single opening with a dam and a weir at Luhan. The weir was used to keep the level of the lake at 2 m above sea level. The canal system used to channel water from the Nile River into the Faiyum Depression consisted of a radial network of relatively high gradient canals. This canal system was unique as compared to the canal systems used in the Nile valley and delta (Said, 1993).

The reclamation project by the Ptolemaic engineers added approximately 325,000 acres of new and fertile arable land to Egypt (Said, 1993). This project along with the wide-spread use of the waterwheel significantly increased the wealth of Egypt and allowed the population to increase to an estimated 4.9 million people, the largest during the long history of Egypt prior to the nineteenth century. This region has archeological sites dating from the Paleolithic to the late Roman and Christian Periods (circa 8000 B. C. – A.D. 641). Most of the surviving archeological remains date to the Ptolemaic and Roman Periods.

7 TODAY'S EGYPT AND THE NILE

Today the hydrology of the Nile plays a major role in the economy and politics of Egypt (Waterbury, 1979, Said, 1993). Irrigated agriculture is still an integral part of the Egyptian way of life. They have continued to use traditional methods handed down through the centuries, including the ancient methods of irrigation, organic manure, and crop rotation. Egypt is still an agricultural country, which survives on water from the Nile. Their history of irrigation and agriculture has made Egypt rich in knowledge. The courage of past Egyptian generations has become the courage and wisdom of the present generation and hopefully for future generations.

7.1 *Aswan high dam*

The first diversion dams (barrages) on the Nile were completed in 1861 as a means of raising water levels to allow for irrigation and navigation. The first dam at Aswan was built in 1902 about 600 miles from Cairo. Construction of the High Aswan Dam began in 1959 and was completed in 1970. The dam was built as a source of hydroelectric power, to provide water for irrigation, and to protect the crops and the people living in the areas downstream. The dam rises 364 feet above the river and is 12,562 feet long along the top of the dam. The Nile River banks today (Fig. 16) are much different looking as compared to those of ancient Egypt. Approximately 97 percent of the present day population lives on about 2.5 percent of the land along the Nile.

Figure 16. The Nile River today in Cairo. (photo copyright by L. W. Mays).

The Aswan High Dam is one of the most controversial of the existing big dams in the world. Economic benefits of the dam have never been in doubt and the dam has been important in Egypt's economic survival. However, the construction of the dam was accompanied with many side effects that are still controversial. These side effects include channel degradation, silt deprivation, dune accumulation, coastal erosion, increased use of pesticides and fertilizers rise of water table and problems of drainage, and changes in water quality. These were problems never faced by the ancient Egyptians. The dam prevents sediment from flowing downstream to the fields and to the Mediterranean Sea as was the natural course. Changes in water quality downstream from the dam include the drop in turbidity, increase in total dissolved solids, higher count of undesirable algae, taste problems, increased density of phytoplankton (Said, 1993). The downstream river is becoming a receptacle of domestic, industrial, and agricultural wastes, with conditions in the delta being even worse because of the reduced velocity of the river, concentration of industrial plants and more intense agriculture. Deterioration of the river has affected the fish population in the downstream river. The real question is, are these mega projects with their environmental consequences sustainable for the future?

7.2 *New projects with Nile River water*

The annual flooding of the Nile has continued into modern times. With the completion of the high Aswan dam in 1988 the flooding is now controllable. Today two major projects include the North Sinai Agricultural Development Project and the Nile River Barrage. The North Sinai Agricultural Development Project (NSADP) is a huge land reclamation project in the North Sinai desert. Flows from the Nile River are combined with irrigation return flows and then transported by the El Salaam (Peace) Canal, under the Suez to the North Sinai for agriculture (see Figure 17). This project has been a major relocation effort to develop agriculture in the North Sinai. Figure 18 shows crops in the north Sinai irrigated by water transported by the El Salaam (Peace) Canal of the North Sinai Agricultural Development Project. How long will this project, which is combining saline irrigation return flows with Nile River flows to irrigate lands in the Sinai desert, be sustainable? This question has not been answered.

The Nile River barrage is located at Naga Hammadi (140 km north of Luxor) in Upper Egypt to divert water for agriculture. The new 330 m wide dam is being constructed some 3,500 m

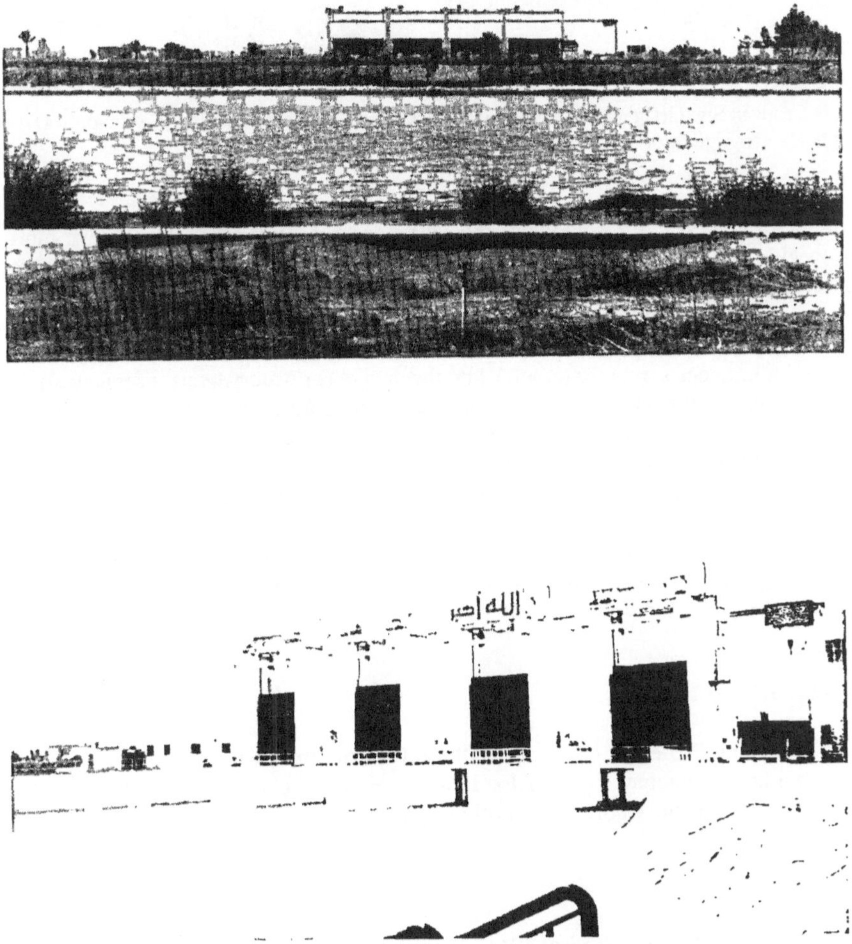

Figure 17. Photos showing location of El Salaam siphon under the Suez Canal El Salaam transfer of water under Suez Canal. (photos copyright by L. W. Mays).

downstream of the existing structure – the reservoir is used to feed a large-scale agricultural irrigation system. The new Naga Hammadi barrage also consists of a 64 MW hydro-electric plant.

8 CONCLUSIONS

Fundamental interrelationships between humans and their environment in the Egyptian floodplain greatly influenced the evolution of land use patterns, the development of irrigation, and the spatial distribution of settlements. The Nile River valley is a seasonally inundated floodplain of the Nile, and being one of the most predictable rivers in the world, the Nile flood is seldom sudden or abrupt

Figure 18. Crops in Sinai irrigated by El Salaam (Peace) Canal of the North Sinai Agricultural Development Project. (photos copyright by L. W. Mays).

and is timely, in contrast to the floods of the Tigris and the Euphrates, which have more abrupt floods. Low Nile floods cause famines. Because of the unpredictability of floods and the production of grains suggest order and stability.

Throughout history the advancements of irrigation of the Nile, starting from natural irrigation and advancing to artificial irrigation and then the development of lift irrigation with the shaduf and then the Archimedes screw (or tanbur), and the saqiya (or waterwheel). From a water management perspective, all evidence known suggests that flood control and irrigation, at the social and administrative levels, were managed locally by the rural population within a basin.

The rise and sustainability of Egypt, with so many great achievements, was based primarily on the cultivating of grain on the Nile River floodplain, without a centralized management of irrigation. What is so unique is that Egypt probably survived for so long because production did not depend on a centralized state. Collapses of the government and changes of dynasties did not undermine irrigation and agricultural production on the local level. "The secret of Egyptian civilization was that it never lost sight of the past" (Hassan (1998)).

ACKNOWLEDGEMENTS

I would like to thank Professor Henning Fahlbusch for the use of his photographs and the great discussions we had on ancient Egypt. Also I have great appreciation for the opportunities that I have received from Enrique Cabrera to travel to Spain.

REFERENCES

Adams, R.M. (1965) *Land Behind Bagdad: A History of Settlement on the Diyala Plains*. Chicago, University of Chicago Press.

Adams, R.M. (1981) *Heartland of Cities: Surveys of Ancient Settlement and Land Use on the Central Floodplain of the Euphrates*. Chicago, University of Chicago Press.

Baer, K. (1960) *Rank and Title in the Old Kingdom: The Structure of the Egyptian Administration in the 5th and 6th Dynasties*. Chicago, University of Chicago Press.

Baer, K. (1962) The low price of land in Ancient Egypt. *Journal of American Research Center in Egypt*, 1, 25–45.

Bell, B. (1970) The oldest records of the Nile Floods. *Geographical Journal*, 136, 569–573.

Bell, B. (1971) The Dark Ages in ancient history. I. The first Dark Age in Egypt. *American Journal of Archaeology*, 75, 1–26.

Biswas, A.K. (1970) *History of Hydrology*. Amsterdam, North-Holland Publishing Company.

Butzer, K.W. (1976) *Early Hydraulic Civilization in Egypt: A Study in Cultural Ecology*. Chicago, The University of Chicago Press.

Butzer, K.W. (1980) Pleistone history of the Nile Valley in Egypt and Lower Nubia. In: Williams, M.A.J. & Faure, H. (eds.), *The Sahara and the Nile*. Rotterdam, Balkema. pp. 253–280.

Caton-Thompson, G. & Gardner, E.W. (1934) *The Desert Fayum*. Two volumes. London, Royal Anthropological Institute.

Cohen, M.R. & Drabkin, I.E. (1948) *A Source Book in Greek Science*, Cambridge, Harvard University Press.

Drowser, M.S. (1954) Water-supply, irrigation, and agriculture. In: *A History of Technology*, Singer, C., Holmyard, E.J. & Hall, A.R. (eds.), New York, Oxford University Press.

Fahlbusch, H. (1996) Ancient Dams in Egypt. *Transactions of the 16th Congress on Irrigation and Drainage*. Vol. 1G, New Delhi.

Fahlbusch, H. (2000) Water in human life: technical innovations in hydraulic engineering in the last 5000 years. *ICID Journal*, 49(4), 1–10.

Fahlbusch, H. (2004) Men of Dikes and Canals. In: Hans-Dieter Bienert & Jutta Haser (eds.), *Men of Dikes and Canal: The Archaeology of Water in the Middle East, International Symposium held at Petra, Wadi Musa* (H. K. of Jordan), 15–20 June 1999), Rahden/Westf, Verlag Marie Leidorf GmbH, pp. 1–16.

Fahlbusch, H. (2004) The Sadd el Kafara – The oldest high dam of the world. In: Hans-Dieter Bienert & Jutta Haser (eds.), (*International Symposium held at Petra, Wadi Musa* (H. K. of Jordan), June 15–20, 1999), *Men of Dikes and Canal: The Archaeology of Water in the Middle East*, Verlag Marie Leidorf GmbH, Rahden/Westf, pp. 365–368,

Garbrecht, G. (1985) Sadd-el-Kafara: The world's oldest large dam. *Water Power & Dam Construction*, July, 71–76.

Hamdan (1961) Evolution of irrigation agriculture in Egypt. *Arid Zone Research*, 17, 119–142.

Hassan, F.A. (1986) Holocene lakes and prehistoric settlements of the Western Faiyum, Egypt. *Journal of Archaeological Science*, 13, 483–501.

Hassan, F.A. (1997) The dynamics of a riverine civilization: a geoarchaelogical perspective on the Nile valley, Egypt. *World Archeology*, 29(1).

Hassan, F.A. (1998) Climatic change, Nile floods, and civilization. *Nature and Resources*, 32(2), 34–40.

Hassan, F.A. (2002) The collapse of the old kingdom: low floods, famines, and anarchy. *Conference on Environmental Catastrophes and Recoveries in the Holocene*, UK, Department of Geography and Earth Sciences, Brunel University, August 29–September 2.

Hillel, D. (1994) *Rivers of Eden: The Struggle for Water and the Quest for Peace in the Middle East*. Oxford, Oxford University Press.

Hughes, J.D. (1992) Sustainable agriculture in ancient Egypt. *Agriculture History*, 66, 12–22.

Issar, A.S. (2003) *Climate Changes during the Holocene and Their Impact on Hydrological Systems, International Hydrology Series*. Cambridge, UNESCO, Cambridge University Press.

Kees, H. (1961) *Ancient Egypt: A Cultural Topography*. The University of Chicago Press.

Kitchen, K.A. (1991) The chronology of Ancient Egypt. *World Archaeology*, 23(2), 201–208.

Lightfoot, D.R. (1997) QANATS in Levant: hydraulic technology at the periphery of early empires. *Technology and Culture*, 38(2).

Maccurdy, E. (1997) *The Notebooks of Leonardo da Vinci*, Vol. 2, new edition, London, Jonathan Cape.

Mark, S. (1998) *From Egypt to Mesopotamia*. College Station, Texas A&M University Press.

Mehringer, Jr., P.J., Peterson K.L. & Hassan F.A. 1979. A pollen record from Birket Qarun and the recent history of the Faiyum Egypt. *Quaternary Research*, 11, 238–256.

Murray, G.W. (1955) Water from the desert: some ancient Egyptian achievements. *Geographical Journal*, 121, 171–187.

Murray, M. (1949) *The Splendour That Was Egypt*. New York, Philosophical Library.

Payne, R. (1959) *The Canal Builders*. New York, The Macmillan Company.

Postel, S. (1999) *Pillar of Sand: Can the Irrigation Miracle Last?* New York, W.W. Norton Company (A Worldwatch Book).

Redman, C.L. (1978) *The Rise of Civilization: From Early Farmers to Urban Society in the Ancient Near East*. San Francisco, Freeman.

Said, R. (1993) The River Nile: *Geology, Hydrology and Utilization*. New York, Pergamon Press.

Schiller, E.J. (1992) *Sustainable Water Resources Management in Arid Countries: Middle East and North Africa, Institute for International Development and Cooperation, University of Ottawa*. Canada, Ottawa, Ontario.

Schnitter, N.J. (1994) *A History of Dams, The Useful Pyramids*, Rotterdam, The Netherlands, A.A. Balkema.

Smith, N. (1971) *A History of Dams*. London, Peter Davies.

Strouhal, E. (1992) *Life in Ancient Egypt*. Cambridge, Cambridge University Press.

Waterbury, (1979) *Hydropolitics of the Nile Valley*. Syracuse, Syracuse University Press.

Willcocks, W. & Craig, J.I. (1913) *Egyptian Irrigation*, London, E. & F.N. Spon Ltd.

CHAPTER 4

Water engineering and management in the classic civilizations

Henning Fahlbusch
Senior Professor, University of Applied Sciences, Lübeck, Germany

1 INTRODUCTION – POLITICAL SITUATION, ATTITUDE TO THE WATER AND NECESSITY OF TRANSFERS

The "classical age" is defined as the time between the Archaic and Roman epoch. The political situation in this phase was characterized in the Greek world (geographically mainly Greece and Asia Minor) by wars between the various city states themselves, the Persian wars and, after the death of Alexander the Great, the power struggle between his successors. West of this region the Roman Empire slowly grew. It developed to become the strongest power in the Mediterranean region after the final defeat of Carthage in 146 B.C. and then step by step conquered the whole Greek World. Nearly no internal wars took place, when the Roman Empire flourished in the 1st and 2nd cent. A.D. It was the so-called *Pax Romana*.

These political conditions influenced the planning and construction of water installations, which mainly were water supply systems, as under the Mediterranean climate no large scale irrigation seemed to be necessary at that time. However, a sufficient water supply was the backbone of each city. When the water demand exceeded the locally available water resources, Greek people consequently installed underground aqueducts. Due to political and military considerations, this avoided detection by an enemy. On the other hand, Roman engineers after the final defeat of Carthage constructed huge bridges for the transport of water above valleys. These bridges signalled the power of the Empire and showed that no enemy had to be feared.

As well as the political conditions, the personal attitude of the population influenced the construction of water supply systems, namely of aqueducts. Greek people often worshipped water. Thorough care for the water and proper maintenance of the various structures characterized their behaviour. As they believed that diseases were spread by air and water they tried to avoid the contact of water with the open air. That resulted in closed pipe-lines for the aqueducts. Nobody could enter the conduits and thus contaminate the water.

The attitude of the Romans was different. Their approach was much more pragmatic. They chose canals for the aqueducts, which were also covered, either by vaults or by slabs. In these comparatively big cross-sections large quantities of water could flow. However, people had to walk in these canals for repair or maintenance purposes, an unthinkable situation for the Greeks.

When founding a settlement in antiquity a sufficient local water supply from springs, wells, rivers, or lakes will probably have existed. People in antiquity looked additionally to the quality of water. Vitruvius[1] describes the methods to judge it in detail. Crouch (1993) pointed out that the Greek settled preferably at places with carstic geology[2]. And it can be shown at the example of Cologne that the Romans also preferred water with dissolved lime. In Cologne water could have been used originating from the catchment area of the river Erft. An appropriate aqueduct would have had a length of about 40 km only. Instead they took water with dissolved lime from the catchment

[1] See book VIII, c. I.
[2] Page 63 ff.

area of the river Urft in the Eifel mountains, which required an aqueduct of about 100 km in length. This obviously was done only because of the water quality (Haberey 1975)[3].

However, since archaic times the increase of the water demand due to the enlarged population resulted often in measures to extend the available water by means of either temporal or local transfers. Temporal transfer meant to store water when the supply exceeded the demand, i.e. the rainy season, in order to use it in the dry season. For local transfer aqueducts had to be built from the source of the water to the place of its consumption. Naturally both transfer methods could also be combined.

In the following section the planning and construction principles for hydraulic structures will be explained for Greek systems, as far as possible using the example of the development of Pergamum's water supply system. This city was chosen because nearly all elements of the Greek supply systems had been applied here and as the development of the system has been well researched and documented. Only the distribution system inside the city will be explained at the example of Priene, as it is not known in detail in Pergamum.

It will be shown that with growing water demand first resources in the vicinity of the capital were used. Than with still increasing population and wealth water resources far away were recruited into the system until in Roman times, in the 2nd cent. A.D., when the city was at the peak of its development, two additional big canals supplied it.

The various elements of Roman systems are afterwards explained in detail at examples from many different places, because they have only partly been applied in Pergamum.

2 PERGAMUM

Pergamum was a city near the western coast of Turkey at the Barkirçay river of today, the ancient Kaikos (Radt 1999)[4]. The castle-mountain has an elevation of about 340 m above mean sea level (MSL). It is bordered by the rivers Ketios in the east, the Kaikos in the south and the Selinos in the west. To the north it is connected to the mountains by a ridge whose deepest point is nearly 200 m below the acropolis.

The empire of Pergamum was founded at the beginning of the 3rd century B.C. by Philetairos, who defended the war treasure of Alexander the Great (900 talents of silver – about 180 t) in his castle. After Alexander's death and the defeat of Lysimachos (Alexander's successor in Asia Minor) he created the Attalids-dynasty. Based on the enormous treasure of money, the new empire grew quickly. Pergamum became a centre of arts and science. According to his testament, Attalos III., the last king, left his Empire in future to Rome. Thus it became the province "Asia" in the Roman Empire. Pergamum joined the rebellion of Mithradates against Rome in the 1st cent. B.C., and this resulted in a certain decrease of its wealth and economic power. This decline was additional supported by an outbreak of the plague. A slow recovery did not begin until the end of the 1st cent. B.C. Then, during the *pax Romana* Pergamum reached its peak of development.

The development of the city started from the castle on top of the acropolis. With the increase of the population the inhabited area was enlarged down the slopes of the mountain. Already in Hellenistic times people settled down west of the Selinus river. When the capital reached its peak in the 2nd cent. A.D. probably most of the estimated more than 100.000 inhabitants lived within reach of the modern town of Bergama, i.e. at the southern foot of the castle-mountain, respectively in the northern part of the Kaikos-plain.

The slow decrease of power and wealth began with the earthquake of 178 A.D., which most of all destroyed Smyrna. This decline occurred in parallel with the slow decrease of the power of the Roman Empire. In Byzantine times it was still a relative strong town which more and more reduced

[3] Haberey explained this in a personal communication.

[4] Page 17 ff.

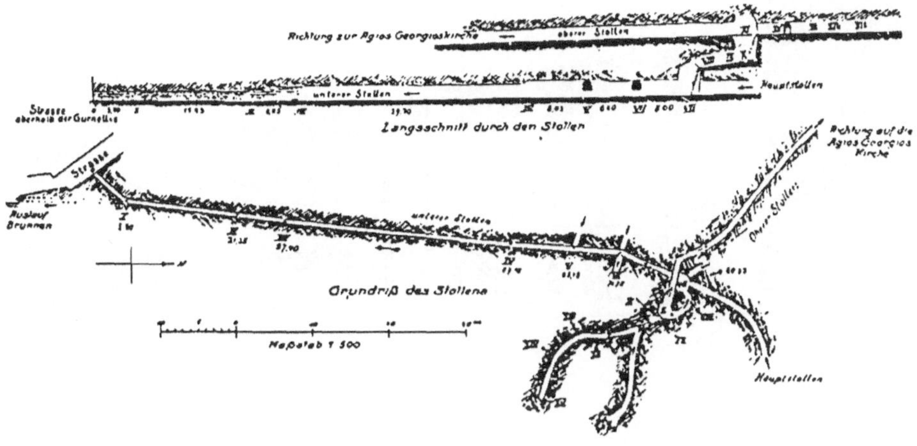

Figure 1. Plan and longitudinal section of the infiltration tunnel at the Gurnellia in Pergamum (Gräber, 1913).

its area and crept again to the top of the mountain in order to fight against Semitic and Persian enemies.

In this paper the development of the water supply system will be explained from the early beginning until the peak of the development. The necessary structures will be explained in the following.

3 LOCAL RESOURCES

3.1 *Fountain-houses: Infiltration-tunnel at the "Gurnellia" and "Agios Stratigos"*

At the Gurnellia, a large square at the southern slope of the acropolis, a fountain-house still existed at the beginning of the 20th cent. A.D. (Gräber 1913)[5], but it is unknown what it looked like and whether this house was an ancient construction.

Gräber described, that the water discharged from a tunnel. At this building especially the method can be shown, how the infiltration was increased in order to collect the maximum quantity of cool, fresh water. Originally the water would have seeped to the surface of the slope probably in a coarse artery or a cleft in the rock. In order to increase the seepage the small cleft was enlarged and finally a tunnel was constructed. Its length into the mountain was more than 100 m. Furthermore this tunnel was extended to another such structure above the first one (Fig. 1). It is a pity that the tunnel is not accessible any longer and that there is no information about the discharge and its development. However the building shows that people understood the principle of groundwater-movement and how to handle and use it.

The "Agios Stratigos" fountain-house (Fig. 2) is a typical example of such local buildings, which have been constructed in many cities of the Greek world like Corinth or Athens (Glaser 1983), to mention just a few. It was constructed at the eastern foot of the castle-mountain in the valley of the river Ketios. There also, little water would have seeped out of a coarse cleft, probably. People realized this and enlarged this cleft to a small infiltration-tunnel, thus increasing the yield into the tunnel. But its extension is unknown, like in the structure at the Gurnellia. At the outside opening of the tunnel in the valley a basin 6.32 m long and 1.6 m wide was constructed, in which the fresh water was collected. From here it could be conveyed to the various parts of the city.

[5] Page 410 ff.

Figure 2. Fountain-house "Agios Stratigos".

Four Doric columns bore a roof above the basin in order to protect the basin and the water in it. As the fountain-house would have disappeared in the lake behind a newly constructed dam, it was demolished and reconstructed besides the road to the top of the acropolis, where it can be visited today.

3.2 *Cisterns*

In the castle area on top of the acropolis there is neither a spring nor a deep well. In order to guarantee the water supply for the inhabitants, especially in case of a siege, cisterns had been constructed to collect rain water in the rainy winter season. These small scale storage structures were hewn into the rock. They were mostly pear-shaped (Fig. 3). At least one layer of hydraulic plaster prevented water losses through the bottom and walls.

The size of the cisterns varied from less than 10 m^3 up to more than 90 m^3 (Table 1). Assuming the same distribution of precipitation in antiquity as today and a daily water demand per person of 8 l in winter and 12 l in summer, Garbrecht (2001) calculated that about 7900 people could have been supplied by the cisterns in case of a siege for a whole year[6].

A cover of the cistern's mouth prevented the contamination of the water by dust and debris. Furthermore no light fell into the structure. In this way the development of bacteria and algae could be avoided thus guaranteeing excellent hygienic conditions. This could be proved successfully at the archaeological works in 1979, when a cistern was cleaned and filled in spring to supply the workers during the campaign in late summer.

Even after the construction of the high-pressure pipe-line, which supplied the acropolis since Hellenistic times the cisterns were still operated as part of the whole water-management-system. This can be concluded from the so-called Astynomen-inscription, which is dated to the 2nd cent.

[6] Page 37 ff.

Figure 3. Broken cistern at the corner of the acropolis of Aspendos (Turkey).

A.D.[7]. According to this inscription the cisterns had constantly to be cleaned and maintained. Penalties were fixed in case of neglect and the responsible persons were indeed punished.

Calculating the available water of the fountain houses (Agios Stratigos and at the Gurnellia) and the cisterns as a constant flow, the discharge probably amounted to about 3 l/s.

4 HELLENISTIC AQUEDUCTS

4.1 *Attalos-aqueduct*

At the beginning of the Hellenistic era quite a number of cities like Athens, Samos or Olynthos already had aqueducts in which fresh water was flowing. Therefore it is not astonishing that the wish arose also in Pergamum to get an aqueduct to supply the population with fresh and flowing water. The necessary money to pay for the construction of such a system was available.

Most probably in the middle or the 2nd half of the 3rd century B.C. the first pipeline was constructed, leading water from a spring in the mountains north of Pergamum to the city. The

[7] See Klaffenbach (1954).

Figure 4. Pipe made of fired clay of the Attalos-aqueduct.

origin of this aqueduct, probably a fountain-house, has not been detected yet[8]. But the line was identified for a length of more than 15 km[9]. The conduit was led down from the mountains with a relatively large incline into the Selinus-valley and then from there with an average gradient of about $I = 0.23\%$ along the eastern slopes to the capital.

The pipes made of fired clay had an inner diameter of 13 cm and a length of up to 60 cm (Fig. 4). This inner diameter does not correspond to the local measure of a Philetairic foot (36 cm) but to the outer one, being roughly ½′. The pipes were laid into an excavated bed a little below the surface of the natural soil. In order to minimize water losses through the pipes they were embedded in a layer of an artificial mixture of clay and sand. The joints between the pipes were also sealed with the same material.

Various stamps imprinted on the pipes show a combination of the letters AB (Fig. 5). According to Schuchardt (1895)[10] these letters are interpreted as "*Attalos basileios*", meaning "belonging to the palace of Attalos". Today this interpretation is doubted by archaeologists and historians. Nevertheless, due to engineering considerations this pipe-line must have been the first one to be constructed in Pergamum and therefore had been certainly installed in Hellenistic times. This happened most probably in the reign of king Attalos I. and therefore this pipeline is called the Attalos-aqueduct.

The bottleneck in the line however was the saddle north of the castle mountain. The Attalos-aqueduct arrived there at an elevation roughly 25 m higher than the crest of a saddle at this ridge. Thus a pressure pipe-line, an inverted siphon, was necessary in order to overcome this difficult spot if a drop in elevation was to be avoided.

---

[8] Examples of Greek fountain houses at the beginning of aqueducts are very rare. In principle their construction was similar to the fountain houses mentioned above. As element of an aqueduct may be referred to the one on the island of Samos, where a church today stands on the ancient structure (Kienast 1977).

[9] See Garbrecht (2001), page 58 ff.

[10] page 393 ff.

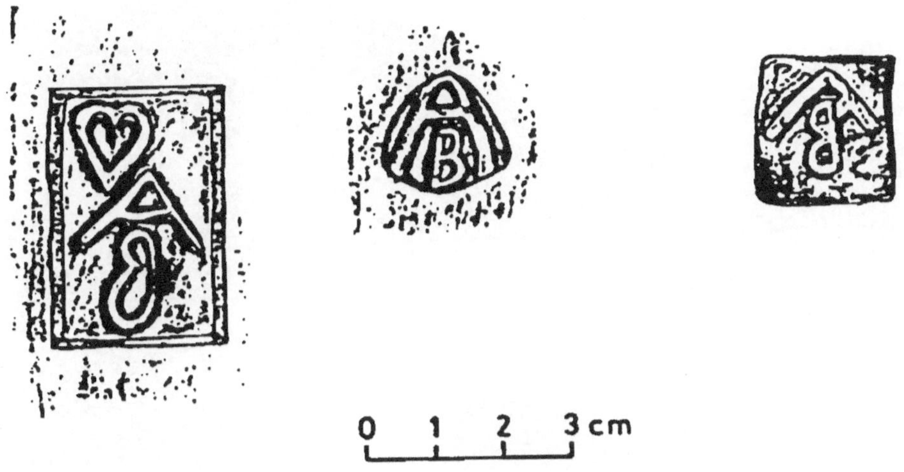

Figure 5. Stamps on pipes of the Attalos-aqueduct.

The transfer from a free flowing pipe-line to an inverted siphon and vice versa was done by means of a basin. The inlet basin of the pressure-pipe-line of this Attalos-aqueduct has not been found, but the outlet basin, was set into a quarried rock. It was certainly constructed from stone slabs. But these ashlars have disappeared and were used later elsewhere.

Some shards of pipes found in the area indicate that the pipes of the siphon consisted of fired clay with a wall thickness of about 6 cm. At certain distances these pipes were fixed in blocks of stones which had been drilled. Four of these blocks could be identified in the vicinity of the line (Fig. 6). The diameter of the drilled holes was just 10–11 cm thus reducing the discharge capacity due to increased energy losses. As it seems this is one of the first large scale applications of a pressurised pipe-line in antiquity[11].

After entering the city the pipes were installed in a separate tunnel, which protected the aqueduct. This expensive type of construction had been used before, for instance in Athens or Samos, also outside the city along the whole length of the aqueduct, whilst in Pergamum only inside the city. Here, in this segment the pipes were embedded in pure sand and the joints between the pipes consisted of lime.

The exact end of the pipe-line is not known. However, the big city well (Fig. 7), located outside of the city walls at the road running to the acropolis, which was constructed of marble plates, lies about 10 m below the last known spot of the aqueduct in the tunnel mentioned above. Therefore it is assumed that it was supplied with water by the Attalos-aqueduct since the 3rd cent. B.C.

Assuming that the cross section of the pipes was half filled, the discharge of the aqueduct would have been in the order of 3 l/s. This was about the same quantity which was already available from the fountain-houses and cisterns. The available supply at that time was therefore doubled by starting the Attalos-aqueduct.

Archaeological findings proved that this pipe-line obviously worked for at least 500 years, therefore even during times when in Roman times the comparatively small discharge seemed to be negligible.

---

[11] A fresco in the palace of Knossos at Crete shows a vertical fountain. As the artists surely will have known an example in reality, a pressure pipe-line must have existed there. And indeed the aqueduct supplying the palace with water from the Mavrokolybo-spring had to cross a small depression in front of it by an inverted siphon with a pressure height of about 8 m. As it looks like this is the oldest known application of the principle of corresponding pipes.

Figure 6. Drilled stone of the inverted siphon of the Attalos-aqueduct.

4.2 *Demophon-aqueduct*

Obviously not many years later, after realising the success of the first aqueduct and its inverted siphon, a second conduit was constructed in more or less the same line as the Attalos-aqueduct but on an elevation about 5 m above the existing pipe-line[12]. Its line is known only on the eastern slopes of the Selinus valley. There the gradient is nearly identical with that of its forerunner. This aqueduct consisted of a twin pipe-line (Fig. 8).

The pipes of this aqueduct had an enlarged inner diameter of 18 cm, which meant ½′ (Philetairic foot) and a length of 50–60 cm, in an average of 1½′. The construction of the pipeline in an earth-bed a little below the surface of the soil was the same as at the Attalos-aqueduct. The pipes often showed two stamps imprinted in the fired clay: ΔΗΜΟΦΩΝ and ΔΙΟΝΥΣΙΟΥ. Referring to the first name, probably the craftsman or factory owner who was responsible for the production of the pipes, the aqueduct was named the Demophon-aqueduct.

The Demophon-aqueduct had to overcome the same bottleneck at the saddle north of the castle-mountain by means of an inverted siphon, but with an increased pressure height of about 30 m. No traces of the inlet basin or the outlet basin have been found, because the later Roman bridge crossed this saddle in the same line and at the same elevation. Most probably the water of the Demophon-aqueduct discharged into the new canal and the old system was abandoned or even destroyed. Therefore only few shards of pipes, made of fired clay, have been found. They indicate that this inverted siphon was constructed of thick earthenware pipes only without any stone-blocks in between them.

Due to the elevation of the aqueduct a second saddle, closer to the castle-mountain, had to be crossed by another inverted siphon, this time with a pressure height of about 5 m only. Here the inlet basin has probably been found. The peculiarity is that it consists of three chambers one behind

[12] See Garbrecht (2001), page 72 ff.

Figure 7. Big city well at the street leading to the acropolis of Pergamum.

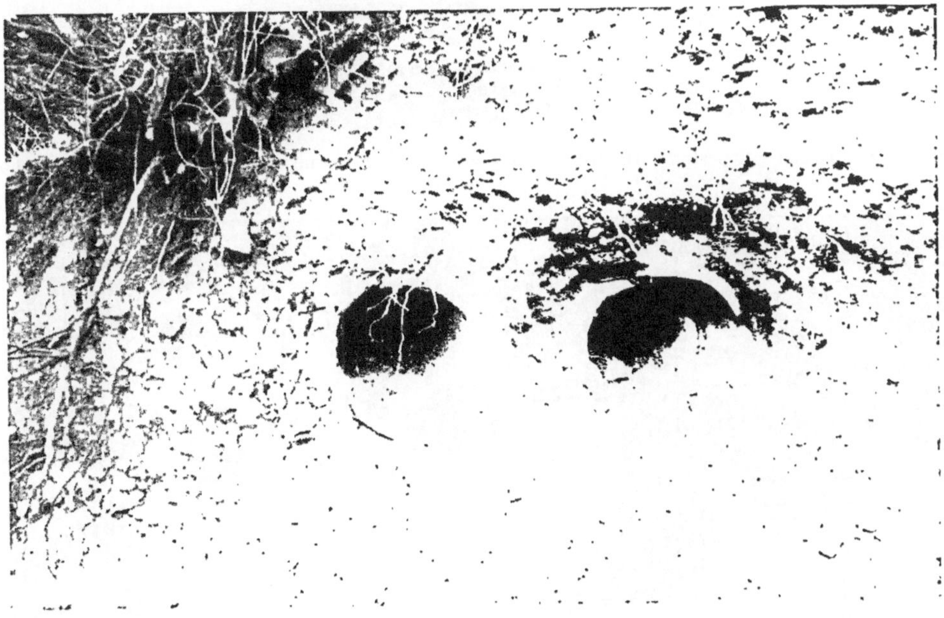

Figure 8. Twin pipe-line of the Demophon-aqueduct.

the other. The dividing walls, made by a stone slab, are perforated by drilled holes (Fig. 9), a very rare construction in antiquity.

The aqueduct then terminated obviously in a fountain well in front of the Demeter sanctuary still outside of the city walls (Fig. 10). The structure to be seen there today belongs to the later Roman Madradağ conduit. However, its excavator Dörpfeld (1910) pointed out, that it was constructed of

Figure 9. Inlet basin with perforated dividing walls upstream of the second inverted siphon of the Demophon-aqueduct.

Figure 10. Fountain well in front of the Demeter sanctuary.

Figure 11. The triple pipe-line of the Madradağ-aqueduct (Foto: Garbrecht).

stones which had been used before in an identical structure[13]. This can only have been the well of the Demophon-aqueduct.

The hydraulic capacity of the pipe-line was calculated to about 25 l/s, but it can be assumed that only half of this discharge arrived at the metropolis because the pipes were most probably only half filled. Even this discharge was more than twice that, which was available in the city before.

4.3 *Madradağ-aqueduct*

After the successful construction of two aqueducts which each included an inverted siphon of 25 m respectively 30 m pressure height the most adventurous project was undertaken. Probably under the reign of king Eumenes II at the beginning of the 2nd century BC the Madradağ-aqueduct was built[14]. It was a triple pipeline of more than 50 km length (Fig. 11). The water from several springs was caught in the Madradağ-mountains – hence the name- and led to the city of Pergamum. The line in the mountainous course often can be easily been found, because the pipes have been robbed in the middle of the 20th century leaving a more or less recognisable ditch. Therefore they can be identified around Bergama being used for many purposes like chimneys or planting pots.

Again the pipes had an inner diameter of ½' (Philetairic foot, i.e. about 18 cm) and a length of 50–60 cm. They were put into a bed of an artificial mixture of clay and sand. Gas chromatographic

[13] Page 355 f.

[14] See Garbrecht (2001), page 89 ff.

Figure 12. Entrance of the tunnel of the Madradağ-aqueduct (Foto: Garbrecht).

analysis of this soil revealed that an organic material with hydrophobic behaviour was added to it, probably olive oil. By this means seepage losses through the porous pipes as well as their joints were drastically reduced. According to calculations, the water losses through the joints were just about 3% of the discharge in the 3 pipelines on the total length of the said more than 50 km.

Various pipes also show stamps, all as monograms. A total of four different stamps have been found so far. But their meaning is still unknown.

In the course of the pipeline a 3 km long protruding hill was crossed by a 180 m long tunnel. The structure most probably had been constructed from both ends[15]. But it was never cleaned.

Therefore this assumption hasn't been proved. Only the entrance was excavated (Fig. 12). As it appears, it seems that this was the first time that a vaulted roof was constructed at a hydraulic structure in Hellenistic times.

[15] The most famous tunnel constructed in this manner is the Eupalinos-tunnel on Samos. This 1040 m long structure is already mentioned by Herodot. It was excavated and cleaned in the 2nd half of the last century. Kienast (1995, page 148 ff) proved that in the southern part at least five independent measurement systems for the elevation existed and another five for the position. The geodetic measurements were prerequisite that both parts of the tunnel, the northern as well as the southern, met in the centre of the mountain. As mentioned already by Herodot the pipeline of the aqueduct was installed in a separate tunnel below the floor of the main tunnel. However, upstream of this Eupalinos-tunnel the pipeline was also placed in a tunnel, which was constructed according to the so-called "quanat"-method. After fixing the line of the aqueduct the positions of shafts were determined and levelled. Then the shafts were dug down to the calculated depth. From the bottom of these shafts the tunnel was then excavated to both sides. Thus it could be worked simultaneously at several places. The distance of the various shafts depended mostly on their depths.

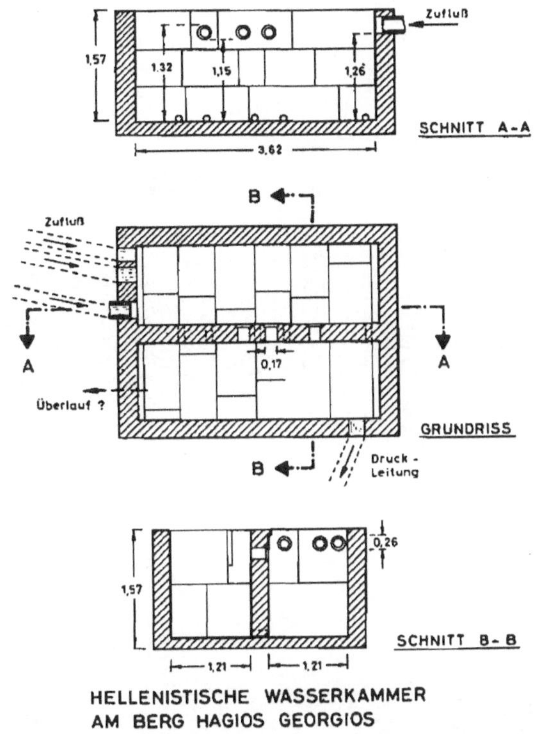

Figure 13. Plan of the inlet basin of the inverted siphon of the Madradağ-aqueduct.

The triple pipeline terminated in a basin at the slope of the next mountain north of the city- the so-called Hagios Georgios on an elevation about 35 m above that of the acropolis. It consisted of two chambers, each measuring 3.63 m × 1.21 m (Fig. 13). These chambers were connected by three holes in the dividing wall. The pipes of the triple aqueduct merged into the north-eastern corner of the first chamber. The water flowed from there through the mentioned holes into the second chamber and from there into a pressure pipeline, which started at the south-eastern corner of the chamber. Due to the enlarged cross sections of the chambers compared to the pipes before reaching the basin the chambers worked as settling tanks.

As already said the inverted siphon started at the south-eastern corner of the second chamber, leading water to the top of the acropolis. The plan shows that this pipeline had a length of about 3.5 km (Fig. 14). According to the longitudinal section the maximum pressure height amounted to about 190 m water column. It took nearly 2000 years before a pipe-line had been constructed bearing a pressure higher than this one in Pergamum.

Nothing has been found from the pipes but their foundation stones (Fig. 15). They consisted of vertical stone slabs with drillings of about 30 cm in diameter. The distance between the slabs varied, the average is about 1.2 m. The pipes were lying between the slabs on the smoothed rock or on especially prepared ashlars. A hole between the vertical slabs and the rock or the ashlars enabled the soldering of the joint between the single lead pipes and their connection-sleeves (Fig. 16)[16].

Nearly all foundation slabs were broken at their tops, obviously in order to take out the pipes. Therefore it could be assumed that they were fabricated of valuable metal. Analysis of soil samples from directly below the former pipe-line, as determined by the foundation slabs, and a couple of

[16] See Fahlbusch (1982), Fig. 42 f.

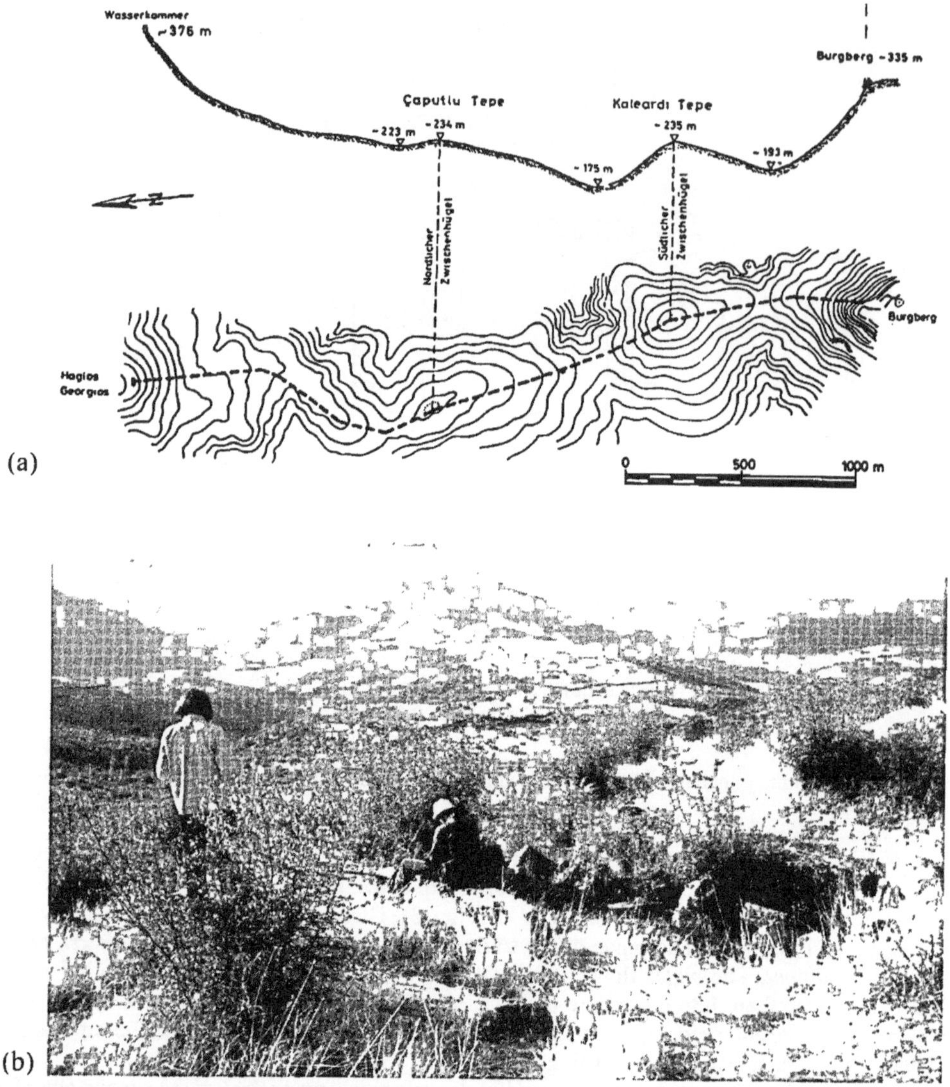

Figure 14. The inverted siphon of the Madradağ-aqueduct: a) Plan and longitudinal section (Garbrecht 2001) b) View from the inlet basin to the castle-mountain.

meters on either side, in order to find traces of copper and tin – as components for bronze – as well as for lead, showed a lead concentration 56 times higher in the line of the pipe-line compared to the outside position. The proportions for the other metals at both places were nearly equal. This fact proved that the pipes had consisted of lead. Obviously the rain water dissolved lead-ions over the centuries, which were then concentrated in the soil.

The exact circular shape of the holes in the slabs indicated that the pipes had been cast, unlike later Roman lead-pipes which were then manufactured from plates that were bent around a cylinder, and then soldered along the seam[17] thus forming a drop-shaped cross-section. Calculations with

[17] Because of this procedure they got a shape of a drop and their walls were much thinner. Vitruvius (VIII,c,vi) describes this method and gives measures about the size and weight of the various pipes, which were thus more or less systematically classified. Frontinus (25 ff) later enlarged this system.

Figure 15. Foundation slabs of the inverted siphon.

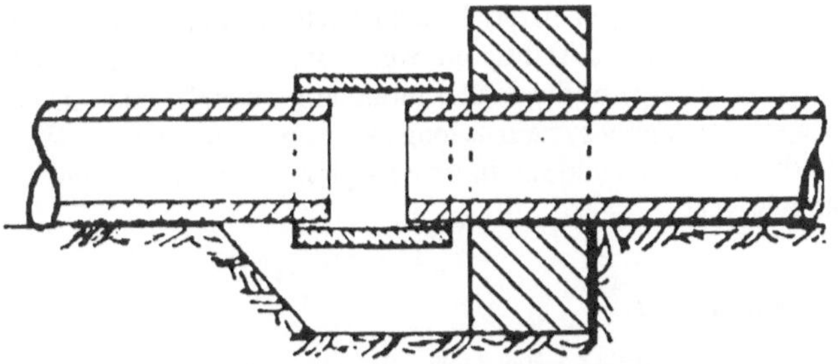

Figure 16. Reconstruction of the connection of the pipes of the inverted siphon (Fahlbusch 1982).

ancient measuring scales and comparisons with other cast lead pipes led to the conclusion that this pressurised pipe-line in Pergamum would have had an inner diameter of 18 cm or ½′[18].

In order to calculate the hydraulic capacity of the Madradağ-aqueduct the friction number of the terracotta-pipes was experimentally determined in the hydraulic laboratory to $\lambda = 0.029$. Based on the gradient of about 0.39% in the final stretch of the aqueduct, the maximum hydraulic capacity of this triple pipe-line was calculated to about 45 l/s. But most of the time the discharge would have been less. The real capacity was determined by the inverted siphon. The hydraulic capacity of this part of the aqueduct was about 30 l/s. It can be assumed that the rulers on the acropolis tried by all means to get this amount of water for as long as possible during a year. Therefore the

[18] Cast lead-pipes have been found at the site of the Artemis-temple in Ephesus (Bammer 1972). One pipe can still be seen in the museum in Selçuk. The various pipes were connected by a fitting made of a drilled stone.

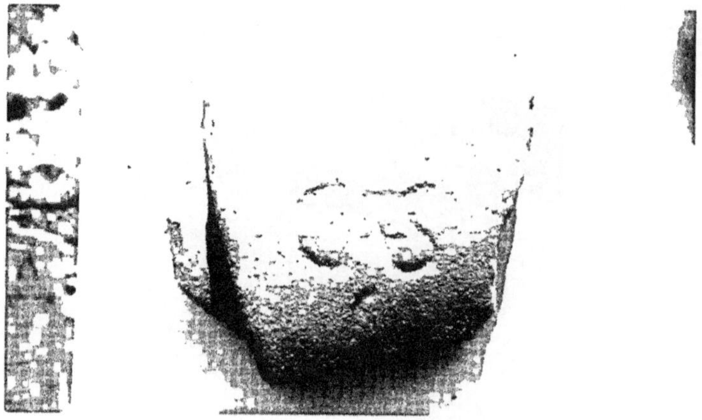

Figure 17. Stamp on a pipe of a twin pipe-line west of the Selinos.

Madradağ-aqueduct increased the water supply from about 17 l/s before to 47 l/s, which was really a huge improvement.

The exact termination point of the inverted siphon on top of the acropolis is not known. Garbrecht (2001) convincingly argued that the pipe-line entered the city through the northern defence wall, ran along a cleft in the rock and terminated inside the palace in a monumental structure, most probably a nymphaeum[19]. The drawing of the excavators shows a pipe draining the room in the palace and still today the mortar of the bed of the tiles can be identified as *opus signinum*, i.e. hydraulic mortar. But naturally nobody knows what the nymphaeum looked like. From the termination structure the water would have flown to the largest cistern besides the later Trajaneum, having a volume of more than 90 m^3. As the water flowed constantly an overflow soon occurred, which was led from one cistern to the next down the slope of the castle- mountain. Thus the cisterns became an integrated element of the flowing water supply system as long as the aqueduct worked. We do not know exactly how long the Madradağ-aqueduct and the inverted siphon functioned. However it was certainly a period of more than 500 years.

4.4 *Other Hellenistic aqueducts*

As mentioned before, people settled west of the Selinus river already in Hellenistic times. The Nikephorium, which is mentioned by Strabon[20], was probably located here. At least three aqueducts supplied this area, two twin pipe-lines and one single pipe-line[21]. The lowest twin pipe-line is similar to the Demophon-aqueduct. The pipes show the same diameter and also two stamped names: NAOY and APOΛΛωNIOY. The aqueduct above this one was constructed similarly and shows one stamp, in which letters were combined into a monogram, which can be interpreted as "CAESAR" (Fig. 17). Should it refer to Gaius Julius Caesar? As nothing is known where the aqueducts came from and what their destinations were further interpretation is impossible.

West of Pergamum the Asklepieion is situated, a very famous spa in antiquity. It got its water by a pipe-line from the Geyikli-Dağ in the west[22]. Few remains could be identified in the mountains, but hardly anything along the course down to the destination, because almost every remnant of the ancient aqueduct has been demolished during the construction of a new modern canal.

[19] Page 123 ff.
[20] Book XIII, chapter IV.
[21] See Garbrecht (2001), page 52 ff.
[22] See Garbrecht (2001), page 43 ff.

Figure 18. Cross section of the Roman Madradağ-canal.

5 ROMAN AQUEDUCTS – MADRADAĞ-CANAL, KAIKOS-AQUEDUCT AND AKSU AQUEDUCT

When Pergamum flourished during the *pax Romana* and got to the peak of its development, two vaulted canals were constructed. One ran more or less parallel to the triple pipe-line from the Madradağ north of the city[23]. Its original clear width was probably about 1½′ (now Roman foot) and its headroom 3′–3½′ (Fig. 18). It was connected to the inlet basin of the inverted siphon for the acropolis at Hagios Georgios. Obviously it was intended to replenish the discharge of the triple pipe-line to the maximum capacity of the siphon in dry seasons to guarantee that the palaces on top of the castle-mountain always got the maximum water possible.

From the basin the line went most probably in a kind of cascade down to the ridge and overcame the saddle on a bridge, which was around 30 m high, obviously taking up the water of the Demophon-aqueduct. Where the Hellenistic pipeline needed a second inverted siphon, again a bridge was constructed, here of about 5 m height. The canal terminated in the already mentioned fountain-well in front of the Demeter-terrace (Fig. 9).

As the water from the Madradağ contains no dissolved lime, no carbonates could have formed a calcareous crust, indicating the depth of water in the canal. However it can be assumed that the discharge was probably in the order of 50 l/s or even much more.

[23] See Garbrecht (2001), page 132 ff.

Figure 19. Cross section of the Roman Kaikos-aqueduct.

The other canal constructed in Roman times was the approximately 50 km long Kaikos-aqueduct, leading water from a spring in the upper Kaikos-valley along its northern rim from the east to the city[24]. Its dimensions were even larger showing an original clear width of 3′ and a headroom of 5′ (Fig. 19). In order to keep the line as short as possible 40 bridges had been constructed in the course of the line, the largest across the river Karkasos is more than 40 m high and longer than 500 m (Fig. 20). It was obviously one of the largest bridges ever built in Roman times. Furthermore five tunnels must have existed, but they have not been excavated. As the difference in height between the beginning and the end of the canal was just roughly 15 m, the extremely small gradient of the slope was kept constant to $I = 0.03\%$.

The calcareous deposits on the wall of the channel indicate the depth of the flowing water. Depending on the friction coefficient used for the calculations the discharge could have been 150–200 l/s.

In 178 A.D. an extremely heavy earthquake destroyed Smyrna and demolished also many buildings in Pergamum. The most severe damage however occurred most probably at the aqueducts. The high bridges of the Madradağ-canal and the Kaikos-aqueduct were probably completely destroyed and not reconstructed. Possibly the bridge of the Madradağ-canal was replaced by an inverted siphon.

For the Kaikos-aqueduct the water of another spring was collected and the 24 km long "Aksu-aqueduct" was constructed as new branch, abandoning the former 11 km of the Kaikos-canal. Thus the reconstruction of two large bridges could be avoided. However in order to omit the bridge across the river Karkasos a deviation was necessary, which extended the line for about 8 km. The bypass thus had a length of about 10 km. The gradient in this new line therefore had to be reduced to 0.01% only and it was constructed indeed in this way. Water did flow through this canal, as is indicated by a calcareous crust.

[24] See Garbrecht (2001), page 228 ff.

Figure 20. Ashlars from the bridge across the Karkasos-valley.

Although it seems that after finishing the repair works the whole water supply system worked again, the slow decline had begun. The first drastic downward step obviously occurred with the earthquake of 262 A.D. and generally continued in parallel with the political loss of power.

6 DISTRIBUTION SYSTEM IN PRIENE

Similar to the situation in Pergamum the aqueduct in the Hellenistic town of Priene, situated at the northern shore of the Great Meandros river, terminated most probably in a basin in the upper part of the city, more than 100 m above the lowest point of the distribution system. From there the water was led in pipe-lines made of fired clay to the various users inside the city (Fahlbusch 2003). The joints of the pipes were sealed with mortar which was based on lime.

The pipe-lines were installed in the streets like today. The rectangular pattern of the streets in Priene followed the proposal of Hippodamus of Miletus. However, this resulted in problems with respect of the aqueducts. Turning a pipe-line around a right-angled corner is difficult as terracotta-pipes can't be bent like those made of lead. Therefore either prefabricated elbows had to be installed (Fig. 21) or the line had to emerge into a small basin and start there again with the next part, but in the new direction (Fig. 22). These basins were primarily made of marble, but later ones made of burnt clay were used, most probably because they could be produced much more cheaply. Both types were used in Priene.

But the basins not only enabled a change of the direction of the pipe-lines but they were a multipurpose device. The division of the water from one main line into several smaller ones was facilitated by these basins as can be seen in Figure 22 where four pipes were fed from this basin.

The small basins were installed at many places at the sides of the streets. They were certainly covered to prevent contamination by rubbish and dust. On the other hand the removal of the cover enabled access to the water, so that it could be taken and used by the people.

At three points it could be shown in Priene that the water in the pipe-lines upstream of the respective fountains must have been under pressure. That meant that if the outflow was closed a

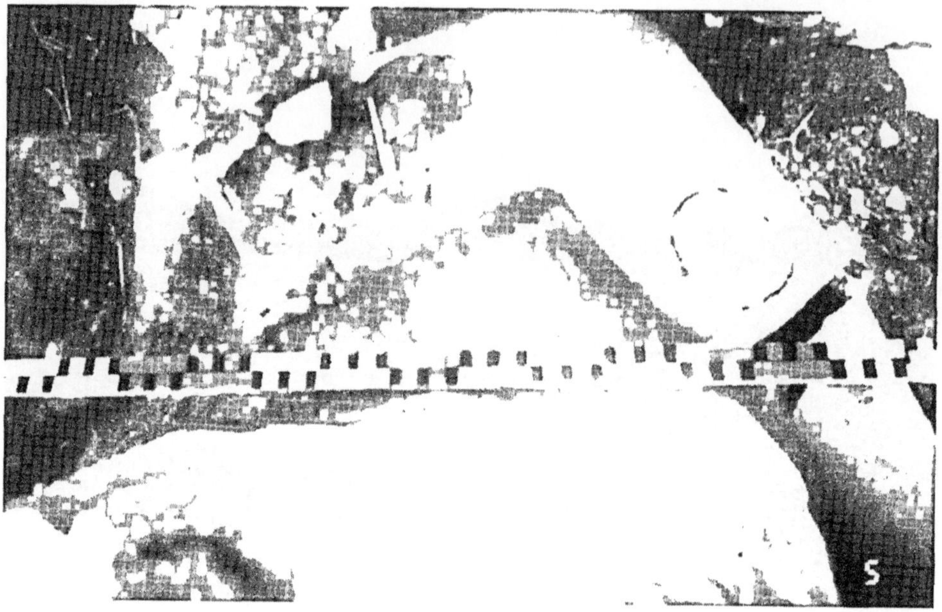

Figure 21. Prefabricated elbow pipe of fired clay in Priene.

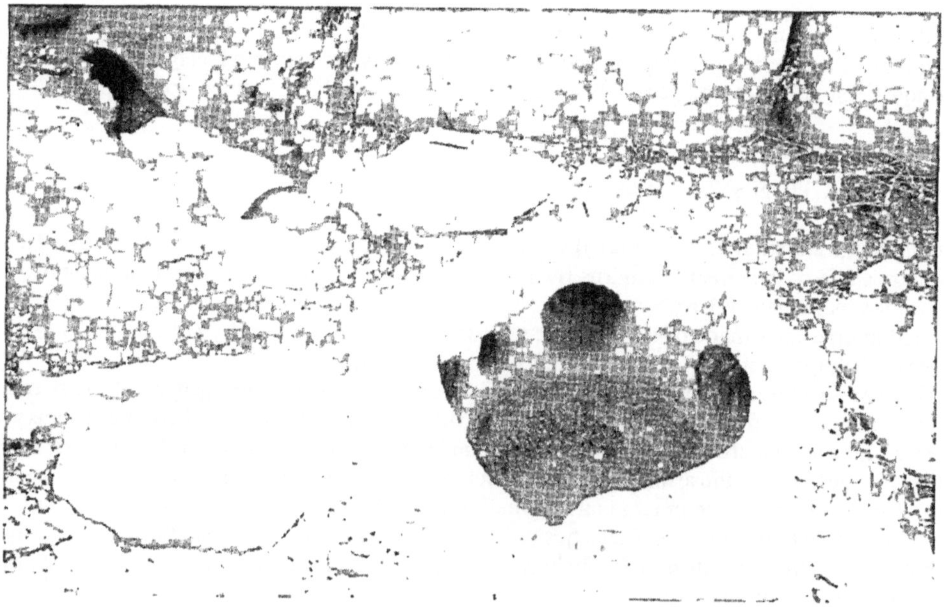

Figure 22. Multipurpose basin in the course of the distribution system of Priene.

huge pressure could build up in the pipes, which threatened to burst them, unless a device was applied to limit the pressure. And indeed the described basins, in which the water had an open surface, limited this pressure. In case of a locked outflow the water would have been backed up to the next small basin, where it flowed out, thus preventing the occurrence of high pressures.

Figure 23. Fountain in the street of Priene.

This purpose was probably the most important. It appears that the difference in elevation between these basins was less than 10 m[25].

The pipe-lines terminated in various fountain structures (Fig. 23) or houses either as a free-flowing pipe-line or as an inverted siphon. From here the people could take the water to the various houses.

7 ROMAN SYSTEMS

As already mentioned Roman engineers planned and constructed water supply systems much more pragmatically then their Greek predecessors. In the following section the various elements of Roman systems will be explained with examples from the canals of Pergamum mentioned before. But contrary to the elements of the Greek systems, which nearly all had been applied in this city, only some Roman elements can be described here. Therefore the various elements of the system will be discussed separately, first of all the structures at the beginning of the canals, i.e. dams,

[25] In principle it was thought, that terracotta pipes could bear very little pressure only. However, Prof. Dr. König (Munich) provided us with a newly produced terracotta-pipe, which also had been produced on a potter's wheel. This pipe was set under air pressure until at least 55 m water column without bursting. But the sealing leaked, that the pressure couldn't be increased further. This test showed that the ability of the pipes to bear pressure seems to be underestimated in the past.

Figure 24. Cornalvo-dam with intake tower in Merida (Spain).

weirs, percolation wells, springs, then the canals, bridges, inverted siphons, tunnels and last but not least the distribution of the water in the cities.

7.1 *Dams*

Every aqueduct is fed by either surface- or groundwater. Contrary to the Greeks, who avoided the use of surface water for the supply of the people as explained above, the Romans used this at least in regions where springs or other groundwater resources did not exist or when their discharge was very small.

In order to collect and store the water after heavy rainfalls dams were constructed. Water shortages in summer could thus be managed by the temporal transfer. Striking examples for these structures are still to be seen in Merida (Spain) where the Cornalvo-dam and the Prosperina-dam are operated still today. As can be seen in Figure 24, an intake tower was constructed on the upstream side of the Cornalvo-dam with intake holes at different heights. This type of construction has remained practically unchanged until today. We still apply Roman technology.

Dams have been constructed in many regions of the Roman Empire but only one in the vicinity of Rome. S. J. Frontinus reports that the water of the Aqua Anio Novus was always brown after heavy rainfalls. Therefore the line of this aqueduct was extended for about 20 km to a newly constructed dam in the Anio valley near Subiaco, from which the water was taken afterwards. The reservoir functioned as a huge settling basin and thus markedly improved the water quality. The Roman *curator aquarum* reports that the water flowing to Rome after the construction of the dam was pure and had a much better quality[26].

7.2 *Weirs*

Rivers are the second source of surface water. In order to lead water into an aqueduct without sediments, often a weir was constructed, backing up the water table in a river. A typical example

[26] Frontinus 93.

Figure 25. Diversion weir in the Rio Acebeda near Segovia (Spain).

for this kind of construction can still be seen today at the Rio Acebeda near Segovia. The plan in Figure 25 shows the weir and the overflow for the water which was not led into the canal. The same construction can be assumed for other diversions of water from rivers, for instance in Trier (Germany), Aix en Provence (France) or even for the Aqua Anio Vetus of Rome.

7.3 *Percolation wells*

Groundwater was often obtained by making it seep into a canal through the joints of a wall, which was constructed without mortar. Splendid examples have been revealed at Kalmuth or the "Grüner Pütz" for Cologne's aqueduct (Haberey 1972)[27]. Figure 26 shows the last, where the canal was dug into the aquifer at the foot of a hill and the water percolated through the joints of the dry wall. Then it flowed to a basin, which functioned as sand trap, guaranteeing that only pure water without sediments started the long journey to the destination at the river Rhine.

7.4 *Adits*

Naturally the discharge of large springs was also used in Roman systems. An example can be shown from the Kaikos-aqueduct in Pergamum. There a spring is situated at the foot of the mountain which delivers abundant fresh water with a high content of dissolved lime. Today the spring is located in a house in which the women of the village do their laundry. Therefore entering the house is prohibited for men.

The water is led from the spring in a small canal into a big open basin in front of the house. Whether this construction origins from antiquity is not known.

[27] Page 56 ff.

Die Leitung aus der Eifel

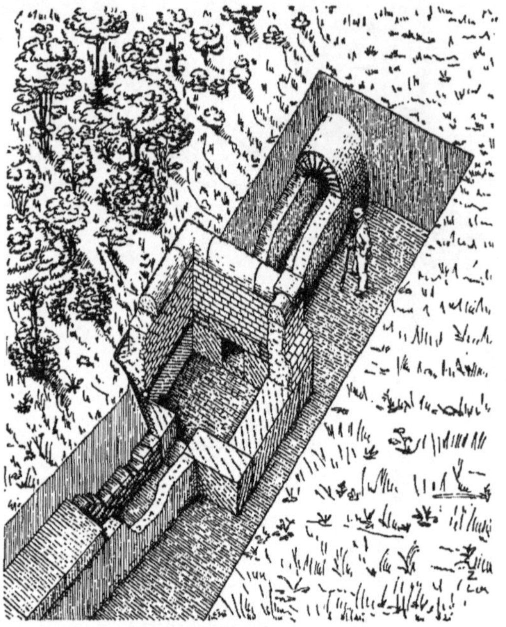

'Grüner Pütz'. Zeichnerische Rekonstruktion der Brunnenstube. Vorne die bergseitig als Sickerleitung gestaltete Quellfassung; hinten die überwölbte Leitung nach Köln (1951). – Maßstab der Längen und Höhen 1 : 100.

Figure 26. Percolation well "Grüner Pütz" of Cologne's aqueduct (Haberey 1972).

7.5 *Aqueducts*

From the adits, percolation wells, weirs or dams the water flowed in the Roman system in general in canals to their destination in the respective *villae*, settlements, towns or cities. Looking at the numerous aqueducts it seems that the following development took place[28]:

- The size of the cross section was chosen according to the estimated discharge which could flow in the canal. However the size varied along the course of the line.
- The cross section was kept constant which enabled manifold uses for instance of encasings especially the soffit scaffoldings for the vaults in a kind of industrialized construction.
- The chosen cross section had to be large enough for people to be able to walk through the canal in order to repair or just maintain it, mainly to remove the calcareous deposits.

A development can also be shown concerning the construction technology. One of the earliest canals, i.e. the Aqua Marcia in Rome, was constructed of big ashlars (Fig. 27) at least in the Campagna near Rome. They were put together without mortar. Probably a thin layer of clay sealed the joints. Later the walls of the aqueduct were carefully built from small stones or bricks. Figure 28 shows the canal of Cologne as a representative example. The careful pointing of the masonry joints is clearly visible. Finally, after the development of Roman concrete made of lime, the so-called "opus caementitium", the concrete of the walls was just poured between casings, which can clearly be seen at one of the canals for Aix en Provence (Fig. 29). There is hardly any difference compared to modern concrete. And at the wall we can still see the imprints of the casings

[28] See Fahlbusch (1982) page 44 ff.

Figure 27. Cross section of the Aqua Marcia near Rome constructed of ashlars.

The choice of gradient depended on the difference in elevations between the origins of the canals and their destinations. If large height differences were available, the slope often varied, being steeper in the mountainous regions (mostly upstream) and flatter downstream near the cities. However, when the available height difference was small and the line long, the gradient had to be reduced. This can be shown for the examples of Nimes (France) or Pergamum's Kaikos-aqueduct with gradients of 0.02% and 0.03%. The gradient of only $I = 0.01\%$ at the bypass at the river Karkasos mentioned above was obviously the smallest. There will hardly be a manmade canal in the world with a smaller gradient until today.

7.6 *Substructures and bridges*

In the lines of aqueducts depression or even valleys often had to be crossed. In order to ensure continued flow in an open channel the canal had to be constructed on a support. This was either a continuous wall when the depth of the depression was small and the construction of arches was almost impossible, or an arched bridge. Depending on the depth and the width of the valley as well as the inclination of its slopes the bridge was constructed with several stories. Low bridges normally show only one row of arches like in Tyana (Turkey) (Fig. 30), even when they were very long like the arcades of the Aqua Claudia and Aqua Anio Novus in the Campagna south of Rome (Fig. 31).

Figure 28. Cross section of Cologne's canal.

The most famous example of a bridge with three rows of arches is obviously the Pont du Gard as part of the aqueduct of Nimes (Fig. 32). As far as is known, it is the highest bridge of antiquity showing a maximum height of 48 m. Engineers obviously constructed inverted siphons in cases where the valley to be crossed was deeper than this.

Looking at the construction type, a development similar to the canals can be observed. Depending also on the material available the bridges were built by using ashlars which were put together without mortar like at the Pont du Gard (Fig. 32) or a bridge in Tyana (Turkey – Fig. 30).

After the development of "*opus caementitium*" this material was also used a great deal for the construction of bridges. It was poured behind the casings which often consisted of a wall of small stones and this was later the visible face of the structure. The problem was to guarantee a good bond between this "face" and the core. Later the stones of the casing walls were replaced by bricks as can be seen for instance at a bridge of Aqua Alexandrina in Rome.

Bridges were often constructed very slim. The most famous example is the bridge in Segovia (Spain) which still exists today to is full height in the centre of the city (Fig. 33). Its filigreed shape greatly impresses every tourist. However such slim structures could get static problems, as can be shown at a bridge of the Forum Julii, modern Frejus (France). Here the bridge was obviously not stable so that piers had to be applied at the side to strengthen the construction (Fig. 34). This measure has proved to be adequate, as the building still stands safely today.

7.7 *Inverted siphons*

Deep valleys were crossed by Roman aqueducts also by means of inverted siphons. The principle was the same as at Greek system with an inlet and outlet basin at the transition points to the free

(a)

(b)

Figure 29. Canal near Aix en Provence made of *opus caementitium*, Roman concrete: a) Cross section b) View from the side with the imprint of the casing.

Figure 30. Bridges made of ashlars in Tyana (Turkey).

Figure 31. Bridge of the Aqua Claudia and Aqua Anio Novus in the Campagna near Rome.

Figure 32. "Pont du Gard" – bridge of the aqueduct for ancient Nimes (France).

Figure 33. Bridge in Segovia (Spain).

Figure 34. Bridge with piers to stabilise the structure in Frejus (France).

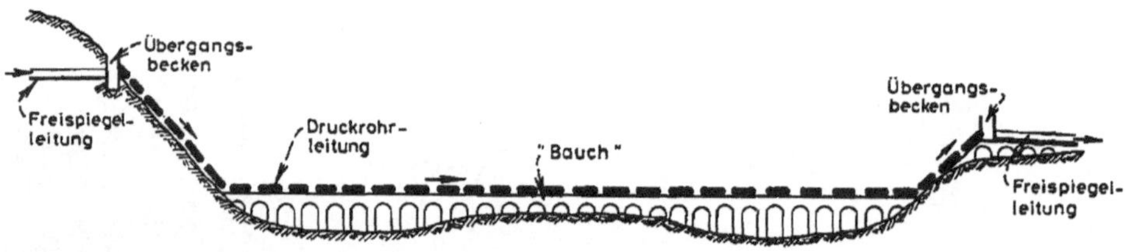

Figure 35. Elements of a Roman inverted siphon.

flowing aqueduct (Fig. 35). However, the pipe-line ran deep in the valley on a bridge. This is described by Vitruvius as follows[29]: "... when the pipe comes to the valley a "*venter*" (stomach) is to be constructed so that the aqueduct runs here nearly horizontally for as long as possible". Obviously the sharp bends in V-shaped valleys should be avoided in order to prevent the forces, which threaten to destroy the pipe-line at these spots.

Excellent examples for the construction of deep inverted siphons can still be seen today in Lyon (France). Each of the four aqueducts needed such a pressurized pipe-line to get to the city. However the most famous examples were constructed in the course of the Gier-aqueduct. In its course four inverted siphons existed with pressure heights of up to 100 m water column. Figure 36 shows the

[29] See book VIII, c. VI.

Figure 36. Inverted siphon at Chagnon – inlet basin.

inlet basin of a siphon near Chagnon and Figure 37 the reconstruction of the respective ramp near Chaponost and what can still be seen today of this structure, and Figure 38 shows the "*venter*" of the same siphon near Beaumont.

Figure 37 shows that the siphon consisted of 9 pipes in total. This fact reveals the important difference to the Greek systems. For instance in Pergamum one pipe-line was sufficient for the whole discharge of an aqueduct. However, due to the increasing population and the growing wealth in Roman times the water demand had risen. This resulted in bigger aqueducts and thus a much bigger discharge. As it was impossible in antiquity to construct huge pipes, several smaller ones, running in parallel, had to be installed to create large discharge rates.

The pipes of the inverted siphons in Lyon consisted of lead. Considering the many inverted siphons and multiple pipes of this city, it is obvious that huge amounts of expensive lead were necessary. The method already mentioned above, of bending plates to manufacture the pipes, enabled the reduction of the wall-thickness to about 10 mm only. Nevertheless the amount of lead available was limited and only very rich communities were able to finance such expensive aqueducts.

Not only capitals of provinces prospered in Roman times but also small settlements developed into respectable towns, which generally also required a good water supply system with well functioning aqueducts. Very often an inverted siphon was necessary in their line which was constructed of pipes made of drilled stones instead of lead. Figure 39 shows the beginning of the twin pipe-line of Laodikeia (Turkey) with the termination point inside the city in the background. The stone-pipes

Figure 37. Ramp of the inverted siphon near Chaponost. a) Remnants in the field; b) Reconstruction (Montauzan).

Figure 38. The "*venter*" of siphon near Beaumont.

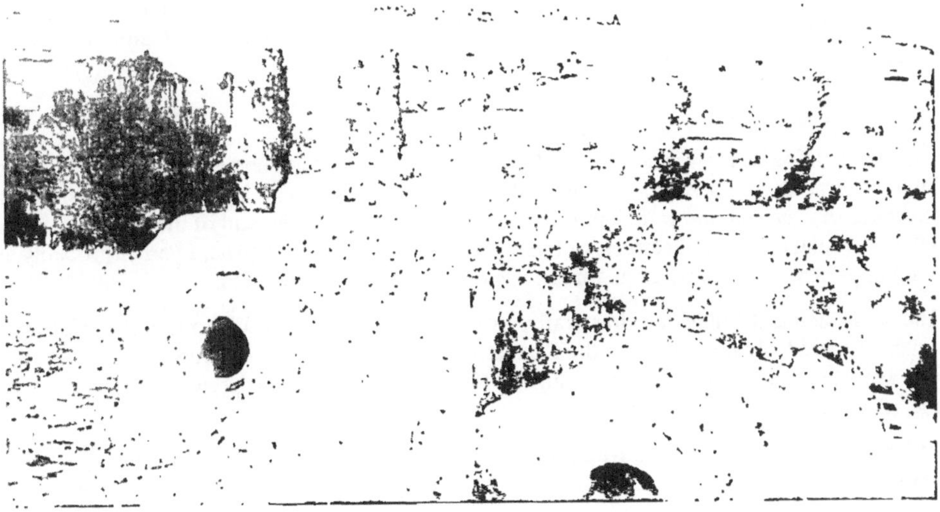

Figure 39. Inverted siphon of stone-pipes in Laodikeia (Turkey).

were assembled according to the collar and socket principle and the joints were sealed with a mortar made of lime.

7.8 *Tunnels*

After depressions in the ground, hills or mountains were the second type of obstacles in the planned line of aqueducts. If the line could not be diverted, or the necessary extension became uneconomically long, a tunnel had to be constructed. There have been two methods for the construction of tunnels, and both were already applied in Greek times[30].

In the first method a tunnel was dug through the hill or mountain from both ends such as the Greek Eupalinos-tunnel at Samos. The famous structure in Saldae is a splendid example for the corresponding Roman structures. A commemorative inscription under the headline "*patientia, virtus and spes*" reports how the surveyor Nonius Datus fixed the line and corrected the mistakes which the miners made during his absence (Grewe 1998)[31].

The second method of constructing a tunnel was the quanat-technique, already mentioned before. The tunnel was excavated to both sides in the direction of the aqueduct from shafts, which had been dug at intervals from the surface. This method had obviously been invented already in the kingdom of Urartu at the beginning of the 1st millennium B.C. and was employed not only in the area of the Roman Empire but in the east as far as China, and in the south as far as the Arabian Peninsula. Recently such a tunnel has been discovered in the vicinity of Cologne, as had been reported in Berlin at the beginning of April[32].

7.9 *Castellum aquae*

An aqueduct very often terminated in a city at a very high point, where a "*castellum aquae*" was constructed to distribute the water to the various consumers and/or townships. The most famous structures are those of Pompeii and Nimes. In February 2006 outlets of a similar structure were cleaned in the Villa Hadriana east of Rome[33].

Ohlig (2001) analysed in detail the *castellum* in Pompeii. He could prove that there orifices had been used to control the discharge to the three townships (Fig. 40). This had become necessary, as the water demand exceeded the available water. Only by such a control system could the water have been distributed more or less according to the demand of the population. The situation in other cities would have been similar.

7.10 *Reservoirs*

When the incoming water was not distributed directly to the various consumers, it was often stored in huge cisterns having a storage volume of many thousands of m^3. Very famous examples are for instance the "*piscina mirabilis*" in Cape Misenum (Fig. 41) at the end of the Serino-aqueduct, which ran around the Gulf of Naples, or the reservoirs in Carthage, Smyrna, Lyon or probably also Pergamum. There were two reasons for the use of these storage structures:

- The water demand fluctuated drastically, as was the case at the piscina mirabilis. Here much water was suddenly needed when the Roman fleet had to be supplied before it sailed.
- The available water fluctuated depending on the discharge of the spring or river as the source of the aqueduct.

At least until Byzantine times, when open reservoirs were constructed at the end of aqueducts for example in Constantinople, the huge cisterns were covered and the water thus protected from

---

[30] See footnote nr. 15.

[31] page 135 ff.

[32] Päffgen B.: Der Quanat-tunnel von Inden; Lecture at the seminar on April 3rd, at the occasion of the seminar on history at "Wasser Berlin 2006".

[33] The report will soon be published in DWhG vol.

(a)

(b)

Figure 40. *Castellum aquae* in Pompeii.

debris and light, similar to the small cisterns in which rain water was stored. The impermeability of the floor and walls was also guaranteed, often by several layers of plaster.

7.11 *Distribution system*

The water distribution system in a Roman city can best be explained by the example of Pompeii. This system is similar to that of Priene, as explained above, but it differs in three points. In Pompeii

Figure 41. *Piscina mirabilis* in Miseneum (Italy).

three main pipe-lines were connected to the castellum aquae, in Priene hardly more than one. Whilst the pipes in Priene were of fired clay those in Pompeii were made of lead. And whilst in Priene several multipurpose basins (Fig. 22) which interrupted the flow in the main lines were installed in the ground, in Pompeii these basins were installed on pillars (Fig. 42) and were made of lead. However, their purposes were the same. But due to its elevation water could not be taken directly from these basins but from fountains placed mostly at the foot of the towers.

8 MAINTENANCE

8.1 *Staff*

We know little about the management and maintenance of the various water supply systems. In many Greek cities the Astynomoi had the task of looking after the various elements, as already described above.

The source of information about Rome is first of all Sextus Julius Frontinus, who was in charge of the whole system as *curator aquarum* under the reign of the emperors Nerva and Trajan and who recorded his experiences. He reports that there were two teams responsible for the aqueducts, one on behalf of the emperor himself and the other on behalf of the public[34]. In every team various craftsmen worked according to the orders of the *curator aquarum*.

---

[34] Para. 116 ff.

Figure 42. Water towers of the distribution systems in Pompeii.

8.2 *Works*

Repair works were necessary in case any component was damaged. There are many inscriptions reporting such repairs as can for instance be seen at Porta Maggiore in Rome . The thorough renewal of the whole of Rome's aqueduct system under the guidance of Sextus Julius Frontinus is described by the *curator aquarum* in his book "*de aquaeductu urbis Romae*", which is a splendid source of information about Roman water supply management and its technology.

However, very often the most important maintenance task was the removal of the calcareous crusts, the so-called sinter, which had precipitated out of the water. Two main factors were responsible for the amount of these crusts: a) the quantity of dissolved lime and b) the turbulences of the water.

At a point of the aqueduct of Aspendos (Turkey) it could be proved that the crusts had been removed at least seven times there. And in Pergamum, where the Aksu-canal had been led down a steep slope to the Yagcili-valley a hydraulic jump resulted in a calcareous crust that filled up the complete cross section of the canal (Fig. 43). The walls, vault and floor of this canal have been removed meanwhile, but the sinter still proves that big amounts of the dissolved lime were deposited there.

If the deposits were not removed it was only a question of time before the cross section was so much narrowed that the character of the canal changed its from an open channel flow to a pressurized flow, as can be shown at the example of Aqua Anio Novus. Figure 43 shows the cross section, i.e. the so-called *specus*, at a spot east of Tivoli. Thick crusts are clearly seen even beneath the vault. Comparing the remaining open space with the original cross section, for instance about 4 km downstream, it becomes clear that less than 30% of the former space was still available for the discharge before the aqueduct was abandoned.

Figure 43. Cross section filled up with calcareous deposits of the Aksu-aqueduct in Pergamum.

The removal of the calcareous crusts was even more important in inverted siphons than at open flow channels, because there the crusts reduced the discharged cross-section from the outset, whilst in canals additional spare space existed. The removal of sinter was fairly easy when the siphon was constructed of lead pipes. During the night, segments of the pipes could be opened, the lead-plates bent outward, the crusts removed, the pipes bent back and resoldered.

However, this procedure could not be carried out, when the pipes consisted of drilled stones, and many of Roman inverted siphons were constructed in this way. Inspecting these pipe-lines nearly all show additional drilled holes from the top, which were usually closed when the aqueduct was in operation. It appears that these siphons could have been treated with boiling vinegar in order to remove the calcareous crust[35]. All elements necessary to carry out such a treatment procedure could be proved at the inverted siphon of Patara (Turkey).

9 CONCLUSIONS

Looking at the water management and the hydraulic structures in the classic civilizations it can be said that people were remarkably able to observe nature and to draw conclusions from their observations. This resulted in the evolution of criteria to evaluate the quality of water and the formulation of hygienic measures to prevent diseases.

For the supply of water local resources were used first. When they were exhausted local and temporal transfers were instituted and the necessary structures built. In this context it is noteworthy that the Greeks knew all necessary physical parameters. The Romans did not add anything to this knowledge. However, new technologies were invented, namely the Roman concrete – the so-called "*opus caementitium*", which enabled the economic construction of even long canals, huge bridges and long tunnels in soft rock.

[35] Fahlbusch (1991).

Well trained staffs were responsible for the supply of excellent water to the inhabitants of numerous cities and the maintenance of the water works for very many decades. Only when the political system collapsed, this brilliant infrastructure of supervision of the construction and maintenance organisation stopped functioning and was finally abandoned.

REFERENCES

Bammer, A. (1972) *Archäologische Gesellschaft zu Berlin 1971/72*, Sitzung am 9.2.1971. In: *Archäologischer Anzeiger*, pp. 714–728.

Crouch, D. (1993) *Ancient Greek City Planning*. Oxford.

Dörpfeld, W. (1910) *Archäologische Mitteilungen*.

Fahlbusch, H. (1982) *Vergleich antiker griechischer und römischer Wasserversorgungsanlagen*. Braunschweig.

Fahlbusch, H. (1991) Maintenance problems in ancient Aqueducts. In: Hodge, A.T. *Future Currents in Aqueduct Studies*. Leeds.

Fahlbusch, H. (2003) Wasserwirtschaftliche Anlagen des antiken Priene. In: *DWhG Wasser-historische Forschungen – Schwerpunkt Antike*, Siegburg. pp. 55–80.

Frontin S.J. (1989) *De Aquaeductu Urbis Romae*; translated by Kühne, G. Munich.

Garbrecht, G. (2001) *Altertümer von Pergamon I, 4*. Berlin.

Gräber, F. (1913) *Altertümer von Pergamon I, 3*. Berlin.

Grewe, K. (1998) *Licht am Ende des Tunnels*. Mainz.

Haberey, W. (1972) *Die römischen Wasserleitungen nach Köln*. Bonn.

Kienast, H. (1977) Der Tunnel des Eupalinos auf Samos. In: *Zeitschrift für Geschichte der Architektur*. München. pp. 97–116.

Kienast, H. (1995) *Die Wasserleitung des Eupalinos auf Samos*. Bonn.

Klaffenbach, G. (1954) *Die Astynomeninschrift von Pergamon*. Berlin.

Monrauzan, G. (1909) de *Les Aqueducs Antiques des Lyon*. Paris.

Ohlic Chr. (2001) *De Aquis Pompeiorum*. Nijmegen.

Radt, W. (1999) *Pergamon*. Darmstadt.

Schuchardt. *Altertümer von Pergamon* VIII, 2. Berlin.

Vitruvius. (1970) *De Architectura*; translated by Granger, F. Vol. II, London.

CHAPTER 5

Water engineering and management in al-Andalus

José Roldán & Maria Fátima Moreno
Area of Hydraulic Engineering, University of Córdoba, Spain

1 INTRODUCTION

Water engineering and management in al-Andalus was limited, almost exclusively, to the hydraulic technology linked to irrigation institutions. In addition to the irrigation systems proper, these institutions included systems for water intake and delivery, legal regulations regarding water distribution and any other aspects related to irrigated agriculture. As Glick (1996) has stated, the history of technology is the history of technical ideas, whether developed through physical objects or by means of social and institutional mechanisms.

Similar to what occurs in other fields of science, historians concerned with the origins of irrigation in Spain are divided into partisans of an Islamic origin theory and those who reject the Muslim influence in this area. In some cases, Christian irrigation has been considered a direct legacy of the Romans, who were, without a doubt, responsible for undertaking large-scale water projects. However, Glick and Kirchner (2000) believe that while some basic components of Roman irrigation systems were most likely emulated by the Christians, the differences between the two cultures regarding patterns for the social distribution of water would seem to suggest that the Romans had a negligible influence on the medieval period. Moreover, the decline in population and the economic crisis of the fifth and sixth centuries suggest that there was little cultural continuity between both periods, although the Roman imprint on the irrigated districts or *huertas* of Valencia and Orihuela, among others, has been widely documented. Box Amorós (1992), for example, cites studies which show that large-scale irrigation works existed in Valencia as early as the Roman period. The presence of place names in areas such as the vega of Lorca demonstrate that irrigation canals bearing Arabic names correspond to the expansion of previously existing irrigated areas.

However, as in other fields, our current knowledge is the result of accumulating and superimposing contributions from indigenous societies to the present day. As Gilman and Thornes (1985) demonstrated, and as Giraldez et al. (1988) subsequently confirmed, irrigated agriculture was practiced in south-eastern Spain as early as the El Algar and Los Millares cultures by means of *boqueras* or earthen dikes, taking advantage of the occasional flooding in dry riverbeds. Thus, irrigation in the latter part of the Middle Ages cannot be attributed to nor was it spontaneously generated from a single culture. In consonance with this, Barceló (1989) stresses the importance of determining the contributions made by indigenous societies in both North Africa and Hispania where the climatic conditions would have necessitated the development of diverse hydraulic technologies for these peoples to adapt to their natural environment prior to the arrival of later civilizations.

Nevertheless, as Glick rightly asserts (1988), it was the Arabs who initiated an era of agricultural revival in the eight century which led to the improvement and intensification of irrigation practices throughout the Islamic world, including al-Andalus. As Trillo (2002) has shown, water was an essential element of Islamic culture to such an extent that the agricultural calendar was modified, which with irrigation, became more continuous. The Islamic civilization was thus one of technological synthesis resulting in the development and refinement of the technical practices of the ancient world.

Much evidence exists to confirm this fact. For example, Latin irrigation terminology was substituted by Arabisms (*acequía*, derived from the Arabic word *sāqiya,* replaced the Latin term *canalis*); hydraulic mechanisms introduced by the Muslims such as the *noria* or Persian waterwheel were widely adopted; irrigated agriculture and extensive irrigated districts or *huertas* were developed around cities with a strong Islamic influence (Valencia, Murcia, Toledo, etc.) or new crops were introduced, many of which had Arabic names and required irrigation due to the climatic conditions (lupin, orange, alfalfa, and cotton, among others). This Islamic inheritance remains evident even today in the place names of areas with medieval irrigation systems such as *huerta de los moros* (Glick, 1988).

Glick indicates the existence of local institutional models which are repeated wherever irrigated agriculture was practiced, from the Code of Hammurabi (18–17 B.C.) to medieval irrigation systems in Valencia. These administrative arrangements were based on the need for the fair and just distribution of water and to prevent conflict. Some of the common principles of these systems were: 1) the proportional distribution of water according to the amount of land cultivated; 2) the individual's responsibility towards the community of irrigators regarding such aspects as the maintenance of irrigation channels, compliance with the established irrigation turns or liability for damages caused to neighboring irrigators and 3) the political autonomy of irrigation systems in which justice was exacted by means of their own self-governing institutions.

Another striking aspect, which in part has given rise to the above controversy, is the lack of documentation from the Islamic period regarding irrigation practices and the construction of irrigation canals, with certain exceptions such as the late twelfth-century work by the geographer al-Idrisi (Carrasco, 1996). Martí (1989) suggests that the scarcity of written documents could be due to problems of a jurisdictional nature as the large-scale water projects were not the competence of Andalusi agronomists, but mathematicians and astronomers.

Owing to the absence of documentary evidence, the institutional and technological aspects of Islamic irrigation can only be examined from subsequent Christian documents or by means of archaeological studies. In order to understand more about the social aspects of water distribution, it is therefore necessary to resort to hypotheses based on land registers, archaeological investigations, place names, litigations, irrigation community regulations, the geographic distribution of irrigation terms and the few extant Arab sources (Glick, 1996).

2 THE SPREAD OF ISLAMIC IRRIGATION. DIFFUSION TO THE NEW WORLD

Like the origins of irrigation, the diffusion of Islamic irrigation and the elements that survived following the Christian reconquest are a matter of great controversy amongst historians of diverse disciplines.

Following the approach by al-Mudayna (1991), we will examine the scope of these irrigation practices through the study of particular geographical areas or hydrographic basins.

Beginning in the north and moving southwards, there is clear evidence of the Islamic contribution to the improvement of water distribution systems and the organization of irrigated lands in the Ebro region. The majority of Islamic irrigation canals have been found in the Alfaro-Tarazona-Saragossa triangle, that is, on the right bank of the Ebro River from Tudela onwards. The most important irrigation canals include those of Canet (Alhama River), Irues (Moncayo mountains), Furón Mayor (Jalón River) and four more in the region of Saragossa (Almozara, Almudafar, Gales and Urdán), which drew water directly from the Ebro and Gállego Rivers. Continuing on to Aragón, Box Amorós (1992) cites the Guadalaviar canal which irrigated the vega near Teruel.

In the Balearic Islands, irrigation systems were largely developed during the Islamic period, leading to the creation of agrarian landscapes that have survived intact to the present day. Two of the most striking examples are the numerous *qanats* or subterranean tunnels for extracting water, which will be discussed in greater depth in Section 3, and the system of irrigated terraces. Specifically, Majorca is the region of al-Andalus with the largest number of *qanats*. These are located in the valleys that cross the Tramontana and Levante mountain ranges and the Puig de Randa. Introduced in the tenth century, the majority of *qanats* were constructed in areas with terraced agricultural systems. In the case of flood irrigation, the terraces are horizontal, while in the case of canal irrigation, the terrace is laid out both longitudinally following the direction of the irrigation canal and crosswise so that water is conveyed to the total area of the terrace. In Ibiza there

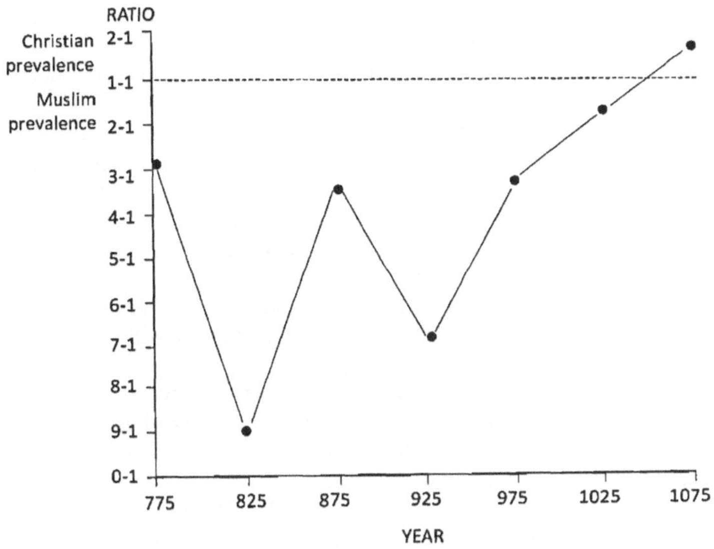

Figure 1. Ratio between Muslim and Christian scientists.
(Source: Glick, 1979)

exists what appears to be a unique irrigation system known as the *feixes* , which will be discussed in Section 4.

The geographer al-Idrisi has left a testimony of certain irrigation structures in Valencia, although it appears that the eight irrigation canals that form the canal network of the Valencian huerta already existed towards the end of Islamic dominion. Although Valencian irrigation is of Roman origin, the Muslims unquestionably contributed to its enormous diffusion. Giner Boira (1997) maintains that the settlers of the Levante region in Spain could not have been of Arab origin as they were unfamiliar with irrigation techniques and did not irrigate their land, but must have been Syrians, Lebanese and Egyptians who had a five-thousand-year long tradition of irrigated agriculture. Boira also holds that the current Tribunal of Waters was created c. 960 AD. Notable examples of Valencian irrigation include the huerta of Valencia and the irrigated districts of Alicante and Elche.

The irrigated districts of Murcia are located in the lower and mid-catchment basins of the Segura River, especially in the area surrounding the capital city of Murcia and Orihuela (which forms part of the Community of Valencia but is located in the lower basin of the Segura), as well as the Campo de Lorca which is irrigated by the Guadalentín River. In the huerta of Murcia it is worth mentioning the *norias*, hydraulic wheels powered by the force of the current that drew water by means of *cangilones* or waterbuckets. The most well-known of these are the *norias* of Alcantarilla and Ñora. In the Islamic period, the water drawn from the river in the Contraparada dam was distributed proportionally (in Murcia, *parada* is equivalent to the Valencian term *rafa*, that is, an obstacle placed in the course of a river in order to raise the level and subsequently divert the water). Two irrigation canals branch off from this dam, the Alquibla canal on the right margin and the Alfujia on the left.

In western Andalusia or the Kingdom of Granada, irrigation differed greatly from the Levante region of Spain given the area's mountainous relief and rainfall regime. In western Andalusia, river and mountain basins were used to build irrigated terraces. One of the most noteworthy aspects of this region (the Alpujarras of Granada and Almería) – characterized by a scarcity of water – is the ingeniousness with which devices were developed to extract water. Here surface water was drawn from permanent or ephemeral streams by means of diversion dams known as *azudes* or *boqueras*, while *qanats* were widely used for extracting groundwater (Hermosilla, 2006). One of the most striking examples of a *qanat* in this region is the *qanat* of Senés in Almería measuring several hundred meters in length. However, due to the small size and wide dispersion of the irrigated districts in this region, we shall not delve further into this subject.

Records of large irrigated districts in the Gualdalquivir Valley do not exist, but there are accounts of irrigated districts located outside urban areas and *almunias*, recreational estates belonging to the nobility, namely those of Jaén, Cordova (la Arruzafa) and Seville (La Buhayra). There is also evidence of irrigated districts on either bank of the Guadalquivir River where water was diverted by means of *norias* (such as the Albolafia waterwheel of Cordova) or dams. For sentimental reasons, and for purposes of comparison, it is worth mentioning the references made by the Arab traveler al-Himyari to the irrigated fields that stretched along the Marbella River outside the town of Baena (Córdoba): "(Baena) is surrounded by numerous gardens, vineyards and olive groves, its land is fertile and well irrigated by running water, located on the banks of an important river called the Marbella River that comes from the south and moves numerous mills" (Cherif Jah and López Gómez, 1994).

Finally, moving beyond the geographical spheres mentioned above, we should not overlook the vegas of Toledo (Huerta del Rey) and Talavera in the Tagus River basin. Given that the Tagus narrows as it passes through the city, it was necessary to develop systems to raise the water; many of which have remained intact for hundreds of years.

The diffusion of Islamic hydraulic technology to the New World via the Spanish conquerors is difficult to establish given that the cultural unity of eight-century Islamic Spain had been broken by the late fifteenth century. However, a close relationship does exist between certain aspects characterizing irrigation water use in the New World and al-Andalus: water rights are linked to the land, water is distributed according to an established system of turns for a given period of time or from a specific canal and the irrigation systems follow the Islamic style. Del Río Moreno (2002) describes the system implanted by Cortés in Mexico to cultivate sugarcane in which water was diverted from the main canal to the *regaderas* (secondary channels) or *apantles* and, from here, to the feeding channels (*tenapantles* or *contrapantles*) located at twelve furrow intervals. Each group of twelve furrows constituted a *tendida* with one man in charge of regulating the flow of three *tendidas*.

The historiography of irrigation in the vast Spanish empire of the New World is scant. According to Meyer (1996), this can be explained by the fact that the largest indigenous settlements were located in areas with abundant water, namely the Andean Altiplano in South America and the central valley of Mexico The Spanish were attracted to these places not only owing to the water to be found, but especially because of the civilizations that had settled there. Later, in the latter half of the sixteenth century, the Spanish gravitated towards areas where water was scarce, reaching territories to the north of Nueva España, specifically Sonora, Arizona, Alta and Baja California, Chihuahua, New Mexico, Coahuila, Nuevo León and Texas. It was to these arid or semi-arid areas – in which the existence of water marked the frontier between desolation and abundance – that the hydraulic technology of the Iberian Peninsula was taken, mainly for purposes of irrigation. Water played a very important role in the conquest of the New World as it became a source of private wealth, capital, income and power, while at the same time changing needs, uses and value systems and heightening the controversies and disputes involving water rights (Meyer, 1996).

3 INTAKE, DIVERSION AND TRANSPORTATION OF WATER

The Muslims likewise played a key role in the spread of technological practices. While in some instances the Hispano-Romans were already familiar with these techniques, they were fundamentally inherited from the great civilizations of the East. This is the case, for example, of the hydraulic wheels mentioned by St. Isidore of Seville in his *Etymologies*, although with all likelihood their origin can be found in the eastern Mediterranean. According to Pavón (1990), Philo of Byzantium (300–200 BC) describes devices used to extract water in his treatise *Pneumatica*, while Vitruvius mentions four types of artifices to raise water in his *De architectura*, although neither of them makes reference to animal-powered waterwheels.

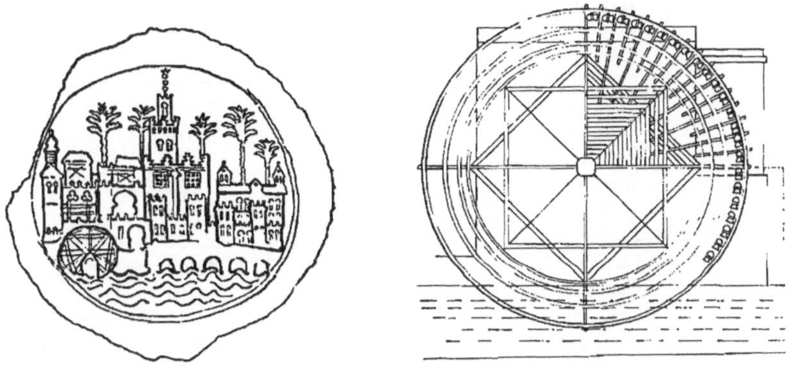

Figure 2. The Albolafia waterwheel of Córdoba. a) Rendering of a 14th-century city seal by J. Caro Baroja. b) Reconstruction by B. Pavón.

The term *noria* comes from the Arabic word *nā'ūra* to refer to any type of wheeled device used to raise water. However, a distinction is made between *norias de corriente* (also known as *noria fluvial* or *noria de vuelo*) and *norias de sangre* (also called *norias de tiro* or *saniya*). Whereas the *noria de corriente* (named *aceñas* in some areas) is a non-geared vertical wheel built in rivers and canals powered by the force of the current alone, the *noria de sangre* is moved by animal power and used to extract well water from a depth of up to 10m. According to Córdoba et al. (2004), the Spanish word *noria* is not a derivation of the Arabic term *saniya* used to indicate these wheels in al-Andalus, but of *naura*, in reference to river waterwheels, though both terms were used interchangeably in Christian Spain. The word *aceña*, a derivation of the Arabism *saniya*, was the name given to vertical waterwheels in the Early Middle Ages.

The Arabic term *naura* appears to be a derivation of the verb *na'ar* meaning "to grunt" in reference to the characteristic grunting sound emitted by the wheels (Pavón, 1990; Córdoba, 2004). In fact, the animal-driven waterwheels in Palma del Río (Córdoba) are known by the name of *chirriones* or "squeakers". According to several authors such as Caro Baroja or Torres Balbás, the famous Albolafia waterwheel of Cordova (see Figure 2) – whose name means "good luck" or "good health" – was dismantled in June 1492 owing to the terrible squeaking sound it made, which bothered Queen Isabel who lay ailing at the Alcázar de los Reyes Cristianos of Córdoba (Córdoba et al. 2004)

The *norias de sangre* (animal-driven waterwheels) are more complex than those moved by current alone as they require knowledge about how force is transmitted by means of a gearing mechanism. For this reason, Caro Baroja (1983) believes these to be a genuinely Islamic invention. As Glick (1979) points out, the Andalusi *noria* is unrelated to the Berber *noria* of North Africa. Most likely, both the wheel itself and the pots are inspired in the Syrian prototype. Likewise, the Andalusis introduced this type of wheel in Morocco and the Christian kingdom through the migration of Mozarab farmers. As Losada (2004) has shown, the Arabs used the *norias* to enlarge the hydraulic area dominated by currents (either in rivers or irrigation canals) in which the wheels were located.

The geographical distribution of waterwheels in al-Andalus serves to shed light on the extent of agricultural development in rural areas. The Libros de Repartimiento (registers of property formerly belonging to Muslims that was subsequently granted to the Christians), however, provide little information regarding these waterwheels, perhaps due to their very abundance. The only existing records, dating from the twentieth century, come from the Ministry of Public Works, which conducted a national survey of waterwheels in 1918. Nonetheless, it is uncertain what procedure was used to tally them, whether or not the survey included all of the waterwheels or only traditional ones and if the methodology varied from province to province (Glick, 1996). The distribution of waterwheels by provinces is shown in Table 1.

The diversion dam or *azud* was yet another device used to extract water resources. The *azud* was built across rivers, making it possible to stop the flow, raise or divert water to the irrigation

Table 1. Distribution of waterwheels by provinces.

City	
Ciudad Real	21,006
Castellón	4,083
Baleares	3,540
Toledo	2,750
Valencia	2,000*
Zamora	1,552
Madrid	1,432
Cáceres	1,010
Valladolid	842
Almería	668
Córdoba	647
Alicante	566
Gerona	505
Murcia	503
Jaén	542

* Surveyor's estimate
(Source: Ministry of Public Works, 1918)

canals. As mentioned above, in the arid south-eastern region of Spain, the use of temporary dikes or *boqueras* (openings or gates through which water is diverted from the irrigation system) in ephemeral streams had been commonplace since pre-Islamic times. Given that the Arabic term *azud* has the connotation of diversion rather than storage, it is more appropriate to use the term "dam" when referring to this type of system.

Among the techniques used for extracting groundwater, the *qanat* is one of the most remarkable. The *qanat* consists of a gently sloped tunnel which transports water from an aquifer to a surface outlet without the need for pumps (see Figure 3). Due to the fact that mines antedated the first *qanats*, Goblot (1979) defines them as a mining technique rather than an irrigation technique consisting of the exploitation of underground water tables (aquifers) by means of drainage tunnels. The tunnels are connected to the surface by a series of wells or vertical shafts separated at a distance of some 5 to 20 meters in order to provide ventilation and remove excavated spoil from the tunnel. A small mound is formed around the shafts to prevent runoff from contaminating the water. The first well or "mother well" is used to localize the aquifer. The tunnel is then constructed back from the surface outlet to the mother well (Argemí et al., 1995). It is precisely this procedure that distinguishes the *qanat* from a mine, since although the latter is comprised of a tunnel that draws water from an aquifer, it is excavated in an inverse manner and does not usually have air shafts.

The physical conditions that determine the use of *qanat* technology can be classified into three groups: climatic, hydrogeological and topographical. Bearing in mind the difficulties involved in their construction, *qanats* are built where surface water resources are scarce, that is, in arid climates. Nonetheless, it is necessary to have abundant phreatic groundwater resources, especially at deep levels. Furthermore, it is important that the system be sustainable, in other words, that it is fed regularly with a sufficient amount of water. The *qanat* therefore requires high elevations in order to receive precipitation from cloud masses, while the tunnels must be built on gentle slopes (between 1% and 2%). Piedmonts consisting of permeable materials located at mountain fronts are the most favorable areas for the construction of *qanat*s. The Iranian tablelands, and in particular debris cones, are the ideal location for these systems.

According to Goblot (1979), the first written records of *qanats* can be found in the chronicles of the Assyrian king Sargon II (722–705 BC) describing the eighth military campaign raged in 714 BC against the Kingdom of Urartu, located northeast of present-day Iran near the Turkish border. *Qanat* technology would later be spread south and eastwards across the Iranian tablelands by the Medes and Persians.

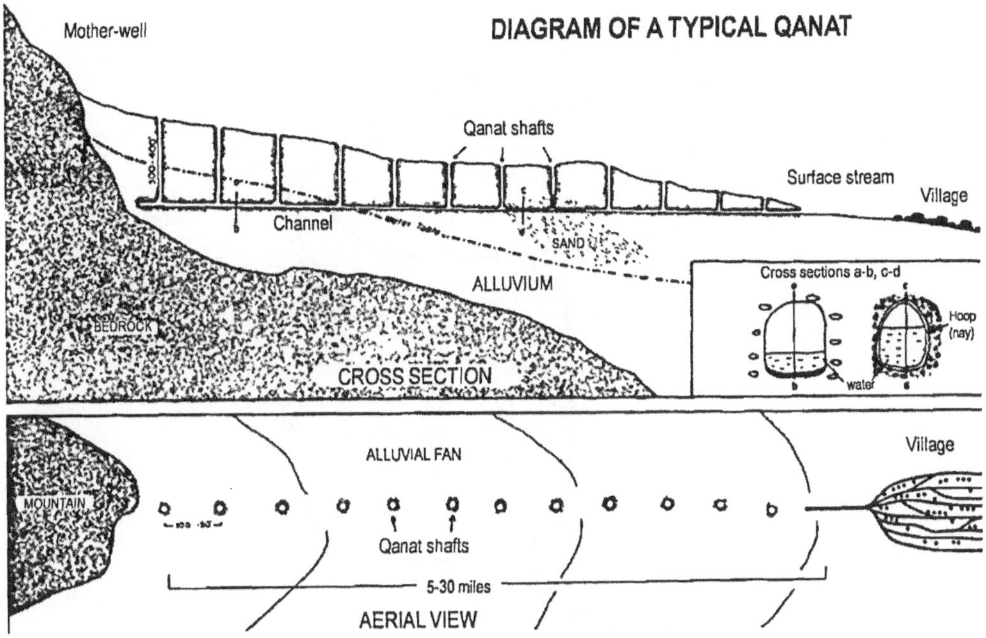

Figure 3. Diagram of a typical qanat.
(Source: Glick, 1988)

The Spanish *qanats* or *viajes de agua* are a direct inheritance of the Islamic and Persian influence in the Iberian Peninsula. The most remarkable example is the *qanat* at Madrid. The city, originally founded as a fortress in 871, owes its existence to the *qanat* system built by Iranians belonging to the Umayyad companies. Later, due to the quality and abundance of the water in the city, Felipe II transferred the capital of the kingdom from Toledo to Madrid in 1561. In a very interesting study describing the *qanat* of Fuente Grande de Ocaña, López-Camacho et al. (2005) affirm that this *qanat*, with 124 km of tunnels, 70 km of groundwater intakes and 54 km of conduits, supplied water to Madrid for a period of ten centuries from c. 900 AD to 1900 AD.

In addition to the Iranian tablelands, and due to its proximity to Spain, it is worth mentioning the Marrakech basin located in the piedmont of the High Atlas Mountain where more than 800 *qanats* measuring from 800 m to 2.5 km in length form a *qanat* network extending over 900 km. This network supplies water to the city and is used to irrigate between 15,000 and 20,000 hectares of land.

Cisterns or *aljibes*, a derivation of the Arabic word *al-yubb* (Pavón, 1990), were essential for supplying the water brought to the city by the *qanats*. *Aljibes* were also used in mosques and households to store the water that ran off the rooftops by means of *atanores* (metal or fired-clay drain pipes) set into the walls.

An *aljibe* still standing in the courtyard at the Mosque of Cordova received water from the courtyard pavement and the rooftops of the building by means of gutters that were designed to quickly discharge the runoff and prevent pools of water from forming (Roldán et al., 2006).

Irrigation water was transported and distributed by means of irrigation canal networks or *acequias*. The techniques used to configure the irrigation canals and distribute water to secondary channels or fields have remained practically intact to the present day. Given that the irrigation canals had to be built on a sloped surface in order to carry out the twofold function of conveying and diverting water, its original configuration is very difficult to modify. The area encompassed by the canal is delineated by an upper and a lower boundary characterized by gravity. Thus the main canal or *acequia* marks the upper boundary above which water cannot be distributed due to

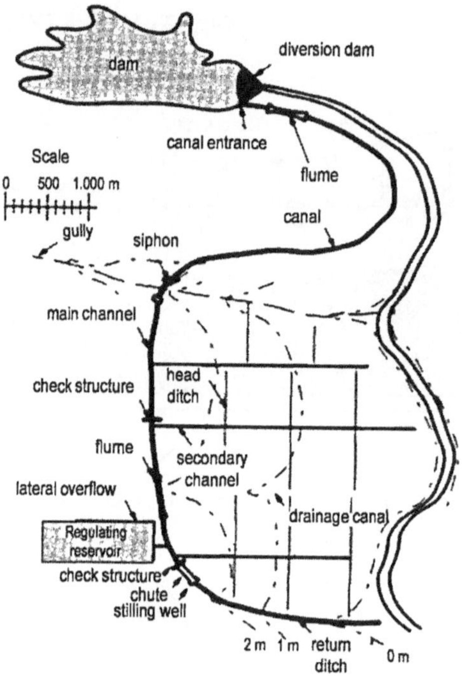

Figure 4. Diagram of a traditional irrigation system (Losada, 2005).

the slope of the terrain, while the lower boundary is marked by the watercourse at the bottom of the valley. The main canal branches out from a diversion dam (see Figure 4) and water is diverted from this canal to secondary channels or fields using wooden or soil divisors to stop the flow.

According to Argemí et al. (1995), *acequías* were built using a stone base set with mortar. The base was then covered with an impermeable surface made of hydraulic limestone and fragments of ceramic material to prevent water loss due to filtration.

The irrigation network is comprised of primary canals (the *acequía*) and secondary channels known as *hijuelas*. The *hijuelas* serve other channels that convey water to the fields. These channels include the *brazal*, which distributes the water taken from the main canal or the *hijuela* to several irrigators and the *regaderas*, which convey the water from a *brazal* to an individual irrigator.

Drainage channels were another important component of the irrigation system as they kept soil free of contamination and prevented water from becoming stagnant. On occasion, this same drainage water was reutilized for irrigation. The drainage system was comprised of *escorrederos*, which received water from one or two irrigators; the *azarbetas*, when water was collected from three or more irrigators or from the *escorrederos* and the *azarbes*, which received water from the *escorrederos* or the *azarbetas* (al-Mudayna, 1991).

4 IRRIGATION TECHNOLOGY

Proportionality was the norm for the distribution of water, that is, each irrigator received water in proportion to the amount of land he held (Glick, 1988). However, the total amount of water that was distributed was not a fixed amount per unit of land but varied according to the discharge of the river. Moreover, the discharge was distributed among the principle channels that took water from it. If little water was available, irrigators could not draw water at will but were obliged to irrigate in a system of turns known as *tanda* or *dula*. This proportional system ensured the equitable distribution of water without measure of time or delivery orifice. An abstract measure called *fila* ("thread of water") was used to represent a share of the total amount of water in a river, spring or canal (Glick, 1988). This concept, albeit with a different system of measurement, is still used today in traditional areas of Spain with gravity irrigation systems (Roldán et al. 1997).

In al-Andalus irrigation, water was distributed in two ways: following the Syrian model, which is typical of huerta agriculture with extensive canals supplied by rivers where water is relatively abundant (water associated to land) or the Yemenite model, which is characteristic of oases where water is distributed according to a time regime and is not associated to the land (Glick and Kirchner, 2000).

As water systems became increasingly more complex over the centuries, irrigation water auctions were set up in certain irrigated districts such as Lorca to sell off portions of water to the highest bidder. Several cultivators would then combine the *filas* they had bought so that they could irrigate their fields with a larger volume of water. In this way, if an irrigator with two *filas* joined forces with two other irrigators that had only one, he could irrigate with the volume of four *filas* in half the time than if he had to do so with just two, while the others were able to irrigate with a fourth.

In the Kingdom of Granada, water was distributed among irrigators by means of *tandas* or irrigation turns. Each *tanda* corresponded to a given time during which a fixed discharge was applied to a field. When the irrigator's time was up, the turn passed to the field below.

In general, water was distributed to irrigators in two ways according to social values and economic priorities; either efficiently or fairly and justly (Glick, 1996). When water was scarce, fairness resulted in a loss of efficiency.

The *fila* is a number that indicates the proportion of the total discharge which a canal takes. In times of abundance, the amount of water that can be drawn depends on the capacity of the canal; while in times of scarcity, water is taken according to a commensurate number of hours. According to Glick (1988), the traditional quotas of *filas* of water are expressed in multiples of twelve where one *fila* is normally the equivalent of one hour of water. The irrigation measurement unit based on hours is customary in the Middle East in places such as Iraq, Yemen or Syria. There the standard water unit is the *qīrāt* and although the system varies from place to place and between canals, it has the connotation of one twenty-fourth part and is normally equivalent to one hour of irrigation.

Argemí et al. (1995) describe certain measurements and proportions used during the Andalusi period whose equivalences are nonetheless difficult to establish. The *abba*, for example, is the equivalent of 24 hours of water or the time it takes to fill a cistern from 6:00 pm to 6:00 am in addition to the 12 hours it takes to empty it, the *azumbre* is equivalent to 3 hours of water, while the *arroba* is a proportion that refers to one-fourth of an irrigation day, or in the case of water distribution, to a measurement ranging from three to four hours. Although both Glick (1988) and Giner Boira (1997) have established the Syrian origin of the Valencian huerta based on the parallelism between the Valencian *fila* and Damascene water measurements, Argemí et al. question this influence, instead emphasizing its Berber origins.

Among the devices used to measure water were holes or flow divisors in which a hole had been made in a stone. Depending on its size, the divisor allowed a fixed number of "threads" of water to pass through, leaving the rest for upstream irrigators. The water from a main canal was divided proportionally between two smaller channels using a divisor that partitioned the flow into two equal parts. For this reason, it was essential that the divisors be accurately designed, leveled, measured and built. In order to measure the time, *clepsydras*, or water clocks, were used. Although they originated in Egypt, the word comes from the Greek term meaning "to steal water". Indeed, the first syllable coincides with that of the word *cleptomania* (McNown, 1976). This mechanism consisted of a bowl-shaped container with markings to represent the passage of hours and a hole at the bottom to allow water to flow out. In order to ensure a linear decrease in water level, the cross-sectional area was smaller at the bottom than at the top. In this way, the decrease in the area of the water surface with decreasing heads tends to compensate for the reduction in outflow rate and, if the cross-sectional area is directly proportional to that rate, the height of the water will fall at a constant rate. Time was also measured by observing the length of shadows; a measurement that corresponded to the time elapsing since sunrise (Cherif Jah and López Gómez, 1994). Thus the time that had passed since daybreak to the moment at which the shadow cast by an irrigator reached a length of eight feet was the equivalent of two hours.

On the basis of the ninth-century treatise *Nabatean Agriculture* by Ibn Wahshiyah, Ibn al Awwan (twelfth century) dedicated specific sections of his book to the signs used to determine if water was near or far from the surface and the manner in which to excavate wells. As regards the former, the existence of certain plant species such as cypresses, brambles and small hawthorns among others indicate the presence of water near the surface. The color, flavor and smell of the surface soil are also good indicators of nearby water since the soil turns moist to the touch and appears to perspire or have dew. As regards wells, Ibn al Awwan refers to their shape (Arabic or Persian), size (to include hydraulic wheels), position (at the highest point of the field to dominate the irrigated area) and the best season for perforating (from August to October).

Ibn al Awwan classified irrigation water according to its source into rainwater, river water, fountain or well water. Rainwater was considered the best type of water and was therefore recommended for the irrigation of horticultural species and for more delicate plants in general. River water was also considered to be of good quality as it is running water, while the water from fountains and wells, which is denser, was preferred for the irrigation of plants with edible roots. Ibn al Awan adds that briny and bitter waters are good for irrigating certain horticultural crops such as purslane, spinach or lettuce. In contrast, he does not recommend salt water for any type of plants.

In consonance with irrigation practice, fruit crops - with the exception of olive trees – were to be irrigated on a frequent basis except when buds are sprouting or during the flowering period. Ibn al Awwan also indicates that it is necessary to water plants with uncovered roots, while delicate plants should not be over-irrigated. Water which lies stagnant for a time is deemed harmful for trees with the exception of fruit trees (Cherif Jah and López Gómez, 1994). Ibn al Awan also gives precise instructions for irrigating fruit trees including localized irrigation (alcorques or small basins), frequent nighttime irrigation (four hours a day from sunset until midnight), fertirrigation (adding manure to the water) and subsaturation (he recommends excavating the soil around the plant, treading upon it lightly and adding manure to retain the water and permit ventilation). Finally, he makes suggestions on the best time of the year to water fruit trees, including olives (traditionally a dry-farm crop), although he concludes that while irrigation can be beneficial for olive trees, the lack of irrigation will not be harmful to them.

Level-basin and graded-border irrigation were the two most common irrigation techniques in the Andalusi period. For this reason, the land was divided into square plots with small channels in the form of furrows or ridges (García Sánchez, 1996). García Sánchez describes the way in which the ridges were laid out; a method which she attributes to the Sicilians. Between every two ridges there was a small channel (*reguera*) which was connected to other channels and the principal canal, which in turn was connected to the plots. The irrigated field was comprised of a series of terraces separated by walls with irrigation plots divided into level basins and graded borders or with small basins at the foot of the fruit trees (Losada, 2004). When discharge was insufficient for the direct application of irrigation water, a reservoir was used to store water. In mountainous zones such as the Alpujarras, the existence of numerous natural springs with small discharges necessitated the construction of regulation reservoirs or storage ponds prior to the distribution of water. In this case, the storage ponds functioned independently, supplying water to relatively small surfaces (Bazzana, 1994).

Land levelling was another fundamental aspect of irrigation systems so that seeds and manure would not be transported by the water from plots at a higher elevation to those on a lower level. The alidade, a topographical device fitted on the back of an astrolabe, was used for levelling the land.

As mentioned in Section 2, a particular irrigation system called *feixes* has been found on the island of Ibiza (Foster, 1952). Here the Muslims excavated a canal network running both perpendicular and parallel to the coastline in a gently sloping floodplain characterized by sedimentary soils. The excavated soil was placed in square-shaped plots (the *feixes*) in order to elevate the soil half a meter above the water level of the canals. Irrigation was subsurface, permitting the filtration of water from the irrigation canals to the subsoil of the *feixes* at a distance of several meters. In order to favor filtration, each *feixa* was crossed by a series of *fibles* or subterranean conduits. One of the *fibles* was placed perpendicular to the coast (longitudinal), while the remaining *fibles* ran parallel to it, although in practice there were few parallel conduits due to the high cost of their construction. A gate was placed at the far end of the longitudinal *fibla* to control the flow of water. It was closed in summer to maintain the water level and opened in winter to discharge excess water.

5 IRRIGATION ADMINISTRATION

Few records exist on irrigation administration in al-Andalus. However, the importance of irrigation in this period is clearly evidenced by the fact that two officials in charge of the administration of irrigation canal water rose to the height of emirs in Valencia and Játiva in the eleventh century.

According to Glick (1988), there existed two types of recognized irrigation officials in al-Andalus: the highest ranking one called the *sāhih al-sāqiya* and a lesser official known as the *amīn al-mā*. From the name of the first derive the words *çabacequia*, *çabacequier* or *sobre-cequiero* meaning "master of the canal". The official called *sâhih* carried out municipal functions that were not provided for by Islamic Law (for example, the surveillance of markets, town policing and the enforcement of norms regarding water use). This officer was entrusted with initiating action on misdemeanors related to the distribution of water. They were therefore not elected by the community of irrigators, but by the governor. In the city of Granada, however, the *acequiero* (mentioned in early sixteenth-century water ordinances of Granada, but of clear Islamic inheritance) held jurisdiction over the mills and tanning sheds.

According to Cherif Jah and López Gómez (1994), the *sāhih al-sāqiya* passed oral sentences as occurs in all administrative procedures under Islamic Law. Given that these were city officials, their jurisdiction could not go beyond the main canal, leaving the organization of the secondary canals to the different tribal authorities. On occasion there existed a particular figure in al-Andalus who specialized in judicial processes related to water known as the *qādi al-miyab* or "official of the water".

Another figure, the *amīn al-mā* ("trustworthy guardian") solely carried out administrative duties, but had no jurisdiction over criminal proceedings. This official was entrusted with the distribution of water and the direction of turns as well as the systems in which the sale of water complicated its distribution, ensuring order in the turns and overseeing water transactions. In Christian times this official was known by the Arabism *alamí* (in Valencian) or *alamín* (in Castilian) or even the more literal name of *fiel de agua* (used in Elche to mean something akin to "the faithful one of the water").

As mentioned above, few records exist from the Islamic period which attest to these administrative processes, However, there are Christian documents dating some years after the reconquest from certain cities such as Valencia that suggest the existence of similar norms and regulations in the Islamic period. This is the case of the *Real Privilegio* of Jaime I in which he orders the *acequieros* to clean the irrigation canals, obliges the irrigators to repair them and prohibits users from returning drainage water to the canals. It also establishes that the irrigators must in turn keep watch on the *acequiero* and denounce him to the water jurists if he is lax in his duties.

Giner Boira (1997) cites a parchment dating from 1223 that gives an account of the sentence passed down on a 20-year-long water litigation between the settlements of Cárze and Torox near Sagunto fifteen years prior to the conquest of Valencia by Jaime I. The sentence passed by the judge of waters, and which finally put an end to the suit, concerned the device used to measure the water – a stone with a hole in it. Due to its deteriorated state, which prevented it from fulfilling the mission for which it was conceived, the judge ordered the stone to be replaced.

In addition to the above, Islamic Law stipulates that the irrigation canals are the property of the community of users that established them, so that solely the community can regulate the matters concerning the canal and has the right to use their water. The irrigators establish their turns but none of them can build a mill or a bridge without the authorization of the others. The Prophet commands that fields must be irrigated in descending order, cautioning that the water should not rise above the ankles in the first plots of land. He also establishes that taxes shall be reduced by half for fields that must be irrigated by extracting groundwater (Vidal Castro, 1995).

6 IRRIGATION TERMINOLOGY

The importance of irrigation in al-Andalus is evidenced by the great number of Arabisms in Castilian related to irrigation and water use in general. Moreover, geographical terms for water derived from the Arabic abound in places where irrigation was practiced or water was distributed; places where hydraulic devices, natural springs, fountains or fluvial currents currently exist or used to exist. Furthermore, it was customary among the inhabitants of al-Andalus to name certain geographical areas according to the characteristics that distinguished them.

Cherif Jah and López Gómez (1994) and Glick (1988) provide ample information on Arabic derivations and toponyms having to do with water. Below we present a glossary of words of Arabic origin related to irrigation and water, although the list is by no means exhaustive and many of the terms included on it may have regional variations with a different meaning than the one given here.

Aceña:	Mill or device to draw water
Acequia:	Irrigation canal
Ador:	Irrigation turn
Alberca:	Irrigation reservoir
Albufera:	Lagoon
Alcantarilla:	Drainage conduit
Alcubilla:	Water box
Alfaguara:	Abundant spring
Aljibe:	A cistern for domestic use usually covered with a barrel vault
Aljofaina:	A container for water
Almenara:	Return ditch
Arcaduz:	Bucket of a water wheel
Atanor:	Drain pipe
Atarjea (o atajea):	Conduit which conveys household water to a drainage system
Azarbe:	Drainage canal
Azarbeta:	Same as above but smaller in size
Azud:	Diversion dam; waterwheel
Canal:	Canal, channel
Cenia:	Wheel moved by the force of the current or by human or animal power depending on the area (Valencian)
Noria:	Wheel to raise water
Rafa:	Board placed across a canal to stop the flow or divert water
Rambla:	Dry riverbed for occasional flooding
Tanda:	Irrigation turn

Some of the most striking examples of water-related place names are those which refer to the storage of water and hydraulic wheels. These toponyms vary from region to region. Although we will not go into great detail here, it is interesting to note the wide range of toponyms derived from the word *noria* in various provinces of Spain: *Arnorias* (Albacete); *Anoria* (Almería); *Nora* (Cáceres, León, Oviedo); *Añora* (Córdoba); *Ñora* (Granada, Murcia); *Naura* (Lleida).

Numerous rivers in southern Spain likewise bear Arabic names. All of them begin with the prefix *Guad*, a derivative of the Arabic *wadi* (river) and end with an appellative that characterizes them. Some of the most well-known of these include:

Guadajoz (Córdoba):	Murky river
Guadalaviar (Teruel):	White river (in Valencia it turns into the Turia River)
Guadalén (Ciudad Real):	River of the fountain
Guadalfeo (Granada):	River of the gully or ravine
Guadalhorce (Málaga):	River of the guardian
Guadalimar (Jaén):	Red river

Guadalmedina (Málaga):	River of the city
Guadalquivir:	Big river
Guadarrama (Madrid):	River of the sand
Guadiana:	River of Ana (a place near Calatrava)

7 CONCLUSIONS

The Islamic contribution to irrigation in al-Andalus is unquestionable. The spread of scientific and technological practices from the East and their subsequent diffusion to the New World has been fully demonstrated. The vast number of place names and terminology having to do with irrigation and water use practices clearly attest to this phenomenon.

The extraction of water by means of *azudes, norias* and *qanats* constitutes one of the most significant contributions of Islamic culture. The principle of proportionality and fairness in the distribution of water are a model that has survived throughout the centuries. Irrigation administration was of notable importance under Islamic Law. The configuration of irrigation canals and techniques to divert water to fields or secondary channels using highly precise divisors still survive today in traditional irrigation networks of Spain.

Finally, as Albarracín and Martínez (1989) affirm, the Andalusis were nature lovers whose poetic evocations of the rural world, including irrigation, are a clear testimony of their affection. As Abu Amir Ibn al- Murabit expresses in his amorous lamentations:

> *"Allí hay un riego hecho con mis lágrimas, ¡oh gacela¡, y una umbría formada por mis costados. Abrévate en esta agua abundante y ven a gozar de esta fresca sombra sin temor a ser rechazada ni asustada."*

REFERENCES

Albarracín, J. & Martínez, J. (1989) El agua y el riego en la poesía árabe andalusí (siglo XI). *I Coloquio de Historia y Medio Físico. El agua en zonas áridas: arqueología e historia.* Almería. pp. 97–119.

Al-Mudayna. (1991) *Historia de los regadíos en España* (… a.C.-1931). Ministerio de Agricultura, Pesca y Alimentación. Madrid.

Argemí, M., Barceló, M., Cressier, P., Kirchner H. & Navarro, C. (1995) Glosario de términos hidráulicos. In: A. Malpica (coord.) *El agua en la agricultura de Al-Andalus.* El Legado Andalusí, Granada.

Barceló, M. (1989) El diseño de espacios irrigados en Al-Andalus: un enunciado de principios generales. *I Coloquio de Historia y Medio Físico. El agua en zonas áridas: arqueología e historia.* Almería. pp. XV–XLVII.

Bazzana, A. (1994) La pequeña hidráulica agrícola en Al-Andalus. In: E. García Sánchez (ed.) *Ciencias de la naturaleza en Al-Andalus. Textos y estudios. III.* C.S.I.C. Escuela de Estudios Árabes. pp. 317–335.

Box Amorós, M. (1992) El regadío medieval en España: época árabe y conquista cristiana. In: A. Gil y A. Morales (coords), *Hitos históricos de los regadíos españoles.* Ministerio de Agricultura Pesca y Alimentación, Madrid. pp. 49–89.

Caro Baroja, J. (1983) Norias, azudes y aceñas. *Tecnología Popular Española.* Madrid. pp. 239–348.

Carrasco, A.I. (1996) La percepción del agua y los sistemas hidráulicos en la obra de al-Idrisi. *Actas del II Coloquio de Historia y Medio Físico. Agricultura y regadío en al-Andalus.* Almería. pp. 57–65.

Córdoba, R. 2004. La noria de tiro en la Córdoba bajomedieval. Elementos y funciones.

En: S. Gómez Navarro (coord.) El agua a través de la historia. Estudios de Historia I. Asociación "Arte, Arqueología e Historia". Córdoba. pp. 79–96.

Córdoba, R., Albendín, A., García Muñoz, J.M. & Ortiz García, J. (2004) Puertos, azudes y norias. El patrimonio hidráulico histórico de Palma del Río (Córdoba). Fundación El Monte. Sevilla.

Cherif Jah, A. & López Gómez, M. (1994) El enigma del agua en Al-Andalus. Ministerios de Agricultura y de Obras Públicas, Madrid.

Del Río Moreno, J. (2002) Influencia de la cultura agraria árabe en la agricultura que implantaron los europeos en América. In: F. Nuez. (ed.) *La herencia árabe en la agricultura y el bienestar de occidente. Universidad Politécnica de Valencia.* pp. 411–424.

Foster, G.M. (1952) The feixes of Ibiza. *Geographical Review*, 42(2), 227–237.

García Sánchez, E. (1996) Cultivos y espacios agrícolas irrigados en Al-Andalus. *Actas del II Coloquio de Historia y Medio Físico. Agricultura y Regadío en Al-Andalus*, pp. 17–37.

Gilman, A. & Thornes, J.B. (1985) *Land Use and Prehistory in South East Spain*. G. Allen & Unwin, London.

Giner Boira, V. (1997) *El Tribunal de las aguas de Valencia*. Fundación Valencia III Milenio. Valencia.

Giráldez, J.V., Ayuso, J.L., García, A., López, J.G. & Roldán, J. (1988) Water harvesting in the semiarid climate of southeastern Spain. *Agricultural Water Management*, 14, 252–263.

Glick, T.F. (1979) *Islamic and Christian Spain in the Early Middle Ages*. Princeton University Press, Londres.

Glick, T.F. (1988) Regadío y Sociedad en la Valencia medieval. Valencia.

Glick, T.F. (1996) Riego y tecnología hidráulica en la España Islámica: consideraciones metodológicas. In: C. Álvarez de Morales (ed.) *Ciencias de la naturaleza en Al-andalus. Textos y estudios. IV*. C.S.I.C. Escuela de Estudios Árabes. pp. 71–91.

Glick, T.F. & Kirchner, H. (2000) Hydraulic systems and technologies of Islamic Spain: history and archaelogy. In: P. Squatriti (ed.) *Working with water in Medieval Europe. Technology and Resource-Use*. Brill, Leiden, Holanda. pp. 267–329.

Goblot, H. (1979) *Les qanats. Une technique d'adquisition de l'eau*. École des hautes études en sciences sociales. Mouton Editeur. Paris.

Hermosilla, J. (ed.) (2006) Galerías drenantes del sureste de la Península Ibérica. Uso tradicional del agua y sostenibilidad en el mediterráneo español. Ministerio de Medio Ambiente, Madrid.

Ibn al Awan. Siglo XII. *El libro de agricultura de al Awan*. Edition and comments on the edition by Banqueri (18th century) by José Ignacio Cubero Salmerón (2001). Empresa Pública para el Desarrollo Agrario y Pesquero de Andalucía, Sevilla.

López-Camacho, B., de Bustamante, I. & Iglesias, J.A. (2005) El viaje de agua (qanat) de la Fuente Grande de Ocaña (Toledo): *Pervivencia de una reliquia histórica. Revista de Obras Públicas*, nº 3451, pp. 43–54.

Losada, A. (2004) Espacios hidráulicos en Al-Andalus. II Simposio Internacional "Repensar Al-Andalus a través del tiempo y el espacio: Agua y agricultura". Córdoba.

Losada, A. (2005) El riego. II. Fundamentos de su hidrología y de su práctica. Mundi-Prensa, Madrid.

Martí, R. 1989. Oriente y occidente en las tradiciones hidráulicas medievales. *I Coloquio de Historia y Medio Físico. El agua en zonas áridas: arqueología e historia*. Almería. pp. 421–440.

McNown, J.S. (1976) When Time Flowed. The story of the Clepsydra. *La Houille Blanche*, 5, 347–353.

Meyer, M.C. (1996) Water in the Hispanic Southwest. The University of Arizona Press, Tucson, Arizona.

Ministry of Public Works. (1918) Medios que se utilizan para suministrar el riego. 2 vols. Madrid.

Pavón, B. (1990) *Tratado de arquitectura hispano-musulmana. I. Agua*. Consejo Superior de Investigaciones Científicas, Madrid.

Roldán, J., Reca J. & Losada, A. (1997) Uso del agua de riego en el valle del Guadalquivir: zonas del Bembézar y de Fuente Palmera. In: López-Gálvez y, J., Naredo, J.M. (eds) *La gestión del agua de riego*. Fundación Argentaria. pp. 99–138.

Roldán, J., Pérez Urrestarazu, L. & Moreno, F. (2006) Canalones Hidráulicos en los tejados de la Mezquita de Córdoba. Al-Mulk, *Anuario de Estudios Arabistas*, 6, 59–67.

Trillo, C. (2002) Regadío y estructura social en al-Andalus: la propiedad de la tierra y el derecho al agua en el reino nazarí. In: Pérez-Embid, J. (ed.) *La Andalucía Medieval. Actas "I Jornadas de Historia Rural y Medio Ambiente"*. Servicio de Publicaciones de la Universidad de Huelva. pp. 71–98.

Vidal Castro, F. (1995) El agua en el derecho islámico. Introducción a sus orígenes, propiedad y uso. In: A. Malpica (coord.) *El agua en la agricultura de al-Andalus*. El Legado Andalusí, Granada.

CHAPTER 6

Hydraulic advances in the 19th and 20th centuries: From Navier over Prandtl into the future

Willi H. Hager
VAW, ETH Zurich, Switzerland

ABSTRACT: The development of hydraulics and hydrodynamics in the 19th and 20th centuries are described. In the first part; the formation of the engineering education is outlined, with a particular emphasis on the advances in France. One the Polytechnics were founded, engineering associations came into life, mainly on a national basis. The impact of hydraulic engineering in particular and engineering in general was largely dependent on these associations, which eventually developed into the modern international associations. The second part introduces some notable hydraulicians of Europe that have contributed to the wealth and the impact of hydraulic engineering as a modern technical science. The paper gives at the end also an outlook into the developments that may be expected in the near future. Given the many issues of the modern world that are related to the water sciences, such as abundance and scarcity of water, and the natural disasters related to rivers, estuaries and the sea, there will be high expectations to our profession from the society in mastering at least partially an equilibrium with water for mankind. The author is convinced.

1 INTRODUCTION

Hydraulics is as old as humans: Because water is a basic need for all life, water supply in particular was developed as soon as larger human settlements were established. In antiquity, the high cultures in China, in India, in Mesopotamia, or in the Americas were aware of this basic need to develop their culture. Later, during the Roman era, water played an essential role and the Roman aqueducts that have survived the past two thousand years still attract our attention. Despite the large water supply schemes developed by the Romans, knowledge eventually was lost due to a less quiet time from roughly 500 AC to 1500 AC. The concerns posed by the water as a basic human infrastructure regained importance only after the Middle Ages around 1500 AC. Waterborne diseases such as pestilence or typhoid fever caused the death of millions. The Renaissance era saw the *New human* that conquered these fates of the dark.

Along with interest in science in general, water received particular interest from the top scientists of the time, including Leonardo da Vinci (1452–1519) and Galileo Galilei (1564–1642). In the 17th and in the 18th centuries, some basic observations were conducted relating towards orifice flow, the understanding of river flow, irrigation techniques and naval engineering. These progresses were slow, however, because a general concept in mechanics and physics was missing. Also, a number of proposals was made for a particular problem and some solutions were possibly accepted in one or the other region. The basic hydraulic problem of weir flow was successfully treated for instance by Giovanni Poleni (1683–1761) only in 1717.

A significant step forward was made by the Swiss Daniel Bernoulli (1700–1782) and Leonhard Euler (1707–1783): Whereas Bernoulli introduced the concept of energy head and provided a relation between pressure and velocity for ideal fluid flow, Euler introduced the equations describing hydrodynamic processes on the basis of a set of differential equations. In both approaches, an essential feature was not accounted for, resulting in two future directions referred to as hydraulics and hydrodynamics. In the latter, processes were described for an ideal fluid in which viscosity was absent, whereas the effect of viscosity was accounted for in hydraulics with an approach that invalidated some of the requirements previously accepted in mechanics. In river flow, for instance, the concept of Bernoulli would predict velocities that were much too large and thus of 'no practical use'; the hydraulic approach introduced by Antoine Chezy (1718–1798) was a relation between the average flow velocity, the hydraulic slope and the hydraulic radius plus a coefficient that accounted in some way for the river roughness pattern. This latter coefficient was subject to criticism from the theoretical school because it was poorly defined, could essentially be 'measured' on a certain river reach and contained time and length as dimensions. "Why should time be included in a measure of roughness?", questioned the scientists, and aimed to solve the problem in vain with a more complex approach. There were similar problems in other branches of fluid flow and it became almost evident that a different technique had to be sought for a successful solution. Also, it was coming true that a new profession had to be defined. The following introduces the age of the engineer and his way of solving problems in the 19th century. His methods have been subject to significant changes in the IT era but his approach has essentially remained identical, namely solution oriented, economic and concise.

2 ENGINEERING AND ENGINEERS

The notion *engineer* originates from the engine, although its current meaning is quite different. The steam machine was invented at the end of the 18th century by James Watt (1736–1819); this development led to a revolution in the technical world because mechanical power was available suddenly at any location. However, how should the power be used? How would the engine work optimally?, and How could this power be used economically? These questions were addressed to these who worked with the engines, namely the engineers. Until the end of the 18th century, engineers were often self-made men who had a flair for mechanical works. It was soon realized that an improved engineering education was urgently needed.

The first important 'engineering school' was founded in Paris. During the reign of Louis XV (1710–1774), the Corps of the Bridges and Roads was established which supported the king in all matters relating to the traffic infrastructure, including military logistics. The current *Ecole Nationale des Ponts et Chaussées* ENPC was thus founded in 1747 with Jean-Rodolphe Perronet (1708–1794) as its first director until 1794. Chézy, previously referred to as great hydraulician, took over the directorate for the last two years of his life. The School was famous for its combination of basic theory and engineering applications. Some of the most famous engineers from the ENPC were Gaspard Riche de Prony (1755–1839), the ENPC director who followed Chézy and who is known for research in open channel flow, Louis Navier (1785–1836) to be presented below, Adhémar Barré de Saint Venant (1797–1886) known in engineering mechanics, Arsène Dupuit (1804–1866) with his works in groundwater flow and Eugène Belgrand (1810–1878) who worked for the water supply of the city of Paris (Divers 1961). Other important former students of the establishment are highlighted by Coronio (1997). Figure 1 shows a view of the original *Ecole* in Paris, serving for more than 150 years for the education of the French leaders in civil engineering. This institution had its peak in the 19th century and felt – as many other French establishments – the terrible consequences of two world wars. Today, a new school serves a much smaller community in the suburbs of the French capital. A legacy of the ENPC is its journal *Annales des Ponts et Chaussées*, founded in 1831 and lasting until 1971. In 1977, a novel form of the journal appeared which is currently successful and often presents historical articles that evidence the impact and power of an old engineering organization.

A second outstanding engineering school was founded during the French Revolution. It was clear to both the French Emperor Napoleon I and the French educating authorities that basic sciences were largely absent in the engineering education. In 1794, the *Ecole Polytechnique* was founded in Paris and served during the 19th century as the model for the idea of technical education throughout the

Figure 1. Interior court of the old *Ecole des Ponts et Chaussées* in Paris; a monument for the deaths and an obelisk are shown in the front (Sébille 1999).

world. The following subjects were and still are taught: Geometry, Calculus, Mechanics, Astronomy, Physics, Chemistry, Architecture, History and Literature, and Economics and Drawing. One is astonished to find here the last four subjects, yet the polytechnic idea was to provide full basic technical and humanistic education (Callot 1982, Belhoste et al. 1994).

Great names in engineering originating from Ecole X, as also referred to, are Jean-Victor Poncelet (1788–1867) one of the inventors of the modern water wheel and director from 1848 to 1850, Gaspard-Gustave Coriolis (1792–1843) known for the Coriolis acceleration, who directed the school during his last four years, the mathematician Charles Bossut (1730–1814) who made notable experiments on fluid resistance, the mathematics professor Augustin Louis Cauchy (1789–1857) with his important works on water waves, the mechanics professors Joseph-Louis Lagrange (1736–1813) with his outstanding book Mécanique céleste, Siméon-Denis Poisson (1781–1840) again known for works in wave hydraulics, Jean-Baptiste Bélanger (1790–1874) with his basic contributions to the hydraulic jump and Maurice Roy (1899–1985), the famous mechanical and aeronautical engineer of the 20th century.

Whereas the *Ecole Polytechnique* provided a basic engineering education, the other so called Great Schools of France such as the *Ecole des Mines*, the *Ecole Centrale* and the ENPC had a more specialized background. There was also a number of professors teaching at several of these schools. This has possibly resulted in the fall of the French educational system because the teachers made not their best out of the profession but were in charge at many locations, obviously for financial reasons. In contrast to ENPC, the *Ecole Polytechnique* had never an outstanding journal: The *Journal de l'Ecole Polytechnique* founded in 1812 and lasting until 1924 appeared not systematically and contained extremely long papers. Most of the French academia published their papers in other

Figure 2. Relief symbolizing the unity of the most important sciences (Pfammatter 1995).

French journals, such as in the *Annales des Mines* founded in 1816, the *Annales de Physique et Chimie* founded in 1798 and lasting until after WW II, in the *Journal des Mathématiques Pures et Appliquées* established in 1836 or in the various journals of the *Académie des Sciences*. Short notes were often inserted in its *Comptes Rendus*, whereas longer articles were submitted to the *Mémoires* of the Academy. Both systems have finally not proofed effective in research and technology because they were hardly adopted from other institutions.

The polytechnic idea spread among European universities first, and then found its way to India and the United States towards the end of the 19th century. Notable institutions were first established in Prague, namely the Czech Technical University in 1806; in Karlsruhe, Germany in 1825; then in Zurich, where the present Swiss Federal Institute of Technology ETH was founded in 1855. Other important schools were the *Politecnico* in Milan, Italy, established in 1863 after the Italian State had been reorganized; the *Escuela de Ingenieros de Caminos, Canales y Puertos* in Madrid founded in 1802; the Delft Technical University in the Netherlands established in 1842; the Imperial College in London founded in 1907; the University of Manchester founded in 1824; and the Polytechnic Institute of Saint Petersburg, Russia, founded in 1899. Within a century, most countries of Europe had set up their technical education and therefore greatly contributed to the advancement of the technical sciences and to the reputation of the engineering profession. Engineering from its original meaning had thus significantly advanced to a novel profession, including civil, mechanical, and chemical engineering. Today, many more branches of the engineering education are taught and therefore add to the advancement of the highly industrialized modern world, but may also complicate the current life due to over-specialization.

3 PROFESSIONAL ASSOCIATIONS

Engineering societies were founded along with the development of the Polytechnic schools. In contrast to the technical world before roughly 1800, the engineering profession organized itself by these Associations and presented itself as a companionship, instead of isolated builders and researchers. In parallel, these associations had their own professional journals in which both the

Figure 3. First Meeting of the International Association of Hydraulic Research, Berlin 1937.

advancement in the profession was presented with papers and the history and the members' biographies of the associations were included. Birthdays or deaths of members were announced and led to a strong interrelationship among the members. Many of these associations have survived until today, but the original confidential contact is currently smaller (Fig. 3). Whereas professional ethics was strongly maintained from the 19th century up to about 1950, professional competition increased and finances for projects reduced such that these formerly strong bands were loosened. It appears that these associations will not survive the 21st century if developments continue as they have from the 1950s. This would be a pity for the profession because many contacts would be lost and the formerly high-valued engineering image would reduce in the public opinion. Time for public affairs has become short, because the modern human being is so much occupied with personal affairs. Electronic media and the general mobility add to this hasty development.

It may be interesting to review the most important journals issued from roughly the 1850s to around 1950 by the European engineering associations. The list of journals is incomplete but reflects those that have had a definite impact in hydraulic engineering. Alphabetically, one might state, therefore, the indications presented in Table 1. It is noted that three countries were and still are particularly rich in engineering journals with a definite impact. These are France, Germany and the United Kingdom. From the 1950s, the English language became important in engineering, such that currently the most important engineering journals are issued in the United Kingdom. This trend will increase in the future, with the USA already keeping the top journals in almost all scientific branches. This will also have a direct impact on the national journals, because their authorship will reduce, and financial problems may lead to a further reduction of journals.

Besides the journals issued from engineering associations, there were in parallel a number of journals published by Academies or Universities. This applied particularly to the countries of Southern European such as Italy. The trend here is also retrograde because young engineers are almost forced to publish in review journals to advance their career. Around 1900, these media had played a significant role next to the standard national journals.

Next to national journals, the engineering community has also issued a number of international media that attract particularly the modern research engineer. The international associations were

Table 1. Main engineering journals of the European countries, (1) Title, (2) Year of foundation to end with – still existing, (3) Journal main topic, and (4) B = biographical information available.

Austria			
Allgemeine Bauzeitung	1836–1918	General engineering	
Ö. Zeitschrift Ing.-Arch. Verein	1849–	Civil engineering	B
Ö. Ingenieur-Archiv/Acta Mechanica	1946–	Mechanics	B
Ö. Wasserwirtschaft	1907–	Hydropower	B
Ö. Wochenblatt Baudienst	1895–1924	Public works	B
Belgium			
Annales des Travaux Publics	1843–1990	Public works	B
Bulletin de Navigation	1926–	Internal navigation	B
CeBeDeau	1950–	Water Resources	B
Revue Universelle des Mines	1857–1973	Engineering	B
Czech Republic			
HDI-Mitteilungen	1908–1939	Engineering	B
Plyn y voda	1921–1966	Water Resources	B
Vodni Hostpodarstvy	1951–	Hydraulic engineering	B
Denmark			
Ingeniøren	1892–	General engineering	B
Tekniske Forenings Tidskrift	1893–1920	Technology	
France			
Annales des Ponts et Chaussées	1831–1971	Public works	B
Annales des Mines	1816–1944	Mining engineering	
Annales des Arts et Métiers	1863–	Public works	B
Génie Civil de France	1880–1977	Public works	
La Houille Blanche	1902–	Hydraulic works	B
Ingénieurs Civils de France	1848–1965	Public works	
Revue des Eaux et Forêts	1862–1948	Waters and forests	
Revue Générale Electricité	1917–	Hydropower engineering	
Travaux	1917–	Public works	
Germany			
Gas-Wasserfach	1857–	Water supply and wastewater	B
Gesundheits-Ingenieur	1878–	idem	B
Polytechnisches Journal	1820–1931	Technology	
Schiffbau	1900–	Marine engineering	B
VDI-Zeitschrift	1857–	Mechanical engineering	B
Wasserkraft	1906–1944	Hydropower	B
Wasserwirtschaft	1905–	Water resources	B
Z. Binnenschifffahrt	1901–	Internal navigation	B
Zentralblatt Bauverwaltung	1881–1944	Engineering projects	B
Hungary			
Hydrologiai Közlöny	1921–	Water Resources	B
Vizügyi Közleneyek	1879–	Water Resources	B
Ireland			
Institution Civil Engineers, Trans.	1845–1935	Engineering	B
Italy			
Giornale del Genio Civile	1863–	Civil Engineering	B
Il Politecnico	1853–1937	Engineering	B
L'Acqua	1923–	Hydraulic engineering	B
L'Energia Elettrica	1924–	Hydraulic engineering	B

(*Continued*)

Table 1. (Continued)

Netherlands			
De Ingenieur	1886–	General engineering	B
Koninklijke Wetenschappen, Verh.	1857–1928	Technology	
Water	1939–1972	Hydraulic engineering	
Waterstaats-Ingenieur	1920–1972	Hydraulic engineering	
Norway			
Teknisk Ukeblad	1858–1967	General engineering	B
Poland			
Archiwum Hydrotechniki	1954–1988	Hydraulics	B
Czasopismo Technize	1880–	Engineering	
Portugal			
Technique	1926–1987	Engineering	B
Romania			
Revue Mécanique Appliquée	1956–	Engineering	B
Studii Hydraulica	1960–1988	Hydraulics	B
Russia			
Bull. Akad. St. Petersburg	1856–1934	Engineering	B
Fluid Dynamics	1966–	Hydraulics	B
Gidro Stroitel'stvo	1931–	Hydraulic engineering	B
Izvestiya VNIIG	1931–	Hydromechanics	B
Spain			
Ingenieria y Construccion	1923–	General engineering	
Revista de Obras Publicas	1853–	Public works	B
Sweden			
Kungl. Tidskrift	1862–	Engineering	B
Tidskrift Teknisk Forskning	1931–1974	Engineering	
Switzerland			
Bulletin Tech. Suisse Romande	1875–	General engineering	B
Revue Polytechnique	1899–	General engineering	B
Schweiz. Bauzeitung	1883–	General engineering	B
Wasser-Energie-Luft	1907–	Hydropower engineering	B
United Kingdom			
British Waterworks Association	1919–1972	Water supply	B
Engineer	1856–	Engineering	B
Engineering	1866–	Engineering	B
Institution of Civil Engineers	1842–	Engineering	B
Institution of Mechanical Engineers	1847–	Engineering	B
Institution of Naval Architects	1860–	Naval hydraulics	B
Institution of Water Engineers	1896–1987	Hydraulic engineering	B
J. Fluid Mechanics	1956–	Fluid mechanics	
Philosophical Magazine	1851–1975	Science	
Royal Society, Proc.	1800–	Science	
Water Power & Dam Construction	1949–	Hydropower	B
Yugoslavia			
Tehnica	1946–1980	Technology	B

Table 2. Main European waterborne engineering associations.

PIANC	1882	Brussels	bi-annual	inland navigation
IUTAM	1922	Eindhoven	four years	mechanics
IAHS	1922	Paris	tri-annual	hydrology
ICOLD	1933	Paris	tri-annual	hydropower
IAHR	1935	Delft/Madrid	bi-annual	hydraulics
ICID	1950	New Delhi	tri-annual	irrigation and drainage

developed from the 19th century and they keep on with more and more specialized groups. One of the earliest water-based engineering associations was the Permanent International Association of Navigation Congresses PIANC founded in 1882 in Brussels, Belgium. The original purpose was the planning and the extension of knowledge among experts in inland navigation. This goal has essentially been retained until today. Table 2 reviews the main international engineering associations in water sciences of Europe. Column 1 gives the official abbreviation, 2 the year of foundation, 3 the seat of the secretariat, 4 the conference rhythm, and 5 the main topic considered. Figure 4 shows the logos of some well known engineering associations.

The papers submitted to an international congress have gone through a process as did in parallel those submitted to the national journals. Until about 1970, both the congresses and the national journals had a high reputation, but a decline developed later on. A number of high-caliber hydraulic engineers have submitted all papers of their career to the same journal, and they were proud because they thought to add to the national engineering reputation. Currently, papers submitted to either national journals or international congresses have lost much of their value, because the latter papers are much lower rated in terms of the impact factor as compared to papers published in review journals. From about 10 years, an even more dramatic development initiated by publishing conference proceedings as CD-ROMs. These papers are therefore hardly read because they are normally beyond a serious review process. Financial constraints are stated to be the main cause for this development; one may ask if this is really true relative to the often low quality and the large quantity of the papers submitted by a 'normal' academic engineer. This trend will hardly be changed in the future because national journals are outdated and considered merely as media with a news section.

Conferences have developed in a similar direction with the meeting of colleagues as the main 'happening' and the presentation of results becoming more and more a must for the specialist. The impact of the international congresses is still large, however, particularly among younger engineers because of the traveling to foreign countries, meeting with friends and sharing new ideas. In addition, competitions among young scientists are popular, such as the J.F. Kennedy Student Paper Award of the IAHR, held from 1993 to honor the late John F. Kennedy, formerly an IAHR president. The IAHR is currently preparing the electronic paper publication to cut further the annual fees for its members mainly of the Third World. One may question how long it will take until the value of an engineering journal is lost.

In the USA, a similar trend may be observed. Take, as an example, the American Society of Civil Engineers ASCE: This large engineering association is one of the few national institutions with an increasing number of members. The ASCE may be regarded as divided into two distinctly different groups; on the one hand there is the national ASCE membership with their thousands of active civil engineers, whereas the international ASCE body originating mainly from academia uses ASCE journals. These are internationally known for their large impact factor due to the global distribution and an excellent reputation. In contrast, the ASCE National Conferences are often below all professional standards, at least for an European.

In summary, one may therefore state that the future of the engineering community, with a stress on the second word, will definitely depend on the development of the engineering associations. These are actually relatively low rated by the common engineer, who sees no direct profit as a

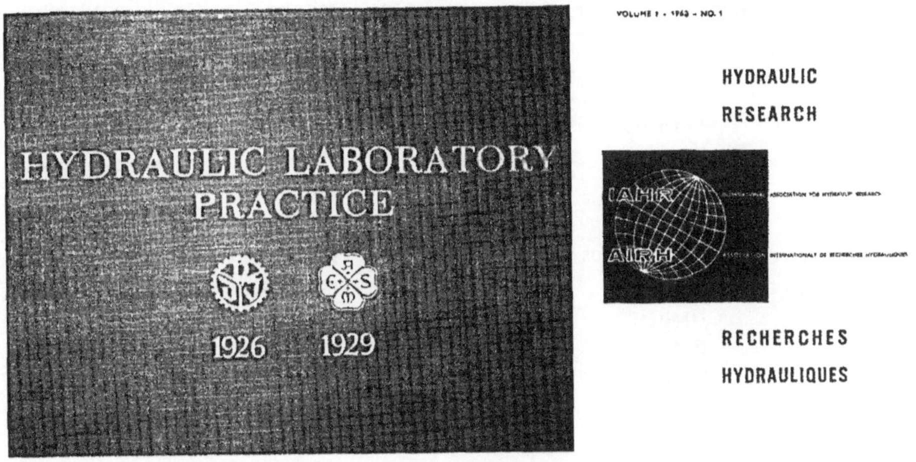

Figure 4. Logos of important engineering associations, namely IAHR, VDI and ASME.

member and therefore does normally not support its activities. Whereas engineering associations were originally important as a national institution, the global spread of the engineering idea has taken lots of its basic values. Young engineers do often not identify with an engineering association because most problems of their profession may be solved without its members. Engineering conferences appear to be still popular because they allow for the immediate exchange of ideas, and is an excellent forum for contacts with colleagues from the national to the global level. Financial problems make these contacts more complicated in the Third World as in Europe and in the richer countries.

Review journals have taken a significant role for the researcher in engineering sciences. Particularly PhD students can hardly publish their research in national journals and in engineering conferences because of their low 'value'. This fact may also be considered the reason for the increasing gap between the practitioner and the researcher, a gap that is currently widening because of a conflict in interests. The Internet may possibly reduce the gap considered because it is a forum open for all engineers, and contacts to any person have become much less complicated. Currently, there is a transition form the classical engineer to what we might call the modern engineer, whose background is not only classical engineering education, standard books and engineering experience, but also a global appreciation of a certain key problem and the international net of engineering knowledge.

4 NOTABLE HYDRAULIC ENGINEERS

The development of the engineering profession from around 1800 to today may be illustrated with selected biographies. In the following, a number of outstanding individuals is highlighted with their curriculum vitae, based on the author's work and volume 2 to be published in 2007 (Hager 2003). Each major European country shall be highlighted with one person who has contributed significantly to hydraulic engineering. Table 3 reviews these engineers based on the following items: (1) Country, (2) Family name, (3) First name, (4) Years of life, and (5) Field of contribution. A short bibliography is added to each individual to obtain an idea of his main writings. The origin of the portrait is indicated with P.

The selection of individuals is based on her or his professional merits in water sciences and the sustainability of technical results. Sometimes it was impossible to find a biographical background

Table 3. Outstanding hydraulic engineers of Europe from 1800 to 2000 (Country selection).

AUSTRIA	Wang	Ferdinand	1855–1917	Torrent engineering
BELGIUM	Denil	Gustave	1865–1940	Fish engineering
CZECH REPUBLIC	Smetana	Jan	1883–1962	Hydrology
DENMARK	Engelund	Frank	1925–1983	Hydrodynamics
FRANCE	Navier	Louis	1785–1836	Mechanics
GERMANY	Prandtl	Ludwig	1875–1953	Mechanics
GREECE	Argyropoulos	Praxitelis	1918–1995	Hydraulics
HUNGARY	Bogardi	Janos	1909–1998	Fluvial hydraulics
IRELAND	Nash	James E.	1927–2000	Hydrology
ITALY	Bidone	Giorgio	1781–1839	Hydraulics
LATVIA	Weinblum	Georg	1897–1974	Naval hydraulics
LITHUANIA	Kolupaila	Steponas	1892–1964	Hydrometry
NETHERLANDS	Dronkers	Jo Johannes	1910–1973	Hydro–informatics
NORWAY	Vogt	Fredrik	1892–1970	Hydropower
POLAND	Matakiewicz	Maksymilian	1875–1940	Fluvial hydraulics
PORTUGAL	Almeida d'Eça	Bento F.	1825–1906	Irrigation
ROMANIA	Coanda	Henri	1886–1972	Fluidics
RUSSIA	Kochina	Pelageya Y.	1899–1999	Groundwater hydraulics
SPAIN	Iribarren	Ramon C.	1900–1967	Harbor engineering
SWEDEN	Fellenius	Wolmar K.	1876–1957	Geotechnics
SWITZERLAND	Schnyder	Othmar	1904–1974	Hydraulics
UNITED KINGDOM	Froude	William	1810–1879	Naval hydraulics
YUGOSLAVIA	Boreli	Mladen	1922–1995	Groundwater hydraulics

for individuals that are known in engineering literature. It was generally simpler to establish a biography for people with an academic background than for engineering consultants, because academia has additional means to highlight the career of their members than does the private engineering sector. This is particularly true for individuals having passed away a century or more ago. Today, a problem in establishing biographies is for instance the exact date of death, because this information is hardly published in 'modern' engineering journals. One university even stated that the career of their members is finished at retirement, and that all the documents are then dropped because of financial and organizational limitations. Modern society has – as already stated – a different attitude toward its members.

The most difficult part of a biography is finding a portrait. Many a user may think that these portraits are available on Internet. This is true for the most important and well known engineers. However, this is not true for many others. One way may be to contact the university of the person, or professional associations in which the person played a role. In many other cases, there is no chance of finding it other than by a detailed telephonic search. I found in volume 1 cited previously the portraits of some 150 individuals only by this approach. The success depends then on the family name. It was for instance simple to find the family of Paul Du Boys (1847–1924), the famous French engineer having established the first theory of sediment transport. It would, however, be extremely time consuming if one would look in Germany or in Switzerland for Hans Meier, given that this is a common name. The success may also depend on the family structure, because only sons keep the name, whereas daughters would not. For some persons, I used years until I finally found what I was looking for, or I stopped....

4.1 *WANG 23.12. 1855 Wien/A – 26.4. 1917 Wien/A*

Ferdinand Wang had first a degree in civil engineering and then the degree in forestry from the then newly founded University of *Bodenkultur* in Vienna. In 1878, he joined the forest services of *Salzburg* County and continued as a torrent engineer with the services in *Villach* first, and in *Brixen* from 1887. In this year he started lecturing on that topic at *Hochschule für Bodenkultur*. He was asked in 1902 to direct the newly founded department of torrent engineering, and was appointed a *Ministerialrat* in 1908. Later, he was a recipient of the Austrian Leopold Order, and was awarded the title of professor at his *Hochschule*.

Wang may be considered the founder of technical torrent engineering, an important issue in steeply sloping catchments where enormous floods may carry not only water but also ice, sediment and wood. Debris flow is a modern issue which was studied by Wang around 1900 in a less scientific way than currently of course. His book *Grundriss der Wildbachverbauung* is rich in terms of experience and presentation by photography and can be recommended for reading to all working in this field. It appears that the combination of civil and forest engineering was particularly successful in Wang's career, because he not only investigated specific disasters but also proposed the means to counter them. The effect of a forest on water retention was thereby fully acknowledged.

Anonymous (1917) KK Ministerialrat Professor Ferdinand Edler von Wang. *Österreichische Forst- und Jagd-Zeitung*, 35(19), 112–114.

Anonymous (1917) Ferdinand Edler von Wang. *Österreichische Vierteljahresschrift für Forstwesen*, 67, 156–157.

Offer (1917) Ministerialrat a.D. Ferdinand Edler von Wang. *Zeitschrift des Österreichischen Ingenieur- und Architekten-Vereines*, 69(28), 418.

Wang, F. (1895) Die Ermittlung der Wasserabflussmengen mit besonderer Berücksichtigung der Verhältnisse in den Wild- und Triftbächen. *Österreichische Vierteljahresschrift für Forstwesen*, 45, 116–124.

Wang, F. (1896). Die Gesetze der Bewegung des Geschiebes und die natürliche Entwicklung des Längen- sowie des Querprofiles – Schuttkegelbildung. *Österreichische Vierteljahresschrift für Forstwesen*, 46, 123–148; 47, 111–135.

Wang, F. (1901) *Grundriss der Wildbachverbauung*. Hirzel, Leipzig.

4.2 *DENIL 2.1. 1865 Namur/B – 7.9. 1940 Bruxelles/B*

Gustave Denil obtained the engineering diploma from Gent University in 1888 and joined the Belgium *Corps de Ponts et Chaussées* in 1889. He remained there during his entire career as an hydraulic engineer contributing to general engineering designs.

Denil is known for his work relating to fish ladders and fish ways. His designs allow various types of fish to overcome differences in elevation by suitable insets into channels connecting two basins. He conducted extensive experiments to study the flow in channels roughened by insets to create zones of quiet and zones of flow such that fish may rest and then continue to swim in the zones of flow. Denil also proposed fish ways along sluices but they were not realized during his career because of high cost. A first paper in 1909 was followed by a large work published between 1936 and 1938, shortly before his death. Today, the Denil fish ladders are a standard design in hydraulic and environmental engineering. Denil also was an early hydraulician contributing to the macro-roughness flow, an issue which is currently a major research topic. Denil eventually gained the position of general director of roads in Belgium, a job that was not much concerned with hydraulics.

Anonymous (1928) Protokoll der Sitzung in Brüssel am 24. Mai 1927. *Zeitschrift des Internationalen Ständigen Verbandes der Schiffahrts-Kongresse*, Brussels, 3(5), 5–8.

Bonnet, L. & Campus, R. (1959) Gustave Denil. *Biographie Nationale de l'Académie Royale des Sciences des Lettres et des Beaux Arts de Belgique*, 13, 330–331. Bruylant, Bruxelles.

Denil, G. (1905) Etude des effets produits par l'ouverture des canaux de navigation sur le régime des eaux souterraines. *10 Congrès Internationale de Navigation Milano SI-C5.*

Denil, G. (1909) Les échelles hydrauliques appliquées à la canalisation et à la régularisation des rivières. *Annuaire de l'Association des Ingénieurs sortis de l'Ecole de Gand*, 5(2), 229–296.

Denil, G. (1909) *Les échelles à poissons et leur application aux barrages de Meuse et d'Ourthe.* Goemaere, Bruxelles.

Denil, G. (1936) La mécanique du poisson en rivière. *Annales des Travaux Publics*, 89(4), 507–583; 89(5), 708–746; 90(1), 70–84; 90(2), 256–284; 90(3), 412–433; 90(4), 610–638; 90(5), 734–763; 90(6), 958–980; 91(1), 131–171; 91(2), 391–411; 91(3), 537–578; 91(4), 783–811.

Humblet, F. Gustave Denil. (Personal communication, 2002).

4.3 *SMETANA 12.5. 1883 Svobodne Dvory/CR – 13.8. 1962 Praha/CR*

Jan Smetana graduated from the Czech Technical College in Prague and was from 1905 to 1911 a Lecturer at the Prague Institute of Hydraulic Engineering. From then to 1919, he was active with both the Bohemian river regulation commission and the hydrographical department of the Prague municipal authority. He headed from 1920 to 1936 the State Hydrological Institute in Prague thereby setting up the engineering research station. Smetana received an habilitation title in 1925 and was a professor at the Czech University of Technology in Prague from 1936 to 1958. Also, he was elected to the Czech Academy of Sciences in 1938, with full membership from 1946. Smetana founded the national Institute for Hydrology and Hydrotechnics *Vyskumny Ustav Vodohospodarsky* VUV and was known for works in hydraulic structures. He developed the free surface profile for flows in rectangular horizontal channels downstream of a vertical gate thereby verifying the Froude similitude. He studied also submerged hydraulic jumps downstream of gates and successfully applied the momentum equation for the sequent depth ratio.

Smetana presided over the technical section of the Czechoslovak Academy of Sciences from 1955; he was a founding member of IAHR and an honorary president of the International Association of Scientific Hydrology IASH. He was awarded the *Klement Gottwald* state prize in 1958, and obtained the Order of the Czech Republic.

Anonymous (1987) Jan Smetana. *Mala Ceskoslovenska encyklopedie*, 4, 708. Encyklopedicky Institut CSAV, Praha.

Kratochvil, S. (1962) Akademik Jan Smetana zemrel. *Vodni Hospodarstvi*, 12(10), 394.

Smetana, J., Pacak, A. & Till, J. (1923) Utilisation des voies navigables pour la production de la force motrice; ses conséquences et ses applications. *13 Congrès International de Navigation* London. 13(1/1).

Smetana, J. (1933) Etude expérimentale du ressaut d'exhaussement. *Bulletin*, 5, 1–32. Institut T.G. Masaryk de Recherches Hydrologiques et Hydrotechniques, Praha.

Smetana, J. (1935) Neue Arten beweglicher Wehre. *16 Intl. Navigation Congress Bruxelles*, 1(2), 1–19.

Smetana, J. (1948) Ecoulement de l'eau au-dessous d'une vanne et forme rationnelle de la surface d'appui de la vanne. *La Houille Blanche*, 3(1/2), 41–53; 3(3/4), 126–146.

Smetana, J. (1948) Etude de la surface d'écoulement des grands barrages. *Revue Générale de l'Hydraulique*, 14(46), 185–194; 15(49), 19–32.

Smetana, J. (1957) *Hydraulics*. NC SAV, Praha (in Czech).

4.4 *ENGELUND 26.11. 1925 Copenhagn/DK – 1.6. 1983 Copenhagn/DK*

Frank Engelund graduated from the Technical University of Denmark, Lyngby, as a civil engineer in 1948 and continued in its hydraulic laboratories as a project engineer, mainly in groundwater flow. In the 1950s he ran his own consulting company with port engineering works in Greenland. He was one of the first Danish professors in hydraulic engineering starting in 1963, and specializing in sediment transport of rivers. His contributions to the stability of sand waves and meanders are known, as also the Engelund-Hansen formula for sediment transport. The latter counts among these that have received general recognition.

By investigating the bed topographies of channel bends, Engelund contributed to stratified flows, dispersion in rivers and turbulent diffuser flow. He always attempted to describe the physical mechanisms that govern hydrodynamic processes. Engelund also released successful Series papers of the Institute of Hydrodynamics and Hydraulic Engineering. He was a member of the Danish Academy of Technical Sciences. His death at an age of less than 60 years was premature. A great personality and successful researcher had disappeared who was the true initiator of hydraulic research in Denmark.

Anonymous (1965) Engelund, Frank Anker. *Who's who in Europe*, 1, 797; 4, 656. de Maeyer, E.A. (ed.). Feniks, Bruxelles.

Anonymous (1983) In memoriam Frank Anker Engelund. *Journal of Hydraulic Research*, 21(5), 391–392.

Engelund, F. & Munch-Petersen, J. (1953) Steady flow in contracted and expanded rectangular channels. *La Houille Blanche*, 7(8/9), 464–474.

Engelund, F. (1966) Hydraulic resistance of alluvial streams. *Journal of the Hydraulics Division ASCE*, 92(HY2), 315–326; 92(HY6), 257–262; 93(HY1), 108–117; 93(HY4), 287–296.

Engelund, F. & Hansen, E. (1967) *Hydraulics*. Teknisk Forlag, Copenhagen (in Danish).

Engelund, F. (1970) Instability of erodible beds. *Journal of Fluid Mechanics*, 42(2), 225–244.

Engelund, F. & Skovgaard, O. (1973) On the origin of meandering and braiding in alluvial streams. *Journal of Fluid Mechanics*, 57, 289–302.

Engelund, F. (1974) Flow and bed topographies in channel bends. *Journal of the Hydraulics Division ASCE*, 100(HY11), 1631–1648; 101(HY9), 1290–1291; 101(HY10), 1367–1369.

Engelund, F. & Fredsoe, J. (1982) Hydraulic theory of alluvial rivers. *Advances in Hydroscience*, 13, 187–217. Chow, V.T. (ed.). Academic Press, New York.

4.5 *NAVIER 15.2. 1785 Dijon/F – 21.8. 1836 Paris/F*

Henri Navier obtained the engineering diploma from *Ecole des Ponts et Chaussées* ENPC in 1808. After having published an important work on bridges, he submitted the work *Canaux de Navigation* in 1816 and practiced engineering until he was appointed lecturer at ENPC in 1819; he took there over as professor in 1831. In the meantime, Navier introduced with Coulomb and Coriolis basic mechanical definitions for work and live force and around 1822 Navier re-published Bélidor's famous Hydraulic Architecture. He became a member of *Académie des Sciences* in 1824. Until 1830, Navier was engaged with engineering projects, such as a 155 m long suspension bridge over River *Seine* which failed because the effect of friction between the chain members had been overlooked. Navier was appointed professor of mathematics and mechanics at *Ecole Polytechnique* in 1830, and vacated in 1832 his other seat for Gaspard Coriolis (1792–1843).

Navier's main achievement in hydraulics are the so-called Navier-Stokes Equations NSE as an extension of the Euler equations by including viscous effects. These equations are currently a basis of most numerical computations and need closure relations for the modeling of turbulence. The NSE were popular only during the last 40 years or so, because no general methods were available to solve them earlier. The contribution of Gabriel Stokes (1819–1903) to the NSE is currently in question.

Chatzis, K. (1997) Economie, machines et mécanique rationnelle: La naissance du concept de travail chez les ingénieurs-savants français, entre 1819 et 1829. *Annales des Ponts et Chaussées*, 82, 10–20.

Lapparent, A. de (1897) Navier. *Ecole des Ponts et Chaussées*, 1, 157–162. Gauthier-Villars, Paris.

Navier, H. (1827) Mémoire sur les lois du mouvement des fluides. *Mémoires de l'Académie Royale des Sciences*, Paris, 6, 389–440.

Navier, H. (1838) *Résumé des leçons données à l'Ecole des Ponts et Chaussées sur l'application de la mécanique*. Carilian-Goeury, Paris.

Poggendorff, J.C. (1863) Navier, Claude Louis Marie Henri. *Biographisch-Literarisches Handwörterbuch*, 2, 260–261. Barth, Leipzig (with bibliography).

Prony, G. Riche de (1837) Notice biographique sur M. Navier, membre de l'Institut royal de France, officier de la Légion d'Honneur, inspecteur divisionnaire du corps des ponts et chaussées. *Annales des Ponts et Chaussées*, 6(1), 370; 7(1), 1–19; 7(2), 272.

4.6 *PRANDTL 4.2. 1875 Freising/D – 15.8. 1953 Göttingen/D*

Ludwig Prandtl studied at the Technical University of Munich. After having submitted a PhD thesis, he joined MAN in *Nürnberg* until being appointed mechanics professor in 1904 at Hannover Technical University. Also in 1904, he moved to the University of Göttingen, with a promotion to full professor in 1907. There he founded a laboratory of worldwide reputation for aerodynamics AVA. He was promoted to director of the Kaiser-Wilhelm-Institute (now Max-Planck-Institute) for fluid mechanics in 1925. Prandtl received an outstanding reputation as a prime researcher and teacher in fluid and aero-dynamics. Together with famous collaborators, he established the Göttingen School of fluid dynamics with a wealth of theoretical results in research and engineering, besides careful experimentation.

Prandtl was the theoretical promoter of the German school of hydrodynamics, which attracted a large number of international top researchers. He initiated the boundary layer and the turbulence theories and considerably developed the theory of wings, the boundary layer control and the hydrodynamic resistance. He obtained honorary doctorates from the Technical Universities of Danzig (1920), ETH Zurich (1930), Prague (1932), Trondheim (1935), Bucharest (1942) and Istanbul (1952). Prandtl's name is connected to a special type of Pitot tube, the equations governing boundary layer flow, the resistance law in rough turbulent pipe flows and the Prandtl number. He is considered the most important hydraulician of the early twentieth century.

Anonymous (1975) *Max-Planck-Institut für Strömungsforschung Göttingen 1925–1975*. Max-Planck-Institut, Göttingen.

Flügge-Lotz, I. & Flügge, W. (1973) Ludwig Prandtl in the nineteen-thirties: Reminiscences. *Annual Review of Fluid Mechanics*, 5, 1–8.

Görtler, H. (1975) Ludwig Prandtl – Persönlichkeit und Wirken. *Zeitschrift für Flugwissenschaften*, 23(5), 153–162.

Oswatitsch, K. & Wieghardt, K. (1987) Ludwig Prandtl and his Kaiser-Wilhelm-Institut. *Annual Review of Fluid Mechanics*, 19, 1–25.

Poggendorff, J.C. (1936) Prandtl, Ludwig. *Biographisch-Literarisches Handwörterbuch*, 6, 2069–2070; 7a, 619–620. Verlag Chemie, Leipzig, Berlin (with bibliography).

Prandtl, L. (1931) *Abriss der Strömungslehre*. Vieweg, Braunschweig.

Schlichting, H. (1975) An account of the scientific life of Ludwig Prandtl. *Zeitschrift für Flugwissenschaften*, 23(9), 297–316.

4.7 *ARGYROPOULOS 21.11. 1918 Athens/GR – 30.6. 1995 Athens/GR*

Praxitelis Argyropoulos graduated from the Technical State University of Greece, Athens, submitting there also a PhD thesis on hydraulic jumps in 1955. He there had taught fluid mechanics and advanced open channel hydraulics from 1943, and was a visiting professor at the University of Kentucky in the mid-1960s. In addition he was a guest lecturer at MIT, before returning to the National Technical University of Greece NTU, Athens.

Argyropoulos' contributions to water sciences are mainly in general hydraulics, hydrology and water resources. His specializations were energy dissipators and hydraulic jumps, with particular investigations in the effect of cross-sectional shape. Also, he presented an early work on the effect of bottom slope for jumps in rectangular channels. Other topics that he considered were the end depth problem, backwater curves, critical flow and stability aspects of supercritical open channel flow. After World War II he was about the only Greek with an international background, given his visits mainly to the USA. Argyropoulos spoke various languages which at that time enabled international contacts. He was a Fellow of ASCE, and a member of various national associations.

Anonymous P.A. (1966) Argyropoulos. *Civil Engineering*, 36(4), 8.

Argyropoulos, P.A. (1958) *Calcul de L'écoulement en conduites sous Pression ou à Surface Libre*. Paris, Dunod.

Argyropoulos, P.A. (1961) The hydraulic jump and the effect of turbulence on hydraulic structures. *9 IAHR Congress*. Dubrovnik, 173–183.

Argyropoulos, P.A. (1962) General solution of the hydraulic jump in sloping channels. *Journal of the Hydraulics Division* ASCE 88(HY4), 61–75; 89(HY1), 251–261; 89(HY2), 165–169; 89(HY4), 219–221.

Argyropoulos, P.A. (1963) Modern conceptions of flood prediction and their correlation to spillway dam design. *10 IAHR Congress* London 2, 153–156.

Argyropoulos, P.A. (1969) General considerations and new data on flow problem under dams. *13 IAHR Congress*, Kyoto, 4(D37), 339–346.

Argyropoulos, P.A. (1970) Note on modern considerations and new data on river roughness. *Journal of Hydraulic Research*, 8(2), 273–276.

Argyropoulos, S. Praxitelis A. Argyropoulos. (Personal communication, 2001).

Christodoulou, G. Praxitelis Achilleas Argyropoulos. (Personal communication, 2001).

4.8 *BOGARDI 11.6. 1909 Torda/YU – 18.11. 1998 Budapest/H*

Janos Bogardi was born in today's Yugoslavia; he graduated from Budapest University in 1931 as a civil engineer. Then he started a private engineering career, and joined the National river engineering service from 1932 to 1940. During this period he followed postgraduate studies at the Academy of laws in *Kecskemet* and at the University of Iowa in 1937/38. He joined the Hydrographical Institute in 1941 to become its director in 1945. Also, he submitted a PhD thesis in 1952. He was the first head of the hydraulics laboratory of the Budapest Water Resources Research Center from 1952 to 1962. Further, he was appointed professor of hydraulics at the Budapest Technical University until being retired in 1979.

Bogardi started as a practitioner and was known at the end of his career as a theoretical scientist, mainly for his contributions to the sediment transport in alluvial rivers. In addition, he was working on scale models and on problems in hydrology. He had excellent international contacts and visited the USA again in the 1960s, the period when publishing his important papers. During the 1960s and the 1970s, he stayed often in Vienna, from where he was awarded the honorary doctorate in 1975. Also, he became a member of the Academy of Sciences of *Padova*, Italy, in 1976. As a practitioner and a scientist, Bogardi was a council member of various societies, such as the IAHR from 1972 to 1977, or the ICID as deputy president from 1967 to 1969. He was awarded the honorary membership of IAHR in 1979.

Anonymous (1999) Janos Bogardi. *IAHR Newsletter* 16(1), 6.

Bogardi, J., Yen, C.H. (1938) Traction of pebbles by flowing water. *Studies in Engineering*. Iowa.

Bogardi, J. (1951) Mesure du débit solide des rivières de Hongrie. *La Houille Blanche* 6(3/4), 108–126.

Bogardi, J. (1959) Neuere Parameter und Invarianten bei der Bestimmung der Geschiebeförderfähigkeit. *Wasserwirtschaft*, 49, 314–320.

Bogardi, J. (1961) Some aspects of the application of the theory of sediment transportation to engineering problems. *Journal of Geophysical Research*, 66(10), 3337–3346.

Bogardi, J. (1968) Bestimmung der Grenzzustände bei der Geschiebebewegung. *Wasserwirtschaft*, 58(7), 205–212.

Bogardi, J. (1972) Fluvial sediment transport. *Advances in Hydroscience*. In: Chow, V.T. (ed.) Vol. 8, pp. 184–260.

4.9 *NASH 9.3. 1927 Sligo/IE – 17.4. 2000 Galway/IE*

James Eamonn Nash graduated as a civil engineer from Galway University in 1949 and started work with the Irish Electricity Supply Board for dam engineering. Right from the beginning, Nash collaborated with the hydrologist J.C.I. Dooge and remained his friend during his lifetime. In 1955, Nash obtained the Master degree of Engineering from Galway University for a work relating to flood propagation in rivers. One year later, Nash moved to HR Wallingford and developed the Instant Unit Hydrograph IUH Method, for which his name became famous. The IUH was represented by the gamma function and linearly relates flows in basic catchments. The paper was awarded the Telford premium by the Institution of Civil Engineers, London, in 1958. The IUH was applied with success to British catchment areas in the early 1960s.

After a two-years stay in Nigeria, Nash returned to Wallingford but soon after took the position of senior lecturer at Galway University. He was one of the founders of the *Journal of Hydrology* in 1963. In 1970 he submitted a PhD thesis and was appointed professor of engineering hydrology. From 1979 Nash organized the International postgraduate hydrology courses, with a large number of students from all continents until 2000. Nash was awarded the honorary doctorate from the University of Nanching, the International Premium of the International Association of Hydrological Sciences IAHS in 1989, and the 1999 Ven Te Chow award from ASCE for his groundbreaking investigations in hydrologic engineering and innovative solutions of hydraulic problems.

Cairnduff, M. (1984) Nash, J. Eamonn. *Who's Who in Ireland*, 148. Dublin, Vesey.

Moisello, U. (2000) J. Eamonn Nash. *L'Acqua*, 78(5), 94–97.

Nash, J.E. (1956) Frequency of discharges from ungaged catchments. *Transactions, American Geophysical Union*, 37, 719.

Nash, J.E. (1958) Determining run-off from rainfall. *Proceedings of Institution of Civil Engineers*, 10, 163–184; 11, 510–521.

Nash, J.E. (1959) The effect of flood-elimination works on the flood frequency of the river Wandle. *Proceedings of Institution of Civil Engineers*, 13, 317–338.

Nash, J.E. (1960) A unit hydrograph study, with particular reference to British catchments. *Proceedings of Institution of Civil Engineers*, 17, 249–282; 20, 464.

Nash, J.E. (1967) The role of parametric hydrology. *Journal of the Institution of Water Engineers*, 21(5), 435–474.

O'Connor, K.M. (2000) J.E. Nash. *Journal of Hydrology*, 234, 113–115.

4.10 *BIDONE 19.1. 1781 Casal-Noceto/I – 25.8. 1839 Torino/I*

Giorgio Bidone graduated from the Turin University, started lecturing in 1803 and became professor of hydraulics in 1815. His working fields were analysis, physics and hydraulics. In the latter he investigated impact effects of flowing water onto a body. This then led to the *salto di Bidone*, actually referred to as hydraulic jump. He described the basic phenomenon including roller flow, but both the small Froude numbers and the adoption of the Bernoulli instead of the momentum equation resulted in mistakes. In 1824 Bidone worked on wave propagation in channels and determined that the celerity varied with the flow depth. Both positive and negative surges were studied, including dambreak flows. He attempted to explain tidal flows as occur on Amazon River. Bidone confirmed the wave theories of Pierre Laplace (1749–1827) and Simeon Denis Poisson (1781–1840) for so-called shallow water flows.

Bidone's interest were also weirs and orifices, where he investigated experimentally the reasons for jet contraction and the relation of discharge to hydraulic head for both free and submerged flow conditions. The discharge coefficient was found to be practically constant, given that accuracies in discharge measurement at that time were limited. His observations were in agreement with French studies published in 1829. His approach was later extended by Giuseppe Venturoli (1768–1846). A particular phenomenon concerning the inversion of a liquid jet along the jet trajectory referred to as the Savart effect was originally investigated by Bidone.

Bidone, G. (1819) Expériences sur le remous et sur la propagation des ondes. *Memorie della Reale Accademia delle Scienze di Torino*, 25, 21–112.

Bidone, G. (1824a) Expériences sur la propagation du remous. *Memorie della Reale Accademia delle Scienze di Torino* 30, 195–292.

Bidone, G. (1824b) Sur la dépense des réservoirs et sur l'accélération et la courbure qu'ils occasionnent à la surface du courant. *Memorie della Accademia Torino Série 1*, 28, 281–330.

Ménabréa, L.F. (1842) Discours sur la vie et les ouvrages du chevalier Georges Bidone. *Mémoire de l'Académie de Turin Serie 2*, 4, 61–84.

Poggendorff, J.C. (1863) Bidone, G. *Bibliographisch-Literarisches Handwörterbuch*, 1, 187–188. Leipzig, Barth.

Tournon, G. (1980) Giorgio Bidone nel secondo centenario della nascita. *Annali dell'Accademia di Agricultura Torino*, 123, 151–162.

4.11 *WEINBLUM 22.1. 1897 Neu-Kalzenau/LV – 4.4. 1974 Hamburg/D*

Georg Weinblum graduated in 1923 as a mechanical engineer from Danzig Technical University after having initiated studies in Saint Petersburg in 1914. He submitted in 1929 a PhD thesis. In 1936, Weinblum was appointed associate professor at Berlin University from where he moved to Danzig University in 1942. After World War II, he spent some time in England and then continued to the USA, where he was a research associate at the David Taylor Model Basin. Weinblum was from 1952 a professor of naval engineering at Hamburg University and a Honorary professor at Hannover University from 1953. Until retirement in 1962 he directed also its newly founded Institute of naval engineering. He was awarded Honorary Doctorates from TU Berlin in 1960, TU Vienna and the University of Michigan, the William Froude Medal from the Institution of Naval Architects, UK, the Medal of Merit from *Association Techniques Maritimes et Aéronautiques* and the 1972 Davidson Medal from the Society of Naval Architects and Marine Engineers, USA.

Weinblum's life was devoted to naval engineering. He was particularly interested in the resistance of naval bodies of which he contributed mainly to the wave resistance. Moreover he investigated the behavior and safety aspects of ships in storm conditions. He was able to put forward a relation between naval shape and the wave resistance. Weinblum was in addition an outstanding teacher and colleague who supported significantly students and stimulated international corporation in ship research. His organizational talents were realized by contributing to associations that promote naval architecture.

Anonymous (1956) Georg Weinblum, *Der Lehrkörper der TH Hannover* 1831–1956, 191. Hannover, Technische Hochschule.

Brard, R. (1974) Georg Weinblum, *Comptes Rendus*, VieAcadémique, de l'Académie des Sciences. pp. 110–114.

Horn, F. (1957) Prof. Dr.- Ing. Georg G. Weinblum, 60 Jahre. *Schiffstechnik*, 4(20), 45–48.

Poggendorff, J.C. (1953) Weinblum, Georg. *Biographisch-Literarisches Handwörterbuch*, 7a, 901–902. Berlin, Akademie-Verlag.

Weinblum, G. (1930) Anwendungen der Michell'schen Widerstandstheorie. *Jahrbuch Schiffbautechnische Gesellschaft* 31, 289–440.

Weinblum, G. (1936) Rotationskörper geringsten Wellenwiderstandes. *Ingenieur-Archiv*, 7, 104–117.

4.12 *KOLUPAILA 14.9. 1892 Tuminiskiai/LT – 9.4. 1964 South Bend/USA*

Steponas Kolupaila studied engineering at Moscow University until 1915 and specialized immediately in hydrometry. His first Handbook on hydrometry dates from 1918. He returned to his home country in 1921 which had just become independent from former Russia. He established a private hydrometric office in *Kowno* and lectured also at the University of Lithuania until 1940. Two productive decades followed until World War II, during which he left his country, first for Germany and in 1948 for the USA, where he was appointed professor of hydrology at Notre Dame University, Indiana.

Kolupaila improved river gauging under the conditions prevailing in the Baltic countries, typically with frosty weather and ice flows. The rivers *Niemen* and *Vilia* received particular attention in his hydrologic studies, even extending to water flow below an ice cover. After Germany had overrun Eastern Europe, Kolupaila first was received by Dr. Ott, the famous velocity meter manufacturer in *Kempten*, and lectured at Munich Technical University. Kolupaila then started the encyclopedic Bibliography on hydrometry containing more than 7000 references and annotations, still a valuable source for detailed historical researches. Kolupaila wrote books on hydraulics and hydrometry and papers on rivers in Lithuania. He was awarded the *Gediminas* Order of Lithuania, and was a *Vasa* commander of Sweden.

Anonymous (1959) Kolupaila, Steponas. *Who's Who in Engineering*, 8, 1381.
Anonymous (1964) Steponas Kolupaila. *La Houille Blanche*, 19(5), 555.
Kolupaila, S. (1951) Modellversuche und Flügelmessungen in schrägen Strömungen. *Wasserwirtschaft*, 41, 147–151.
Kolupaila, S. (1960a) Water measurements in hydraulic structures and power plants. *La Houille Blanche*, 15(4), 344–363.
Kolupaila, S. (1960b) Early history of hydrometry in the United States. *Journal of the Hydraulics Division*, ASCE 86(HY1), 1–51; 86(HY4), 131–132; 86(HY6), 117–119; 86(HY7), 33–37; 86(HY9), 125; 87(HY3): 175–181.
Kolupaila, S. (1961) *Bibliography on hydrometry*. Notre Dame, Notre Dame University Press.
Kresser, W. (1964) Kolupaila, S. *Österreichische Wasserwirtschaft*, 16, 236.
Sperling, W. (1964) Kolupaila, S. *Bulletin IAHS*, 9(4), 127–129.

4.13 *DRONKERS 24.5. 1910 /NL – 20.2. 1973 s'Gravenhage/NL*

Jo Johannes Dronkers started working after his studies at the *Rijkswaterstaat*. He submitted a PhD thesis on tidal hydraulics to the University of *Leyden* in 1939. After World War II he specialized in the computational approach of estuarine hydraulics, with early applications of computational hydraulics in the 1950s. Following the flood disaster in 1953 which resulted in extensive damages to the Netherlands, Dronkers was at the forefront of simulating other cases that might occur in the future for decision strategy. Several of his papers submitted to international journals focused the efforts made in the Netherlands. His 1964 book is one of the first in the domain of tidal computation and a successful combination of experimental and computational knowledge. It was awarded the Conrad Medal and had a large input on hydraulic engineering, especially in the USA. At the end of his career, Dronkers was appointed head of the hydraulics department of the Delta Service, Rijkswaterstaat, The Hague.

Dronkers was a mathematician with an engineering background able to present the powerful methods of computational hydraulics. With some French and Dutch colleagues, he may be considered a personality having demonstrated that computational hydraulics is an alternative or even a method by itself to solve complex hydraulic problems. He was awarded the title Officer of the Dutch Order *van Oranje-Nassau* in 1959 for his outstanding contributions to tidal computations.

Dronkers, J.J. & Veen, J. van (1949) Calcul de la marée. *17 Congrès Internationale de Navigation*, Lisbonne, S2–Q1, 159–177.
Dronkers, J.J. & Schönfeld, J.C. (1955) Tidal computations in shallow water. *Journal of the Hydraulics Division*, ASCE 81(714), 1–50.
Dronkers, J.J. (1960) Delta project. Tidal computations in coastal areas. *Transactions, ASCE*, 125, 1281–1289.
Dronkers, J.J. (1964) *Tidal Computations in Rivers and Coastal Waters*. North-Holland, Amsterdam.
Dronkers, J.J. (1968) Closure of estuarine channels in tidal regions. *De Ingenieur*, 80(12), B127.
Dronkers, J.J. (1969) Practical aspects of tidal computations. *13 IAHR Congress*, Kyoto, 3, 11–20.
Ferguson, H.A., Leendertse, J.J. & Prins, J.E. (1973) Dronkers, J.J. Ter herdenking. 1910–1973. *De Ingenieur*, 85(44), 473–474.

4.14 *VOGT 23.12. 1892 Kristiania/N – 26.1. 1970 Oslo/N*

Fredrik Vogt is known for his contributions to surge tank computation. He was educated at *Trondheim* Technical University where he obtained the civil engineering diploma in 1914, continued as an assistant at its hydraulics laboratory and submitted a PhD thesis in 1923. His main results were then published in German journals pertaining to the optimum design of damping devices to reduce the extreme water level elevations. From 1927 to 1929 he was a post-doc student at the Bureau of Reclamation, Denver, USA. Upon returning to Norway, he was appointed lecturer at Norway Technical University NTH. In 1943 Vogt visited England and stayed briefly with the Farnborough institution. Upon returning again to Norway in 1945, he became NTH dean until 1948. From then, Vogt was the director general of the Norwegian power plants until 1960.

Vogt wrote a book on the hydropower plants of his country. He thus directed an important institution for Norway's hydropower development besides university activities. He was of course involved in various projects, mainly as a consultant, and thus was an applied scientist, a practitioner and an administrator. His papers treat dam engineering, next to surge tanks. He was awarded the Royal Norwegian Scientific Prize in 1927 for his PhD thesis, was a member of the Norwegian Academy of Sciences from 1936, and of the Norwegian Technical Academy of Sciences from 1955. He received honorary doctorates from the Universities of Stockholm and Helsinki, both in 1949.

Devik, O. (1971) Minnetale over Generaldirektør, Dr. techn. Fredrik Vogt, Det Norske Videnskaps-Akademi i Oslo. *Årbok*, 130–137.

Ludin, A. & Nemenyi, P. (1930) *Die nordischen Wasserkräfte – Ausbau und wirtschaftliche Ausnutzung*. Berlin, Springer.

Selberg, A. (1977) Vogt, F. *Norsk Biografisk Leksikon*, 18, 132–134. Oslo, Aschehoug.

Vogt, F. (1923) *Berechnung und Konstruktion des Wasserschlosses*. Stuttgart, Enke.

Vogt, F. (1924) Die hauptsächlichsten Gesichtspunkte für die Anlage von Wasserschlössern. *Deutsche Wasserwirtschaft*, 19(4), 113–116.

Vogt, F. (1933) Temperature straining in thick concrete dams. *1 Congrès des Grands Barrages*, Stockholm, 3, 293–309.

Vogt, F. & Solem, A. (1968) *Norwegian Hydro-power Plants*. Oslo, Ingeniørforlaget.

4.15 *MATAKIEWICZ 27.6. 1875 Niepolomice/PL – 3.2. 1940 Lemberg/UA*

Maksymilian Matakiewicz originated from the former Austrian possession *Galicia* in the later Poland, located in today's Western Ukraine. He received engineering education from the Technical University of *Lemberg*, today's *Lwiw*, and submitted there a PhD thesis in 1905. Until his appointment as associate professor at his university in 1908, he contributed to the national public works' service. Matakiewicz was promoted to full professor of hydraulic structures in 1911. In 1929/1930 he was the public works minister of the Republic of Poland and presided over the Academy of Sciences in Warsaw from 1931 to 1933. He was also a member of the *Masarykova* Academy in Prague from 1932, and a member of the Technical Committee of the Polish Ministry of Transport from 1933.

Matakiewicz greatly contributed to open channel flow by presenting relations between the average velocity, the surface gradient and the river width, without recourse to a roughness coefficient. This approach is based on regime considerations and on the relation between slope and size of the bed material. His equations, though complex, were popular in the early 20th century. Hubert Engels (1854–1945) generalized the approach of Matakiewicz by additional observations of his students. In 1932, Matakiewicz extended his former velocity formula to bed slopes larger than 1% and thus may be considered an early contributor to flows in high-gradient streams.

Engels, H. (1910) Geschwindigkeitsformel von M. Matakiewicz. *Zentralblatt der Bauverwaltung*, 30(99), 646–647.

Matakiewicz, M. (1905) Versuch der Aufstellung einer Geschwindigkeitsformel für natürliche Flussbette. *Österreichische Wochenschrift für den öffentlichen Baudienst*, 11(51), 767–774; 12(21), 317–324; 12(30), 445; 12(36), 504–505.

Matakiewicz, M. (1911), Empirische Untersuchungen über den Zusammenhang der Bewegungselemente bei natürlichen Flussbetten. *Zeitschrift Gewässerkunde*, 10(2), 97–125.

Matakiewicz, M. (1932) Geschwindigkeitsformel für natürliche Betten und sehr grosse Gefälle. *Zeitschrift des Österreichischen Ingenieur- und Architekten-Vereines*, 84(17/18), 85–86.

Szklarska-Lohmannowa, A. (1975) Matakiewicz, M. *Österreichisches Biographisches Lexikon* 1815–1940. In: E. Obermayer-Marnach, (ed.) Wien, Verlag der Österreichischen Akademie der Wissenschaften. 135.

University Library (2001) Maksymilian August Matakiewicz, Inzynier, Dr. NT, Professor. Warsaw, University.

4.16 *ALMEIDA D'EÇA 17.10. 1825 Esgueira/P – 2.2. 1906 Lisboa/P*

Bento Fortunato de Moura Coutinho de Almeida d'Eça graduated in 1853 from the Lisbon Army School. He was admitted as a civil engineer in 1854 to the ministry of public works and mines, after having followed courses also at *Ecole des Ponts et Chaussées* in Paris, and stayed there all over his career. He managed to climb up the military hierarchy to brigadier in 1893, and general of division of the engineering corps in 1897. During his career he was involved in the improvement of river *Tejo* running into the bay of Lisbon and then into the Atlantic. The river was tamed with a number of levees and bank erosion was reduced with appropriate river engineering works. Some years later, Almeida d'Eça was asked to improve the conditions of rivers in the *Alentejo* Province, south of the *Tagus* River, and to set up dams for irrigation and flood protection purposes. He may be considered a successful practising hydraulic engineer who improved the water conditions of his country.

Almeida was awarded various prestigious distinctions, such as the Grand Cross of the Order of *Aviz*, Commander of the French *Légion d'Honneur*, and Commander of *S. Tiago da Espada*. He acted also as a first class inspector in the corps of public works. As was typical in his time, he was further active for railway development, such as for the *Beira-Alta* line.

Almeida d'Eça, B. (1866) *Memoria acerca das irrigaçoes na França, na Italia, Belgica e Hespanha*. Lisboa, Imprensa Nacional.

Almeida d'Eça, B. (1877) *Memorias acerca do regimen do Tejo e outros rios apresentadas ao Ministério das Obras Publicas nos annos de 1867 e 1872*. Lisboa, Imprensa Nacional.

Almeida d'Eça, B. (1883) Plano geral das obras que convira fazer para melhorar o regimen do Tejo e beneficiar os seus campos adjacentes. *Revista de Obras Publicas e Minas*, 159/160, 58–121.

Almeida, d'Eça, B. (1885) *Memoria acerca do aproveitamento das aguas no Alemtejo*. Lisboa, Imprensa Nacional.

Loureiro, A.B. (1906) *Elogio historico de Bento Fortunato de Moura Coutinho d'Almeida d'Eça. Associaçao de Engenheiros Civils Portugueses*. Lisboa, Imprensa Nacional.

Quintela, A. de Carvalho & Miranda, J.C. (1993) *Biblioteca da direcçao – Geral dos recursos e aproveitamentos hidraulicos 1985*. Lisboa, Ministerio do Ambiente e Recursos Naturais.

Quintela, A. de Carvalho, Bento Almeida d'Eça. (Personal communication, 2000).

4.17 *COANDA 7.6. 1886 Bucharest/RO – 25.11. 1972 Bucharest/RO*

The Coanda-effect discovered in 1934 is the spectacular phenomenon that can be observed by holding a finger against a tap water jet: The jet adheres to the finger and is deflected towards the finger. Similarly, the flame of a candle can be deflected by an air stream. That effect is a result of the combined action of boundary curvature and fluid viscosity. The Coanda-effect is applied in aviation for circulation control of aerofoils and jet-deflection devices. Other applications are in cooling of circular cylinders and swirl atomizers. Recently, the effect was also applied in civil engineering at the intake of final settling tanks.

Coanda studied in Bucharest until 1905, and took additional education at the universities of Berlin, Liège and Paris. He then started as an applied engineer in aerodynamics in France, was associated with the Bristol airplanes up to World War I and returned to France. He developed the principle of jet turbines as early as in 1910. In 1919 he introduced the first hovercraft which came into public use in the 1950s. Also, he extensively tested the effect of wing thickness for advanced flying characteristics. A notable scientist in physics, chemistry and biology, he was awarded the title Commander of the *Ordre du Mérite pour la Recherche et l'Invention* in 1960. Towards the end of his life in 1967, Coanda returned to Romania. A symposium on the Coanda-effect was organized to mark his outstanding achievements.

Anonymous (1994) Coanda, Henri. *Inventeurs et scientifiques*: Paris, Larousse, 151.
Carafoli, E. (1968) Henri Coanda, pionnier de l'aéronautique mondiale. *Revue Roumaine des Sciences Techniques*, 13(3), 389–398.
Coanda, H. (1911) Etude de la résistance d'air par la chronophotographie. *L'Aérophile*, 19, 253–256.
Coanda, H. (1932) Procédé de propulsion dans un fluide. *Brevet Invention Gr.* Cl. 2, No. 762688, République Française.
Felder, A. (1993) *Untersuchungen zum Coanda-Effekt – Mögliche Anwendung im Bauingenieurwesen. Mitteilung*. München, Institut Hydraulik und Gewässerkunde, Technische Universität, 55.
Gheorghiu, C.C. (1979) *Romanian Inventions and Priorities in Aviation*. Bucarest, Albatros.
Métral, A. & Zerner, F. (1948) *L'effet Coanda. Publications Scientifiques et Techniques du Ministère de l'Air 218*. Paris, Service de Documentation et d'Information Technique.

4.18 *KOCHINA 13.5. 1899 Astrakhan/RU – 3.7. 1999 Moscow/RU*

Pelageya Yakovlevna Kochina (born Polubarinova) graduated from Petrograd University as a mathematician in 1921 and started work at its geophysical observatory already in 1919. From 1927 to 1934 she was a Lecturer at Leningrad University and a staff member of the Institute of Civil Aviation Engineering. In 1935 she moved to Moscow's Steklov Mathematical Institute and left for the institute of mechanics of the USSR Academy of Sciences in 1938. Submitting a doctoral thesis in mathematical and physical sciences in 1940, she was an associate of that institute until 1957. From 1958, she directed the Department of Applied Hydrodynamics in Novosibirsk. In 1970 she returned to Moscow to direct the section of mathematical methods in mechanics at Moscow University.

Kohina is known for her fundamental contributions to the theory of flows in porous media. She in particular developed a general method for solving two-dimensional seepage problems in homogeneous soils. Kochina's research was characterized by a deep and well-organized link with practice, a subtle attention to the physical essence of the phenomena considered, an exact mathematical formulation of the relevant physical problem, and by a brilliant mastery of mathematics. She was awarded the Stalin Prize in 1946, full membership of the USSR Academy of Sciences in 1958, Hero of socialist labor in 1969, and the Order of the Friendship of Nations in 1979.

Anonymous (1959). Y. Kochina, *Izvestiya Nauk*, Moscow 3, 6–14 (with bibliography).

Anonymous (1999). Kochina, P.Y. *Journal of Applied Mathematics and Mechanics*, 63(2), 149–160.

Cooke, R. (2000). *Pelageya Yakovlevna Polubarinova-Kochina*. Available from: http://www.physics.ucla.edu.

Kochina, P.Y. (Polubarinova) (1938a) On an integral flow equation for tanks of constant depth. *Izvestiya*, Akademii Nauk SSSR, ser. Math. 2, 249–270 (in Russian).

Kochina, P.Y. (Polubarinova) (1938b) Application of the theory of linear differential equations to some cases of motion of groundwater. *Izvestiya*, Akademii Nauk SSSR, ser. Math. 2, 371–398; 3, 329–350; 3, 579–602 (in Russian).

Kochina, P.Y. (Polubarinova) (1952) *Theory of Groundwater Flow*. Moscow, Nauka. translated by de Wiest, J.M. in 1962. Princeton, University Press.

Vronskaya, J. (1989) Kochina, P.I. *A Biographical Dictionary of the Soviet Union*, 1917–1988, 187. London, K.G. Saur.

4.19 *IRIBARREN 15.4. 1900 Irun/E – 21.2 1967 Madrid/E*

Ramon Iribarren Cavanilles graduated in 1925 from *Escuela Especial de Ingenieros*, Madrid, to continue with the *Gerona* Roads' Department. He was appointed director of ports of the *Guipuzcoa* group in 1929, a position he held until his tragic car accident nearly 40 years later. In 1939, Iribarren was appointed professor of maritime ports at *Escuela de Caminos*, Madrid, a similar institution to *Ecole des Ponts et Chaussées* in France, with a stress on civil engineering education. He was a Member of the technical committee in the port department of the public works in Spain, and thus was at the forefront of almost all larger port projects in Spain. In 1944, he founded the port laboratory of Spain and thus conducted an enormous amount of works which were mainly published in the proceedings of *Congrès de Navigation*.

Iribarren introduced the so-called *Guipuzcoa* wave method by which the maximum wave characteristics in deep water conditions may be computed. In the 1949 paper, the coefficients of wave run-up on sloping and vertical beach walls are predicted and a second order theory for transverse wave expansion is outlined. Definite results relating to dike design were presented in 1953. A beautiful summary of Spanish harbour works as presented in an exhibition is available. Iribarren was awarded the Grand Cross of Alphonso X, the public merit and the aeronautical merit. He was a cavalier of *Légion d'Honneur*, a member of the Spanish academy, and the French marine academy. He was the Spanish representative of the *Commission Permanente de l'Association Internationale des Congrès de Navigation* PIANC. Iribarren was an international consultant of the University of New York and the US Beach Erosion Board, and was a council member of the 3rd Coastal engineering congress in Boston.

Centro de Estudios y Experimentacion de Obras Publicas CEDEX (2000) *Iribarren – Ingenieria y mar*. Madrid, Ministerio de Fomento.

Iribarren Cavaniles, R. (1941) Obras de abrigo de los puertos. Planos de oleaje. *Revista de Obras Publicas*, 91(1), 13–25.

Iribarren Cavaniles, R. (1949) Planos de oleaje en segunda approximacion. *Revista de Obras Publicas*, 99(11/12), 519–534.

Iribarren Cavaniles, R. & Nogales y Olano, C. (1953) Nouvelles conceptions sur les digues à parois verticales et sur les ouvrages à talus servant: (a) Pour la protection des ports, (b) Pour la protection des rivages. *18 Congrès Intl. de Navigation*, Roma, SII-Q1, 45–66.

4.20 *FELLENIUS 10.9. 1876 Viksberg/S – 2.9. 1957 Stockholm/S*

Wolmar Knut Axel Fellenius obtained the civil engineering degree from the Stockholm Technical University in 1898. He was then a building inspector, a chief of the engineering department at *Härnosönd* from 1903 to 1905, and the chief of the *Göteborg* harbor department until 1911. After having been a lecturer in hydraulics at Chalmers Technical Institute from 1906 to 1911, he was appointed professor at Stockholm Technical University, and founded a hydraulics laboratory in 1920. As a consulting engineer, Fellenius designed the harbors of various Swedish cities and later was a member of governmental commissions for that purpose. He retired in 1942 from Stockholm University.

Fellenius might be better known in soil mechanics than in hydraulics for the stability calculation of a sloping plane. He was a predecessor of what was later developed by Karl Terzaghi. Fellenius contributed to scale effects during the 15th Congress of Navigation at Venice in 1931, and studied bank erosion of lakes due to waves. He was awarded honorary membership of *Fridericiana* University, Karlsruhe in 1921, obtained the honorary doctorate from Darmstadt Technical University in 1936, was a corresponding member of the Norwegian Engineering Society from 1924 and in 1926 presided over the Division of highway construction and hydraulics of the Swedish society of engineers. From 1935 to 1948, he was the first president of the International Association of Hydraulic Structures Research, today's IAHR.

Anonymous (1944) Fellenius, Wolmar Knut Axel. *Svenska män och kvinnor*, 2, 505–506. Stockholm, Bonniers.
Ekelund, B. (1957) Wolmar Fellenius, *Teknisk Tidskrift*, 87, 743–744.
Fellenius, W. (1927) *Erdstatische Berechnungen*. Berlin, Ernst & Sohn.
Fellenius, W. (1927) Loss of head in protecting racks at hydroelectric plants. *Ingeniörs-vetenskap-Akademiens*, 79. Stockholm, Svenska Bokhandelscentralen (in Swedish).
Freeman, J.R. (1929) W.K.A. Fellenius. *Hydraulic Laboratory Practice*. pp. 517–558. New York, ASME.
Reinius, E. & Cederwall, K. (1987) Wolmar Fellenius and the hydraulics laboratory. In: Garbrecht, G. (ed.) Stockholm. *Hydraulics and Hydraulic Research* – A Historical Review. Rotterdam, Balkema. pp. 293–296.
Sichardt, W. (1951) Prof. Wolmar Fellenius 75 Jahre. *Bautechnik*, 28(10), 245.

4.21 *SCHNYDER 25.3. 1904 Kriens/CH – 28.10. 1974 Rüschlikon/CH*

Othmar Schnyder obtained the degree of mechanical engineer from ETH in 1926, and submitted a PhD thesis on the static design of regulation rings in 1928. He then joined the *Von Roll* iron works in *Klus*, close to *Solothurn*, for nearly twenty years thereby advancing to vice-director and in parallel contributing to the engineering computation of water hammer. Starting around 1936, he combined his job and 'hobby' by developing emergency machinery for hydropower plants, such as ring valves or leaf gates. By 1947 he was allowed to install a hydraulic laboratory at *Von Roll*, but Schnyder was forced to leave research in favor of administration. In 1950 he founded his own consulting office and in 1954 moved to *Malters* to found *Hydro-Progress*, an engineering design factory for hydraulic machinery. Mainly working for Swiss and Austrian plants, Schnyder was a successful business man in a specialized company. He had a special sense in matters of mechanics, and collaborators would not ask him *why*, but only *how* to do it.

Schnyder was the first to present a general engineering tool for computing water hammer waves as occur in pipelines due to temporal changes at their boundaries. He thereby accounted for boundary friction, changes of pipe diameter or pipe wall thickness in a basic approach that was later also proposed by the Frenchman Louis Bergeron (1876–1948). The Schnyder-Bergeron graphical method lost its popularity once computers came available in the early 1960s.

Hager, W.H. (2001) Swiss contribution to water hammer theory. *Journal of Hydraulic Research*, 39(1), 3–11.

Schnyder, O. (1929) Druckstösse in Pumpensteigleitungen. *Schweizerische Bauzeitung*, 94(22), 271–273; 94(23), 283–286.

Schnyder, O. (1932) Über Druckstösse in Rohrleitungen. *Wasserkraft und Wasserwirtschaft*, 27(5), 49–54; 27(6), 64–70; 27(8), 96.

Schnyder, O. (1935) Über Druckstösse in verzweigten Leitungen mit besonderer Berücksichtigung von Wasserschlossanlagen. *Wasserkraft und Wasserwirtschaft*, 30(12), 133–142; 30(14), 172.

Schnyder, O. (1937) Comparisons between calculated and test results on water hammer in pumping plants. *Transactions, ASME* 59(13), 695–700.

Schnyder, O. (1943) Druckstösse in Rohrleitungen. *Von Roll Mitteilung*, 2(3/4), 1–56.

4.22 *FROUDE 28.11. 1810 Dartington/UK – 4.5. 1879 Simon's Town/ZA*

William Froude received his education at Oxford University and became an assistant to Brunel, the great English engineer. In 1846, Froude retired to devote his time to theoretical investigations. In 1855, Brunel asked Froude to investigate the phenomena of ship rolling, and Froude presented a paper thereon in 1861. To explain rolling Froude developed the trochoidal theory of deep-water waves and introduced bilge keels to diminish the angle of roll, a method first applied for the warship *Greyhound*.

Froude's work on ship resistance is outstanding. Based on his similarity law according to which the ratio between speed and square root of length of a ship should be equal in model and prototype, he was able to upscale observations to prototype. Froude did not realize that the presently called Froude number was introduced about 30 years earlier by the Frenchman Ferdinand Reech (1805–1880), who did not proceed to an experimental verification, however. Froude was the first to use model families for ships and verified his model law, although he detected minor differences due to viscous effects. It was then that the hydraulic modeling was first successfully applied and recognized as a key approach for designing hydraulic structures. Froude was asked to conduct observations for the Admiralty, and his laboratory in *Torquay* was opened in 1871. The experiments relating to the warship *Greyhound* were the first to convince experts of the enormous advantages of hydraulic modeling. Following these observations, model testing of ships became a must. Froude was also able to separate the effects of skin friction, wave resistance and form resistance. He passed away during a trip to the Cape in South Africa.

Abell, W.S. & Gawn, R.W.L. (1955) *The Papers of Froude*. London, Institution Naval Architects.

Anonymous (1880) William Froude LL.D, F.R.S. *Proceedings of Institution of Civil Engineers*, 60, 395–404.

Day, L. & McNeil, I. (1996) William Froude. *Biographical Dictionary of the History of Technology*, 274–275. London, Reference.

Froude, W. (1861) On the rolling of ships. *Transactions, Naval Architects*, 2, 180–229.

Froude, W. (1872) Experiments on the surface-friction experienced by a plane moving through water. *British Association for the Advancement of Science*, 42, 119–124.

Froude, W. (1875) Address to the mechanical section. *British Association for the Advancement of Science*, 45, 221–239.

Wall, K. (1993) The Froude number – South African connection. *Civil Engineering SA*, 1(3), 16–18.

4.23 *BORELI 31.8. 1922 Zadar/HR – 5.3. 1995 Beograd/YU*

Mladen Boreli graduated in 1950 as a civil engineer from the Belgrade Technical University, after having been involved in war activities. He was then an assistant at its hydraulic institute and profited from a research grant which brought him to the University of *Grenoble*. Boreli submitted a PhD thesis on groundwater flow in 1954 before returning to Belgrade, where he was promoted to associate professor of hydraulics in 1960, and to full professor in 1966. Boreli retired in 1987, after having been one of the outstanding scientists of Yugoslavia.

Boreli had an international reputation for groundwater hydraulics. He was also the organizer of the 1961 IAHR Congress held in *Dubrovnik*, and was involved in the organization of several workshops relating to groundwater flow. Boreli was the author of various books on general hydraulics which were extensively used in the former Eastern Europe, with translations into several languages. He was the recipient of the Scientific National Award in 1968, after having received the October Award in 1964, and the Order with Golden circle in 1965. He was presented the Honorary Diploma from the *Jaroslav Cerni* Institute in 1994.

Anonymous (1991) Mladen Boreli. *Ljetopis*, 94, 433–444 (with complete bibliography).

Boreli, M. (1954) Contribution à l'étude des milieux poreux. *Thèse* présentée à la Faculté des Sciences. Grenoble, Université.

Boreli, M. & Bruk, S. (1957) Critical analysis of river similarity. *7 IAHR Congress*, Lisboa, 2(D4), 1–22.

Boreli, M., Batinic, B. (1961) Considération sur la loi non-linéaire de filtration – Répartition des vitesses de filtration au voisinage du point de sortie. *9 IAHR Congress*, Dubrovnik, 506–515.

Boreli, M. (1980). *Hidraulika*. Beograd, Naucna Knjiga (in Serbian).

Kravtchenko, J., Sauvage de Saint-Marc, G. & Boreli, M. (1955) Etude d'une singularité dans les écoulements plans des liquides pesants en milieux poreux. *La Houille Blanche*, 11(4), 533–542.

Muskatirovic, J. (1995) In memoriam: Prof. Dr. Mladen Boreli. *Transactions*, Jaroslav Cerni Institute, 42, 353–356.

5 RETROSPECT AND OUTLOOK

5.1 *19th century*

Around 1800, hydraulics was a science based on some principles that were derived both from experimentation and from theoretical principles. Whereas Isaac NEWTON had laid the foundations in mechanics, the true initiators of theoretical hydraulics are Daniel BERNOULLI (1700–1782) and Leonard EULER (1707–1783). They presented equations currently referred to after their names based on the energy and the momentum conservation laws. However, the two approaches were not welcomed by the hydraulic engineers because it was soon realized that they were in disagreement with hydraulic experiments. The inclusion of viscosity in the Eulerian formulation as presented by Louis NAVIER (1785–1836) was a significant addition to hydrodynamics from the modern point of view, yet it confused again most of the hydraulic engineers of the 19th century because its solutions were – and still are in many cases – beyond engineering capacities. Presently the Navier-Stokes equations NSE are considered the most general formulation of one-phase flows describing fluid flow with a differential approach. Their solution requires special techniques and normally a considerable computer capacity. Currently, the NSE are rarely used in hydraulic engineering, therefore.

Figure 5 shows typical engineering work in the 19th century. Because of the then modern infrastructure of railways, river valleys suddenly were narrowed and the river course had to be 'corrected'. Within few years from the 1840s, the river thus was in a company with the railroad systems and later with roads, once automobiles were available. In addition, communities profited from the until then mostly deserted river reaches and added parts of its industries close to the water course to produce hydro-electricity. Whereas most of the human settlings until the early 19th century were located well away from flooding zones, damages increased toward 1900. The reason was often attributed to the forest management of the earlier times and the erosion associated with the lost soil structure. France was the country that took the first actions, and forest engineering as a part of agriculture developed shortly later also in Austria and in Switzerland. To protect the population against floods, dikes along the rivers were erected but the consequences were hardly accounted for, because of the lack of hydraulic models. The most simple notions of river engineering were then often not established, and all these works were often designed without any hydraulic background. The gap between the hydraulic engineer and the applied designer became evident already then, and this gap has from then remained an essential problem in engineering.

The 19th century disposed of two basic approaches to solve problems in hydraulic engineering: (1) Prototype experimentation, and (2) Analytical computations. Some flow situations were solved

Figure 5. Hydraulic engineering in the 19th century, from Minor, H.-E. and Hager, W.H., eds. (2004). *Flussbau in der Schweiz*. Stäubli, Zurich.

rather accurately, including those with a frictionless fluid such as weir, gate or orifice flows in which the so called contraction coefficients were experimentally determined. Further, the wave theory was developed essentially in the 19th century, including the sinusoidal, the cnoidal and the trochoidal wave theories. Advances in this matter were particularly made in the United Kingdom, with applied mathematicians of the caliber of George Biddell AIRY (1801–1892), John MCCOWAN (1863–1900), William John Macquorn RANKINE (1820–1872), John Scott RUSSELL (1808–1882), George Gabriel STOKES (1819–1903), John William STRUTT (1842–1919) the later Lord Rayleigh and William THOMSON (1824–1907) the later Lord Kelvin.

Other topics that were mastered in the 19th and the early 20th centuries involved two-dimensional inviscid flows that were particularly investigated in Germany. Notable researchers with a contribution to these flows are Heinrich BLASIUS (1883–1970), Hermann von HELMHOLTZ (1821–1894), Gustav Robert KIRCHHOFF (1824–1887), Heinrich Gustav MAGNUS (1802–1870) and Wilhelm Eduard WEBER (1804–1891). They have advanced the vortex theory in inviscid fluids, the potential flow theory with applications also in aerodynamics and the conformal mapping approach to solve for two-dimensional flow fields.

Whereas these topics were relatively well covered up to 1900, others had received practically no attention. These include groundwater flow, unsteady pipe flow, sediment transport in rivers and estuaries, two-phase air-water flows and dam hydraulics with the features of high-speed flows. From a conceptual point of view, viscous flows posed a particular problem because both Gotthilf HAGEN (1797–1884) and Jean POISEUILLE (1797–1869) observed independently two different flow regimes. For viscous fluids of relatively small speed, the so called laminar flow regime pertains, whereas high-speed almost inviscid flows result in the turbulent regime. The latter pipe flows were particularly analyzed by Henry DARCY (1803–1858) and Julius WEISBACH (1806–1871). A more systematic analysis of the two regimes was experimentally investigated by Osborne REYNOLDS (1842–1912) thereby introducing the Reynolds number.

The effect of viscosity complicates the mathematical analysis of viscous fluid flow considerably. From around 1850, particularly the Frenchmen known as tough mathematicians contributed to the understanding of these flows. Outstanding names in this branch until the 1950s are Auguste BOULANGER (1866–1923), Joseph BOUSSINESQ (1842–1929), Marcel BRILLOUIN (1854–1948), Pierre DUHEM (1861–1916), Joseph PÉRÈS (1890–1962), Adhémar BARRÉ DE SAINT-VENANT (1797–1886) and Henri VILLAT (1879–1972). They have contributed to the understanding of the effect of viscosity in fluid flow but have not really made a break-through with a lasting effect up until today. It needed rather a physicist or an engineer with a theoretical background than a mathematician to solve the enigma of turbulence, as Boussinesq once remarked while making his basic contribution to the turbulence exchange coefficient in the 1870s.

5.2 *20th century*

Until 1904, the currently significant Navier-Stokes equations NSE had hardly been considered as an alternative to other basic equations in hydraulics and hydrodynamics. Ludwig PRANDTL (1875–1953) introduced a revolution with a paper presented during the 1904 Mathematical Congress held in Heidelberg, Germany. Instead of considering the entire fluid flow bounded by a rigid boundary, he proposed to separately investigate the flow close to the boundary referred to as the boundary layer, and the flow away from the boundary where flow conditions are almost perfectly reproduced by the equations of inviscid fluid. Prandtl realized that the effect of fluid viscosity is essentially confined to a thin layer, such that the NSE could be drastically simplified and solved with a basic approach. This concept has had far reaching consequences for problems in hydrodynamics and particularly in aerodynamics, whose origins are also in 1904 by the first powered flight of the Wright brothers in the United States. The boundary layer concept proved to be so successful because it once again bridged the gap between the fluid dynamicist and the hydraulic engineer in providing to both an approach of relevance in solving problems at least approximately. It also evidenced the important role of fluid viscosity in fluid dynamics, an aspect that was hardly quantified earlier.

Prandtl may be considered the outstanding scientist in applied hydrodynamics of the 20th century, therefore. His Göttingen School was the center of developments from 1910 to roughly 1935,

Figure 6. Hydraulic Laboratory of ETH Zurich, then directed by Eugen Meyer-Peter.

when the shadow of World War II appeared over Germany. Some of the great names of this era include Albert BETZ (1885–1968), Adolf BUSEMANN (1901–1986), Henry GÖRTLER (1909–1987), Theodor VON KARMAN (1881–1963), Richard VON MISES (1883–1953), Johann NIKURADSE (1894–1979), Hans REICHARDT (1901–1977), Ludwig SCHILLER (1898–1970), Hermann SCHLICHTING (1907–1982), Walter TOLLMIEN (1900–1968) and Karl WIEGHARDT (1913–1996).

Engineers and scientists of the 20th century had elaborated their research investigations with the addition of a highly important approach, namely laboratory experimentation. Despite such observations were conducted from ancient times, the modern hydraulic laboratory was created mainly in Germany by the well known engineers Hubert ENGELS (1854–1945) at Dresden University, Theodor REHBOCK (1864–1950) at the University of Karlsruhe and Alexander KOCH (1852–1923) at the University of Darmstadt. Whereas practicing engineers could hardly believe that processes in prototypes could indeed be modeled with scale models in the first decades of the 20th century, they were overwhelmed by the close predictions made with this approach from around 1920, such that this technique became from then a standard (Fig. 6). It had become evident from then that results from scale models had to be carefully checked against so called scale effects; normally, the scale of the model was too large such that the basic processes modeled in small did not reproduce all major effects that actually occurred in the prototype. Today, all observations collected from scale models are known to be limited against these deficits.

From World War II, hydraulics expanded considerably because a number of problems previously formulated were systematically attacked. These related to environmental hydraulics for instance, to the assessment of water quality and to wastewater hydraulics. All these flows essentially involve a two-phase current made up of the fluid and the solid phases, and complicating significantly a one-phase flow consideration. Other problems tackled in this era related to two-phase air-water flows, such as high-speed flows in dam engineering, or again water quality problems. During this era, the huge research capacities of the United States became obvious, and most of the developments were initiated in the country that had taken leadership also in the political and the economical platforms. No names will be presented here because of the great number of people involved in the modern hydraulic research process.

The research activities were further drastically changed by the addition of a forth basic means to investigate fluid flow. Whereas the three tools previously listed include (1) prototype, (2) mathematical analysis and (3) hydraulic laboratory methods, the advance of computers added the forth element. The modern hydraulic engineer disposes a number of independent approaches that have to be carefully selected for a particular problem. These days, the hybrid approach in hydraulic engineering is often selected as the optimum, corresponding to a combined usage of any of the four approaches previously outlined.

5.3 *21st century*

At the start of the new century, it is difficult to make predictions on what will be important in hydraulic engineering over the coming years. Two basic types of problems may be mentioned, namely the 'microscopic flows' often relevant in mechanical engineering, with a flow domain covering just a volume of some liters of water or even less. Such problems are posed for instance in medical engineering, where an artificial heart is considered and tested for certain flow conditions. Similar conditions apply also to combustion motors for a variety of applications in practice. Optimizing the combustion process needs a detailed look at the various motor elements from a hydro-mechanical view and requires both experimental tests and spatial computations based on the NSE.

(a)

(b)

Figure 7. (a) Impact of granular slide onto water body with corresponding PIV vector field during initial impulse wave generation, (b) subsequent solitary wave formation, from VAW research project.

The second case of interest in the future may be described with large surface computations as typically occur with geophysical flows. In hydraulics, such configurations result for scenarios in natural hazards. Consider as an example rock slides from mountains that block a valley, such that an artificial lake forms which may eventually overtop the 'dike'. These flows endanger the downstream valley and require a detailed look into both the sediment and the water flows. Another case of note is the generation and the propagation of impulse waves resulting from slides into lakes or fjords and the displacement of large water volumes (Fig. 7). In the latter case, a three-phase flow has to be modeled requiring the combined use of experimental and computational infrastructure. The modern world has become so complex and so interacted that almost an indefinite number of scenarios may be considered. Most of the hydraulic problems require a detailed analysis of not only the hydraulic engineer, but also the attention of geotechnical engineers, of geologists and experts in material sciences, and of geophysicists and hydrologists. While our daily infrastructure is rapidly expanding, professional means must be defined to follow these developments. The future will become much more linked and will therefore be more complex to assess. From one side, young people may be frightened from this perspective; on the other hand, life becomes more dynamic with more changes and adoptions to be made within a short time interval. I am personally convinced that these essential questions have a significant impact on our profession. Let's cope with these challenges and add to the wealth of the global community: Long life hydraulic engineering, therefore!

6 CONCLUSIONS

Hydraulic engineering is a fascinating field with a number of complex and breath taking problems. These were solved from the beginning of mankind, first in view of water supply, then in view of water disposal and energy production, and today to conserve a valuable basic resource. The development of hydraulic engineering from around 1800 was reviewed in terms of the engineering profession, the engineering education and the engineering community. An outlook to the near future demonstrates that the position of the hydraulic engineer in the modern society will be increasing as compared to the past because questions in water sciences are more relevant in the global environment than in the past.

ACKNOWLEDGEMENTS

I would like to thank Prof. Enrique Cabrera for the organization of the 2006 Conference held in Alicante. It was a great opportunity for exploring different fields in the history of hydraulics, shared by a number of colleagues with whom the entire span over the centuries of hydraulics were considered. Prof. Cabrera has had the difficult task to propose a theme of interest and he has succeeded in doing so. I would also like to acknowledge some documents in this paper originating from my PhD students at ETH Zurich.

REFERENCES

Belhoste, B., Dalmedico, A.D. & Picon, A. (eds.) (1994) *La formation polytechnicienne 1894–1994*. Paris, Dunod.

Callot, J.-P. (1982) *Histoire de l'Ecole Polytechnique*. Paris, Lavauzelle.

Coronio, G. (1997) 250 *ans de l'Ecole des Ponts et Chaussées en cent portraits*. Paris, Presses de l'Ecole Nationale des Ponts et Chaussées.

Divers (1961) Ecole Nationale des Ponts et Chaussées. *Regards sur la France*, 5(14). Paris, Service de Propagande.

Hager, W.H. (2003) *Hydraulicians in Europe 1800–2000*. Delft, IAHR.

Pfammatter, U. (1995) 1795: Geburtsstunde der modernen Architekten- und Ingenieur-Ausbildung. *Schweizer Ingenieur und Architekt*, 113(45), 1030–1034.

Sébille, R. (1999) Les ponts et chaussées. *Universités et grandes écoles à Paris*, 164–168. Paris, Bussière.

Part C

The great challenges of water in the 21st century

CHAPTER 7

Water, history and sustainability, a complex trinomial hard to harmonise in Mediterranean countries

Concepción Bru
University of Alicante, Spain

Enrique Cabrera
ITA. Universidad Politécnica de Valencia, Spain

ABSTRACT: Analysing some of the main indicators of the water policies currently in force in European Union member states leads to conclusions which, though apparently surprising, can be explained when the analysis is carried out from a historical perspective. On the one hand, per capita water consumption is low in the Northern countries despite the fact that their water resources bear much less pressure, especially in terms of quantity. The explanation surely lies in the high prices which are paid for water; after all, they actually recover all the costs incurred. On the other hand, we find the countries in the South where, due especially to irrigation, water stress is much higher and, however, they keep subsidised rates and, consequently, their infrastructures are in a much worse condition. Therefore, those countries where water is in short supply are the ones which manage it the worst. The present paper tries to find the reason for this paradox, which is not so much of a paradox if the facts are observed from a historical perspective, and identifies the actions that it will become advisable to adopt gradually in order to redress this peculiar situation where the weight of history represents a great burden.

1 INTRODUCTION

No wonder the great civilisations of Antiquity, most of them on the shore of the Mediterranean, kept a close relationship with water. The guarantee of a regular supply beyond the basic subsistence needs permitted the appearance of the first major cities on the river banks. Mesopotamia (which means 'between rivers', in Greek) was the cradle of civilisation. Like the Ancient Egypt by the Nile, the Mesopotamian civilisation was built around the rises in the levels of the Tigris and the Euphrates which provided natural irrigation for the crops. In both civilisations, agriculture could go beyond flood plains thanks to the construction of canals which transported water and permitted its regulation. Especially worthy of mention are the generous and fertile floodings of the Nile, where level meters (the well-known 'nilometers') determined the taxes that citizens paid in kind at the time (Kerisel, 1999). At the opposite end, we find famines caused by the periodical droughts which Mediterranean countries have suffered from time immemorial. This is reported both in the Genesis (12:10 and 12:21) and in the Bible (Bruins, 1993).

The challenge of dominating water has always appeared among the first goals of mankind. And disputes have broken out in those places where water was in short supply. It is worth remembering that the word 'rivals' comes from the Latin term riva – 'river bank'. It should consequently not surprise us to check that the history of water-related engineering is so closely linked to the development of mankind. From Roman aqueducts to Arab irrigation ditches, human beings have always striven to prevent water shortage from limiting their development. However, despite their efforts, the seasonal lack of water has always meant a problem which mankind has had to confront, above all in areas like the Mediterranean, where water is scarce. And it is precisely water scarcity that, making a virtue of necessity, has led to write the most brilliant pages of this story.

Man very soon realised that crops became more abundant thanks to irrigation. Thus, the earliest irrigated systems, all of them located in Mesopotamia, are more than 30,000 years old (Bonnin, 1984). Among all of them, stand out those built by the Sumerians on the banks of the Tigris and the Euphrates more than 7,000 years ago. However, the North of Europe was never conditioned by the lack of water. The appearance of more uniform rainfall regimes and more severe weather patterns numbed the irrigation culture. It consequently came as no surprise when, in the late 1980s, the former president of the International Association of Hydraulic Research, IAHR, the Norwegian T. Carlstens, told everyone that he had first seen an irrigation system in a trip to Israel. He was more than thirty years old by then. That is why, for the better and for the worse, the water culture has always been and, above all, is linked to the areas in which it is not abundant. This is the conclusion that can be drawn after reading the main written works devoted to this story. The studies by Bonnin (1984), IAHR (1987) Schnitter (1994), Levi (1995), Viollet (2000), Evans, 2000 or Blackman and Hodge (2004) dedicate at least 80% of their contents to the Mediterranean culture and its marvellous hydraulic works.

Spain, a genuinely Mediterranean country, is a paradigmatic example. The still active thousand-year-old Tribunal de las Aguas [Water Court] in Valencia also gives proof of it. Founded by Abderraman III back in the 10th century (Giner Boira, 1997), it meets every Thursday to settle the disputes which water generates in the irrigated lands of Valencia. The saying El agua embriaga más que el vino [Water intoxicates more than wine] summarises the relevance of water in the Mediterranean, a popular feeling collected by the best Spanish chronicler of all time, Miguel de Cervantes. Inspired by the severe drought witnessed during the early years of the 17th century, in chapter 52 of the first volume of Don Quixote, he relates the following: "It was the case that during that year the clouds had denied their dew to the land and processions, rogations and disciplines were being made everywhere, asking God to open the hands of his mercy so that it could rain for them; and for that purpose the people from a village located nearby came in procession to a devout chapel which stood in a slope of that valley". Water shortage and prayers, so closely linked to the religiosity of past centuries, have made it possible to reconstruct the droughts occurred in Spain between 1500 and 1900 (Domínguez-Castro et al., 2007).

The close connection between water, culture and farming will reach its peak in Spain at the end of the 19th century when, in the context of an isolated and depressed country so different from today's Spain, Joaquín Costa, one of the most brilliant politicians of his time, saw the solution to the crisis in Spain in irrigation and consequently in hydraulic works. "If you want a more prosperous future, irrigate your lands as much as you can. So did the Arabs and they are still remembered for the hydraulic works that they built" (Costa, 1911). This context of nineteenth-century Spain happened to coincide with the rise of civil engineering driven by the invention of reinforced concrete in that same second half of the 19th century. The result is well-known. The number of large reservoirs (over 1,200 in total) per inhabitant in Spain is the highest in the world (Cobacho, 2000). The irrigated land area, nearly four million hectares at present, experienced its greatest growth between 1950 and 1990 coinciding with the construction of most of these dams (Corominas, 2009). And Spain is, together with Italy (EEA, 2009), the country with the largest irrigated land area in Europe, although closely followed by the rest of Mediterranean countries which have gone through a similar historical process (Fig. 1).

Nevertheless, in the last few decades and especially since the publication of the Brundtland Report which the United Nations asked its Environment Commission to elaborate (Brundtland, 1987), society has started to become aware of the environmental impact generated by reservoirs. As a result, their construction has been, if not paralysed, at least slowed down, above all in developed countries. This means that irrigation now can only develop by resorting to more efficient techniques with which it is possible to irrigate more land with less water, a demand which heightens the growing need for urban water. This is clearly visible in the comparison of columns 3 and 4 in Table 1. But, of course, nothing in this world is free, because we alleviate the tension suffered by the water resource increasing that suffered by the natural environment. Indeed, from the water point of view, the irrigation systems are more efficient but consume more energy. Water is distributed through pipes which require pumping stations that did not exist in the past. And

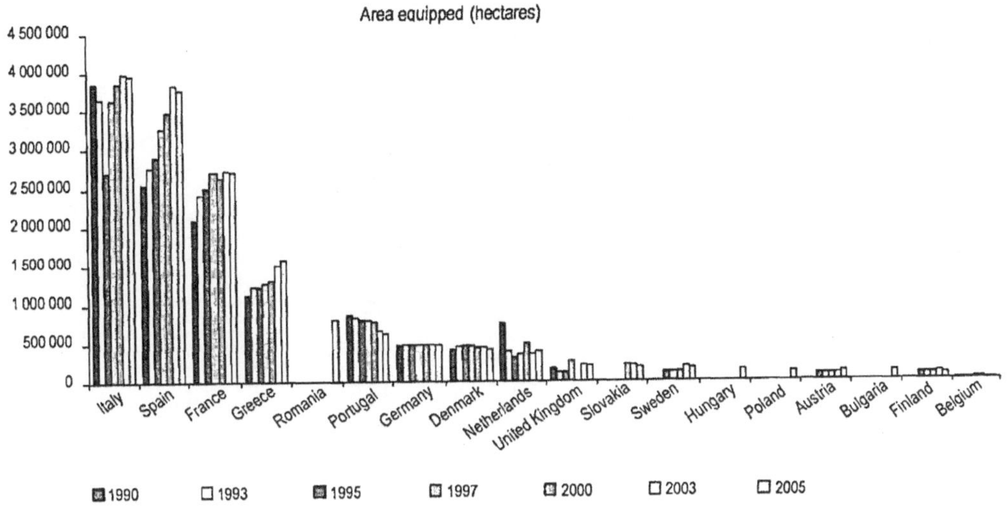

Figure 1. Evolution of the irrigated land area in Europe (EEA, 2009).

Table 1. Evolution of irrigated systems in Spain (Corominas, 2009).

Year	Area, 103 Ha	Water required Hm3	Water used Hm3	Energy consumption (Gwh/year)
1900	1,000	9,000	5,400	0
1930	1,350	12,250	7,594	182
1940	1,500	12,750	8,288	191
1950	1,500	12,375	8,553	309
1970	2,200	17,600	12,320	1,056
1980	2,700	20,925	14,648	2,093
1990	3,200	24,000	17,400	3,480
2000	3,410	23,870	18,499	4,893

consuming energy means emitting greenhouse effect gases. We will refer to this in great detail below.

The new impacts do not only have an impact on the air. The soils are affected as well. The localised supply of water to the plant does not facilitate soil washing, as a result of which the problems acquire a global dimension. And, therefore, what is convenient from one point of view can become inconvenient from another. It is thus necessary to undertake global analyses which incorporate the pros and cons of the different solutions, not an easy task due to the fragmentation of competences in real practice. It should not be forgotten that water and energy are generally issues which concern different ministries.

In any case, the growth of irrigation systems is not only limited by the impossibility of having more water but, above all, by the economic decline into which agriculture has been falling in recent years. In fact, agriculture employed 24.34% of the population and contributed to 10.50% of the GDP in 1973 (Escudero, 1986). Twenty years later, these percentages have decreased dramatically. The land gives employment to 8% of the active population and its contribution to the GDP is situated just below 5%. And, however, it still consumes nearly 80% of the water available (MIMAM, 1998). The figures are in a constant free fall as shown in Figure 2. In 2005, the contribution made by the farming sector to the Spanish economy only reached 2.5%. And regardless of the circumstances specific to each country, most of the Mediterranean countries have followed parallel paths, Greece

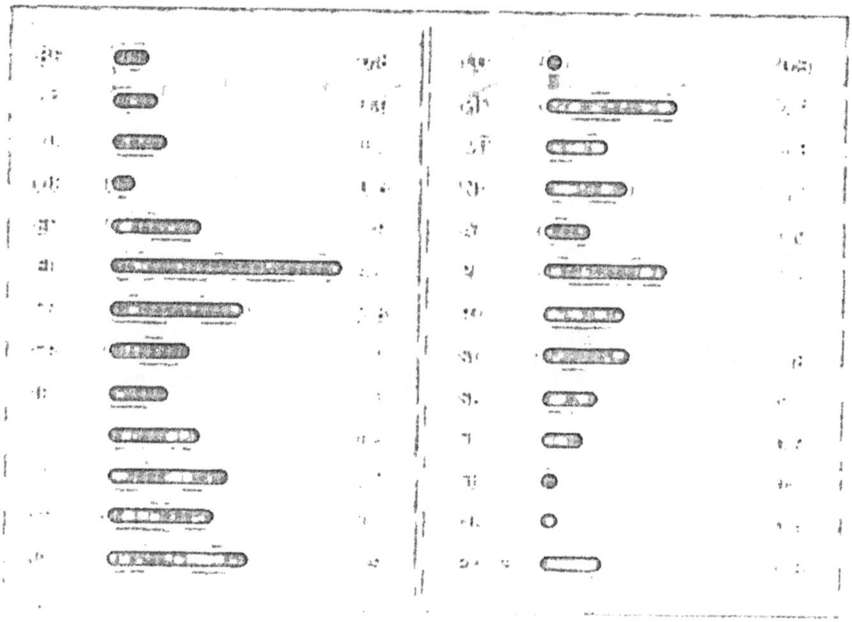

Figure 2. Contribution of agriculture to the 2005 GDP in EU member states (EU, 2007a).

having made the highest contribution with a poor 4.8%, when all these countries were eminently agricultural only fifty years ago.

Two facts can explain what has happened:

- The mechanisation of the land permits to reduce labour maintaining production.
- The growth of wealth is helping minimise the weight of agriculture in the economies of the most advanced countries.

The farming tradition of these countries, where the State has always assumed all the costs associated with such expensive infrastructures, has become a burden in the culture of citizens, who refuse to pay neither for the investment not for the maintenance itself, a responsibility which logically corresponds to those who use these works. On the other hand, the rights of use and administrative licenses, both for surface and ground waters (OECD, 1999), have been kept until the present day without having made any adaptations whatsoever to the present situation. This is a way of thinking which has naturally extended to the drinking water supply since urban and agricultural uses share many of these infrastructures after all.

Summing up, therefore, the water culture in Mediterranean countries has grown stronger throughout history and has arrived practically unscathed to the present day. And not only culture hinders changes. As a result of the complex system created, the legal issues are the most complicated ones. In fact, the Code of Hammurabi already included seven articles related to water law four thousand years ago. Article 56 in particular dealt with an old problem which has survived until today: how to sanction a farmer ('irrigator') who has flooded his neighbour's field. In Spain, the regime for the concession of surface water dates back to the Middle Ages and the first Water Act is from 1879, a law which was not modified until 1985, when ground waters were declared as a public property. In short, culture, legal rights, institutional fragmentation and, ultimately, a set of vested interests shape a maze from which it is hard to get out. Disentangling such a great mess through the adaptation of the rules of the game to the current context is as necessary as complex.

2 WATER CULTURE IN COUNTRIES WITH AN AGRICULTURAL TRADITION

The Spanish case is not unique. And even though the weight of culture depends on history, it is worth mentioning that the water policy of agricultural countries in the 20th century is very similar. It happened in the west of the United States, as was narrated by its National Academy of Sciences (NAS, 2004). In particular, and under the title Era of Development, the history of civil hydraulic engineering started in 1902 with the approval – promoted by Roosevelt – of the Reclamation Act which assigned to the Federal Government the starring role and the responsibility for promoting the great hydraulic infrastructures. A second impulse was going to be added to the preceding one. It was the great depression of the late 1920s. Public work construction largely helped the country's economic recovery and increased employment levels so that, at the end of the 1930s, 40% of the budget allocated to investments in public works by the Federal Government was dedicated to hydraulic works. Among all those works stands out the Hoover reservoir, which not only gave supply to the city of Las Vegas but also made it possible – including a transfer – to supply water to California (in 1921, the US Congress allocated 60% of the water from river Colorado to California). Since then, the Colorado has no longer been that wild river which was able to excavate the most famous canyon in the world. It is nothing but a stream with level rises that 'laminate' the reservoirs distributed along its course.

However, although the undeniable political appeal of these investments represented an invitation to continue with this dynamics, it was the environmentalist groups that formulated the first objections in the early 1950s. On mostly economic grounds, they questioned the Echo Park reservoir at the National Monument to the Dinosaur. This fact was going to mark a 'before' and 'after' dynamics. It is necessary to justify the profitability of those works without ignoring other criteria such as water quality protection. The hydraulic work is no longer good and acceptable by definition. And objections start to appear. They relate to the previously non-existent environmental impacts, since the action of man on the natural environment until the 20th century did not alter it to a significant extent.

The process of water policy revision which started in the 1950s culminated fifteen years later with the beginning of a new era in the late 1960s, the Epoch of Protection, which will ultimately lead to the creation of the EPA (Environmental Protection Agency) in 1970. The EPA soon enacted a set of provisions meant to protect water quality among which stand out the Water Pollution Control Act and the Safe Drinking Water Act. They were both published in 1974.

What happened in US water policy during the last decades of the 20th century does not matter so much from the perspective of the present paper. And, furthermore, it can be consulted in the paper which has inspired this summary (NAS, 2004). In any case, it deserves to be underlined that the United States held back the hydraulic overdevelopment policy a few decades earlier than Spain, amongst other reasons because the policy reached its peak before in the US. Its powerful economy made possible a higher growth rate for hydraulic works. But regarding the desires to bring water from places where it is abundant to those in which it is in short supply, the Mediterranean countries have been and will be pioneers. After all, they had those wishes much earlier. And even an expansive water policy is still in force in some of them, the least developed ones. The stages in the process described above must definitely be fulfilled.

Therefore, the story repeats itself because the problems are always the same too. A powerful civil engineering which, thanks to its great works (above all dams and transfers), solves the problems related to water supply and permits to transform dry areas into irrigated land must necessarily enjoy great prestige. After all, it makes possible both economic development and population growth in previously poor rural areas. Reservoirs were the emblematic works and the banners of development all over the world. But, ultimately, the passing of time has demonstrated that all these actions take their toll (there is no free lunch, Milton Friedman's well-known maxim) and has put things in their right place. The impact caused by those large projects, both on the natural environment and on the population living in the surrounding areas may end up being enormous, so much so that many of the mid-twentieth-century great projects would be unthinkable nowadays in the wealthiest countries because of the impact that they would generate.

Indeed, the last feat of civil engineering, the Three Gorges dam in China, is going to house the largest hydroelectric power station in the world. The 22,500 Mw that they hope will be active shortly (by 2011) will largely exceed the current record of Itaipu, on the border between Brazil and Paraguay (14,000 Mw). It is history repeating once again. The Three Gorges dam exemplifies the eternal debate between the pros and cons of these projects. It permits to regulate the floodings of the river and generate a huge amount of clean electricity for China, the country which emits the most greenhouse gases and the only one which does not admit any control over its emissions. In fact, it does not appear in the United Nations database (Table 2). However, the project has also had an enormous environmental and social impact. The ecosystems in the immediate environment have been irreversibly affected and more than one million people had to leave their homes during its construction. The project would have raised much less controversy when it was originally conceived in 1919, but today's society is much more concerned about environmental problems than it used to be. After all, it is just one disaster after another in that wretched country.

Nevertheless, despite all these challenges, water policies in many places of the world still follow the same patterns which characterised those of the 20th century. Spain is a clear demonstration in this respect once more. Most of the solutions provided have a common denominator: using more resources. One misses the economic and environmental dimensions of the problem which make it necessary to keep rather than promote, to manage the demand rather than boost the supply. And again the cycles repeat themselves, though with their corresponding times. After all, Burgi, who was in charge of research at the Bureau of Reclamation, the organisation which developed the great reservoirs in the United States, including the Hoover dam, already wrote the following in the 19th century: "As public values have moved away from the emphasis on the development of water resources to water management, the Bureau's research program has also changed from the development of the resource supply to the management of its demand" (Burgi, 1998). This clear message has also reached Spain. But, for the time being, it is only words, the facts are radically different.

3 THE TURNING POINT

During the first half of the 20th century, man used the territory as well as its resources in a sustainable way but things changed after the end of World War II. Industrialisation, the widespread use of chemical fertilisers to increase field productivity, the unstoppable growth of the population and, what is worse, its concentration in urban areas, started to degrade the natural environment to a considerable extent. So much so that, from a global perspective, what has happened to our rivers is almost an anecdote. And when the untreated spills began to degrade the water masses located near the cities, the citizens' concern about preserving the natural environment grew at the same rate as the spectacular economic development experienced during the second half of the last century. Among the three types of pressure – industrial, agricultural and anthropic – the third one is the most easily visualised (Fig. 3).

But, of course, the citizens' perception of these pressures varies depending on the place – whereas the industrial countries in the North will perceive them first, those more oriented toward agriculture in the South will become aware of them years later. Two facts explain this: firstly, the higher economic growth resulting from a more rapid and sound industrial development; and secondly, the fact that industry generates a punctual kind of contamination (the water is spilt in specific places) which is perceived earlier. In the agricultural countries, agrochemical fertilisers, together with irrigation water, seep into the subsoil and only after a few years do the aquifers end up becoming contaminated. It is a process which takes its time and that is why it takes citizens longer to become aware of it. This is a story which started nearly at the same time as industrial pollution. The economic boom experienced by the world favoured the arrival of the new phase. If the land is profitable, especially if agriculture exports its produce towards the rich countries in Northern Europe, what matters is fruit quality and size, which favours the massive introduction of

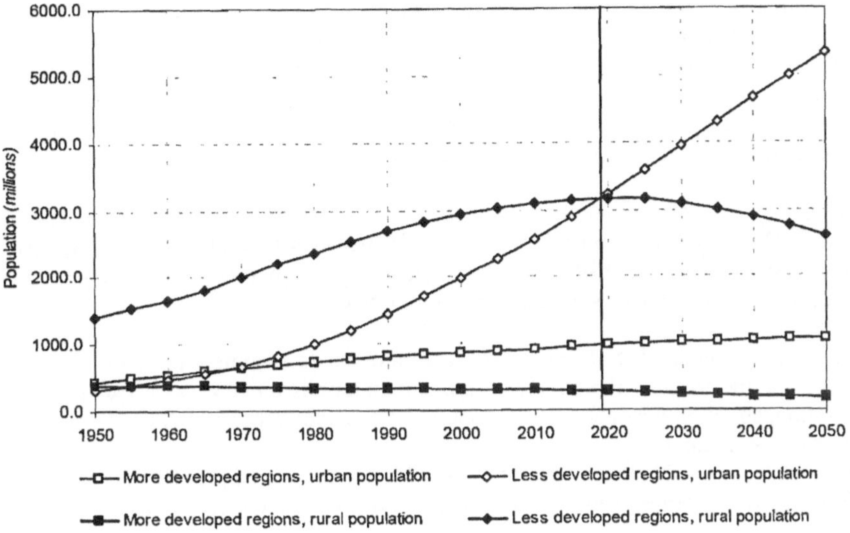

Figure 3. Evolution of the population since 1950 (UN, 2008).

agrochemicals. The development of agricultural machinery coincides with the decline of traditional farming work, well represented by the ploughing of the fields.

It is consequently industrialisation that favoured the mechanisation of agricultural work and that is why changes in agriculture began one decade later, coinciding with the use of agrochemicals. Based on the discovery of DDT in 1939 (its discoverer – Muller – received the Nobel Prize), they would have to wait until the chemical breakthroughs made it possible to synthesise pesticides and herbicides from DDT. These were aggressive and hardly selective chemical products which not only did away with the plagues which typically affected fruit and vegetables but also attacked all the ancillary fauna, thus altering the existing ecological balance in the system. The use of most of those plaguicides, pesticides and herbicides is forbidden nowadays (UE, 2009b). It was a period during which nitrogenated fertilisers were administered generously. Due to the lack of knowledge about the process, a large part of those fertilisers did not get to the plants and, dragged by the irrigation water, they ended up contaminating the ground waters (Agustí, 2000).

It was the time when chemistry was introduced into agriculture. The fertilising was done following a 'calendar' (based on the fruit flowering and growth periods) and it was believed that the tree needed the fertiliser more. At present, it is widely known that the tree uses nitrogen in a more sustained way over time and that this is consequently the right way to supply it. Other minerals which the soil needs (such as phosphorus and potassium) are fixed by the soil itself and, as they do not reach the aquifers, they do not contaminate (Agustí, 2000).

Nevertheless, the side effects of these farming practices obviously had to become visible sooner or later and by the mid-1980s, the measurements in the ground water that give supply to many towns revealed nitrogen concentrations exceeding 50 mg/l, the upper limit for water to be classified as drinkable (BOE [Spanish Government Official Gazette], 2003). And, at the same time, people began to talk about ecological agriculture understood as the one which makes an optimum use of natural resources and does not use synthetic chemical products neither for fertilising nor for fighting plagues, thus obtaining organic elements while simultaneously keeping the fertility of the land along with the environment. As far as sixty years ago, nobody ever spoke about ecological agriculture as only one type of agriculture – the traditional one – existed.

So much pressure on the natural environment (we have only referred to agricultural pollution but, as has already been said, urban and industrial pollution have caused more serious impacts) generated a great deal of social unrest. And the reaction did not take long to appear. It is summarised by the expression 'sustainable development', which is not only present in all political speeches but even in the title of political initiatives and projects: the last one, the Sustainable Economy Act. It is well-known that everything has to be sustainable now because the society knows too well that nothing has been sustainable so far. But when the moment of truth arrives and the time comes to materialise actions, the facts are different. It is clearly shown in the clarification of its meaning provided below.

In the prologue to the above-mentioned document elaborated by the United Nations' World Environment Commission, its president and also former Prime Minister of Norway Gro Harlem Brundtland, explained that, in 1983, the UNO General Assembly requested from that commission "A global program for change" summarised in four objectives:

- Proposing environmental strategies to reach sustainable development in 2000.
- Ensuring that the concern about the environment materialises in a greater level of international cooperation, above all between the developed countries and those which have not achieved that degree of development.
- Exploring the most suitable strategies to deal with environmental problems.
- Defining the most common environmental sensitivities and feelings.

The result is collected in the book Our Common Future (Brundtland, 1987) where the term Sustainable Development appears for the first time, defined in the report as "the process of change in which the exploitation of resources, the orientation of technological evolution and the modification of the institutions are in keeping with and also increase the current and future potential to meet the human needs and aspirations". Or, more colloquially, "Sustainable development is the one which guarantees the continuity of human progress by means of actions that respect the resources of future generations".

In any case, although measures have been adopted and considerable progress has been made in some aspects, the objectives seem to be further and further away. And that makes sense. The destruction of the environment goes ahead, on average, much faster than the solutions implemented. A spectacular example can be found in CO_2 emissions, which have never stopped growing so far, as shown in Table 2 (UNFCCC, 2009). The failure of the recent summit on climate held in Copenhagen does nothing but confirm this reality. It is thus advisable to underline that only one of the three changes requested, that related to technological evolution, has registered a significant improvement which, however, still does not suffice to offset the clear delay regarding sustainable resource exploitation. As for the adaptation of institutions to the present-day needs, the situation is now undoubtedly worse, at least in Spain. It is true that the *Estado de las Autonomías* [State of the Autonomous Regions] has increased decentralisation but, unfortunately, it has equally contributed to scatter competences, particularly those on which the interest of this paper is focused, the ones linked to water. This represents a clear backward step from the perspective of integral and sustainable resource management.

The concept of sustainability which is being examined takes its full meaning when applied to the planet's most important renewable natural resource: water. And not only referring to the resource itself but also in relation to the works and infrastructures that make it possible to manage it. Apart from being maintained so that they continue to deliver their services, they cannot impact on the environment. Instead, we must properly introduce into it what certainly constitutes a change of paradigm. The idea which prevailed during the first half of the 20th century was the domination of nature for the benefit of man. However, as a kind of defence instrument, the environment has shown numerous side effects which were ignored or insufficiently valued when the works were built. And precisely those side effects were the reason why the UNO commissioned the Brundtland report with which Environmental Engineering in general and the concept of sustainable development in particular were born.

Favoured by the evidence of its need, this concept has become deep-rooted in society. But the environment continues to deteriorate day after day, because the concept is still more at the

Table 2. Evolution over time in the emission of Greenhouse Effect Gases during the 21st century (Tm CO_2) (Greenhouse Gas Inventory Data, UNFCCC, 2009).

	2000	2001	2002	2003	2004	2005	2006	2007
Australia	404,392	427,079	791,075	630,040	328,461	596,239	551,058	825,888
Austria	64,104	65,420	71,106	75,807	74,425	75,679	74,352	70,835
Belgium	143,568	142,046	140,600	144,078	144,502	141,549	135,552	129,827
Bulgaria	60,314	60,166	58,339	64,833	63,294	64,241	65,146	68,991
Canada	636,781	627,327	801,492	796,890	858,081	772,380	759,545	792,495
Croatia	20,675	18,941	19,927	23,649	21,925	22,707	23,279	26,082
Czech Republic	138,661	140,443	136,401	139,215	139,849	138,541	144,654	149,103
Denmark	70,797	70,059	68,034	72,881	68,381	65,063	71,624	66,965
Estonia	16,920	14,400	14,232	14,290	11,578	12,077	10,246	14,116
European Community	3,847,717	3,873,345	3,833,662	3,865,827	3,888,484	3,845,217	3,828,150	3,792,548
Finland	51,126	53,239	54,331	62,014	57,184	40,390	47,727	53,080
France	515,697	508,040	491,713	490,546	488,400	487,510	475,510	463,433
Germany	976,066	1,004,433	984,843	988,088	973,938	947,043	964,433	939,985
Greece	124,673	123,255	122,506	126,278	126,232	126,837	123,014	128,203
Hungary	77,188	77,174	76,245	76,744	75,703	75,766	74,756	71,806
Iceland	5,085	5,043	5,051	4,997	5,018	4,945	5,462	5,694
Ireland	69,093	70,726	68,614	68,335	68,408	69,768	69,188	68,220
Italy	470,279	462,335	460,096	443,608	481,975	478,349	473,178	481,862
Japan	1,265,360	1,239,702	1,262,668	1,268,399	1,263,430	1,272,256	1,260,385	1,292,903
Netherlands	216,939	218,331	217,812	219,031	220,387	214,562	210,909	210,041
New Zealand	50,626	52,758	53,419	54,775	50,200	51,901	53,722	51,714
Norway	36,280	35,652	29,592	28,304	28,831	25,781	30,932	29,168
Poland	364,775	360,911	341,715	352,956	349,891	351,234	358,787	358,384
Portugal	75,732	78,304	82,720	90,158	84,651	88,949	82,701	79,517
Romania	97,525	101,471	110,125	117,353	119,901	112,199	116,640	116,068
Russian Federation	2,368,009	2,117,555	1,488,310	1,268,219	1,489,482	1,997,884	2,208,089	2,005,776
Slovakia	46,038	44,882	43,760	45,376	45,755	48,525	45,909	43,754
Slovenia	13,736	14,506	14,562	14,426	14,407	14,947	15,837	14,948
Spain	359,515	359,294	376,069	383,067	398,626	413,735	405,725	414,325
Sweden	32,555	36,621	35,126	36,670	37,685	38,061	41,283	44,952
Switzerland	52,399	52,701	51,878	50,922	53,291	54,530	54,250	50,617
Turkey	212,398	189,974	201,815	218,719	221,498	242,888	256,739	296,364
Ukraine	338,093	344,217	363,110	364,483	373,018	382,655	401,528	392,549
United Kingdom	676,829	680,226	658,171	663,055	660,014	654,230	649,663	638,493
United States	6,290,721	6,150,324	5,913,602	5,715,743	5,770,372	5,985,872	6,000,560	6,087,487
TOTALS	**20,190,667**	**19,820,899**	**19,442,721**	**18,979,777**	**19,057,277**	**19,914,511**	**20,090,534**	**20,276,193**

discourse level than at the action level. Continental fresh waters – surface as well as ground waters – are increasingly exhausted and also contaminated, which compromises the well-being of future generations, One of the strongest pieces of evidence, as a result of the action-and-reaction principle, is the increasing social prominence achieved by environmentalist movements.

Summing up, the United Nations made an accurate diagnosis in the late 1980s which still remains valid today. However, the actions which have been undertaken are at the very least insufficient and, what is worse, those who have the responsibility for reorienting policies seem to be unwilling to do it. How else can we interpret the words uttered by Mehan, in June 2003, within the framework of the World Water and Environmental Congress of ASCE in Philadelphia? During the opening lecture (Mehan, 2003) "Water Challenges in the 21st century" he went so far as to say that "The challenges posed by water in the 21st century will require a type of engineering which has nothing to do with that used in the times of the Hoover reservoir. Because our finite water sources are the object of a greater demand, and thus suffer, 21st century engineering will be more concerned about managing the demand than about increasing the supply; it will be interested in the micro rather than in the macro level and will equally be oriented toward a specific basin rather than toward a multi-state region, and more concerned about soils and trees than about steel and concrete". This shows that, fifteen years after the publication of the report elaborated by the Brundtland Commission, many of those who make decisions in this field in the USA have still not understood or perhaps have not wanted to understand. This will be clearly illustrated by the facts presented below.

4 THE FACTS

After describing the causes which led to the publication of the report Our Common Future and in the light of such a clear diagnosis, it is worth analysing the reactions in the different countries. And, once again, we can draw a distinction between Northern and Southern Europe. The circumstances of both regions are very different and so are their reactions, of course. Thus, with a short water engineering story, a very visible punctual contamination (caused by untreated industrial and urban spills) will become the main driving force of change. This is expressed as follows in a report elaborated by the German Environment Ministry (BUNR, 2001) "In the reconstruction years following the end of World War II, neither East nor West Germany were able to integrate the efficient use of water into the expansion of industrial activities, as a result of which water pollution at the end of the 1960s reached levels which generated great social alarm".

Indeed, efficiency is a key strategy when it comes to use water in a sustainable way. Detracting less water from the natural environment, we increase the guarantee of supply; we favour biodiversity (the other species will have more water at their disposal) and, ultimately, less pollution is caused. This was reminded by the German Environment Minister Trittin during the Fresh Water Conference in Berlin which served as a preparation for the Earth Summit in Johannesburg (Trittin, 2001). According to Trittin, sustainable policies must de supported on four main pillars:

- Efficient water management is the key element to fight against poverty and reach sustainable development.
- Efficient water use largely depends on the creation of efficient management structures.
- The best way to achieve efficient water use is decentralisation because local users are the most interested in maintaining the availability of water resources in the long term.
- All the agents involved must actively participate in the process.

The Northern European countries, especially Germany, Denmark, the Netherlands or Switzerland, quickly realised that water use needed to be as efficient as possible. And the best way to achieve that aim is to force the user to recover all the costs entailed by its sustainable exploitation. In that way, managers will make an effort to improve their management, and users will strive to contain their expenses. We can start highlighting the first part – i.e. the more cost recovery the more efficiency – relating the cost recovery principle to network performance (Figure 4). On the one hand, the downward curve represents the annual costs incurred in maintenance and renovation of the network and of the measuring systems. The more expense the more efficiency, and the fewer losses. On the other hand, the upward straight line visualises variable water costs (energy, reagents or, should there be any, environmental costs). The sum of both provides the total cost curve, the minimum level of which is the optimum loss economic level. Figure 4 (Cabrera et al., 2009a) shows that if we increase the slope corresponding to variable water costs (among which stand out the energy-related and environmental ones) the optimum point moves to the left. In other words, efficiency is favoured. And the opposite happens if water becomes cheaper.

The direct relationship between price and efficiency does not only become evident in urban use. It is also evident in agricultural use. Comparative studies between productivity levels in irrigated lands depending on the provenance of the water (which determines the price paid by the farmer) have proved this. Indeed, while an Andalusian farmer paid 0.01 €/m^3 for surface water around the year 2000, he paid ten times as much for ground water (due to the energy cost). And the higher the price of water, the higher the productivity achieved. Table 3 summarises some of the main conclusions drawn from a study carried out in Andalusia (Corominas, 2000), a study that is

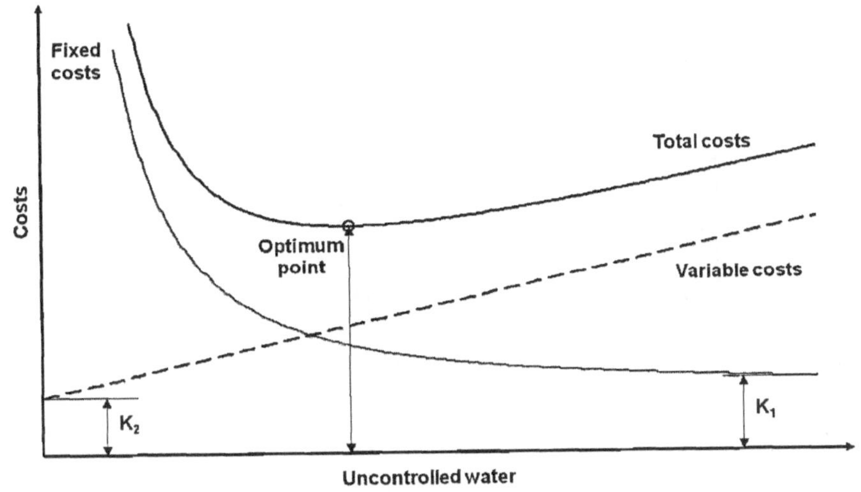

Figure 4. Concept of optimum loss level.

Table 3. Agricultural productivity in Andalusia according to water origin (Corominas, 2000).

Water origin	Productivity (€/m^3)	Efficiency (m^3/Ha)	Water/work unit (m^3/WU)
Surface	0.60	4,360	15,189
Ground	2.42	4,854	43,407

as interesting as it is unusual, which confers a special value on the results presented in detail in Table 3.

It is obviously impossible to draw comparisons between countries in the North and those in the South for irrigated lands, as can easily be done in relation to urban use. That is so because irrigation hardly exists in Northern European countries. However, it becomes clear once again how important it is to pass on all the costs in order to improve efficiency. And should water be subsidised, this must be done in a way that favours its rational use.

The need to be efficient along with the evidence that the passing-on of costs favours that efficiency justifies why, coinciding in time with the publication of the Brundtland report, the countries in Northern Europe started to apply this principle and levied an environmental tax on water. This is proved by the case of the city of Copenhagen (Fig. 5). From 1987 (the year in which the report was published) onwards, the price of water has experienced a dramatic increase.

This behaviour was imitated by most Northern European countries. In fact, one year before the publication of the Water Framework Directive (WFD), which refers to the cost recovery principle in its article No. 9 (EU, 2000), the situation in Europe regarding the price of drinking water is the one described in Figure 6, which shows the Water Rate (WR) and Complete Cost Recovery (CCR) percentages in relation to the average income. Thus, for example, the water rate in Spain meant 0.4% of the income, a percentage that has kept decreasing as can be seen. Obviously, cost recovery in the Nordic countries (with a higher per capita income) represents a lower income percentage. But that is where the paradox lies: the more scarcity the lower price.

Nothing has been done in Spain to bring the price of water closer to the cost that its sustainable management entails during the ten years elapsed since the publication of the WFD (EU, 2000). At present, Spaniards still dedicate ca. 0.4% of their income to it. In fact, according to the last publication of the Instituto Nacional de Estadística [National Statistics Institute] (INE, 2009) – data

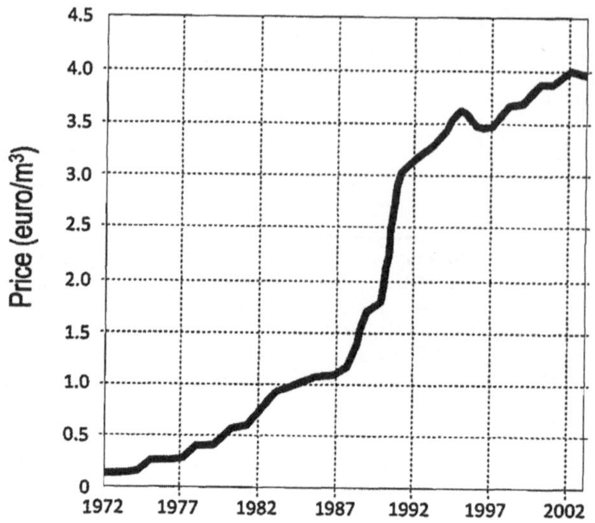

Figure 5. Evolution in the price of drinking water in Copenhagen (Napstjert, 2002).

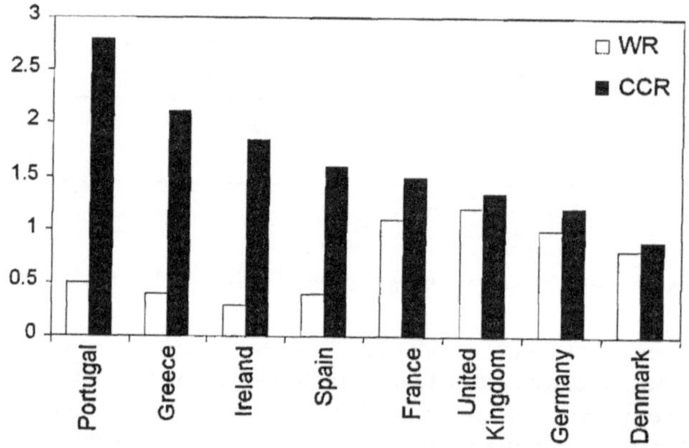

Figure 6. WR and CCR with respect to the average family income (Merkel, 2003).

corresponding to 2007 – the average water consumption of Spanish citizens (Table 5) was 157 l/inhabitant and day, 1.29 €/m^3 being paid for the whole cycle, which represents an annual cost of 74 € per person. As Spain's average per capita income in 2007 was 23,396 €, the percentage dedicated to water meant 0.32%. This provides further evidence that recovering water costs in Europe is a cultural rather than a social issue. The percentage considered socially acceptable in developing countries is 5%, about 15 times the percentage currently registered in Spain.

There are even some countries which take cost recovery to the extreme, dividing drainage rates in two blocks. You pay for the volume of water to be treated, as that volume determines the size of the treatment plant and the area which waterproofs the dwelling. Table 4 (BUNR, 2001) specifies the two rating systems applied in Germany (1999 prices): the habitual one (a single payment) and the split-rate one that has already been explained. It can be observed that, as a result of the greater renovation of systems undertaken in the East, on which practically no investments had been made during the communist period, rates are higher in those Eastern regions. It is worth mentioning that once the costs have been recovered, prices hardly change. It only proceeds to change them

Table 4. Year 1999 urban drainage rates in Germany (BUNR, 2001).

	Drainage rate (divided in two blocks)		Drainage rate
	Sewage (€/m^3)	Rain water (€/m^2 year)	(a single block) €/m^3
Germany	1.79	0.77	2.28
German Federal Republic	1.72	0.78	2.23
Eastern Germany	2.39	0.59	2.54

according to inflation rates. In fact, the prices applicable in Germany in 1999 were similar to those valid in 2003 (DVGW, 2005). The 1.79 €/m^3 in Table 4 corresponding to water treatment go up to 2.14 €/m^3 whereas the payment for waterproofed surface area hardly changes (0.77 €/m^2 and year in 1999 and 0.82 €/m^2 and year in 2003).

At this stage, it can be of interest to compare the drainage and treatment rates valid in Spain. According to the INE, the total average cost per cubic meter was 0.86 €, which resulted from adding the water supply, 0.64 €/m^3, and 0.22 €/m^3 for drainage and treatment (INE, 2005). This means that Germany paid up to ten times as much as Spain for treatment in 2003 (2.14 €/m^3 as opposed to 0.22 €/m^3) a cost to which must be added the cost derived from waterproofing the territory (0.82 €/m^2 year). This difference is much higher than that registered for water supply in that same year (1.72 €/m^3 as opposed or 0.64 €/m^3, only three times as much). The conclusion is clear: the whole urban cycle is subsidised, but this especially applies to drainage and treatment. It must be remembered that treatment plants and sewers have been financed with European funds during the last 15 years and the repayment of that 'loan' has not passed on to the user's bill.

The German drainage and treatment rates are additionally bringing about the reappearance of rain water urban deposits. Their use is justified by economic, environmental and operational reasons. From the economic point of view, they are profitable because thanks to them it is possible to save water and rationalise drainage networks. In fact, they are small storm deposits which laminate the peaks of the entry hydrograms generated by Mediterranean rainfall patterns (a lot of rain in little time). When rates include all costs, also the environmental ones, these systems entail significant savings in the user's bill. Two facts justify them from an environmental perspective. Because they function as small storm deposits, they reduce the pollution in the first rain water discharges, to which must be added that the water savings mean detracting less water from the natural environment. The advantages in operational terms are obvious too, above all in the rural areas of developing countries (Fewkes, 2006).

In Spain, the same as all over the Mediterranean, these deposits were widely used until the urban water networks made them fall into oblivion in the early decades of the 20th century. Nevertheless, they have been experiencing a spectacular revival during the last few years in countries where urban water rates justify them. Thus, continuing with the case of Germany, over 100,000 units have been installed during the last decade. It should not surprise anybody. These deposits reduce the rate both within the limits of the waterproofed surface area which they drain and in that proportional to consumption (the stored water is reused). Their repayment period is situated between 12 and 19 years (Fewkes, 2006).

Particularly convenient in single-family detached, semidetached or detached houses – a typology which is increasingly abundant – they were born on the shores of the Mediterranean and Herodotus (Bonnin, 1984) already reports how they were used in the ancient Tyre more than 2,600 years ago. They were well known by our farmer ancestors and now they are experiencing a rebirth in the North of Europe favoured by the rate systems applied there. So much so that their presence explains the recovery of unitary consumption in Germany in years of drought, just the opposite of what happens in Spain. Figure 7 (DVGW, 2005) shows a slight increase of unitary packages due to the drought in 2003. It can also be appreciated that the decrease in the unitary demand starts with

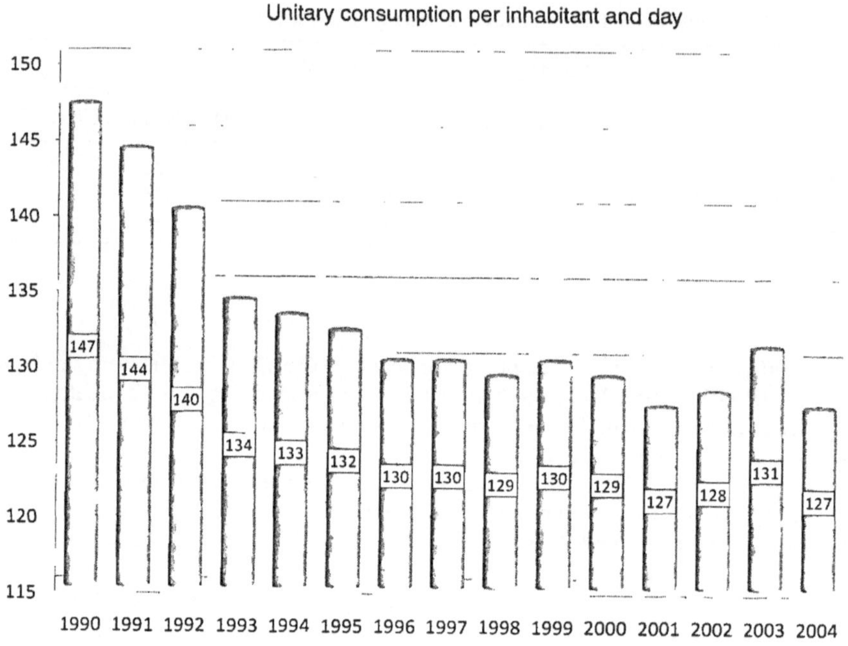

Figure 7. Evolution of unitary consumption in litres per inhabitant and day in Germany (DVGW, 2005).

the application of the cost recovery principle, the reaction from the countries which have shown more receptiveness to the Brundtland report.

This is not the case of Spain. Judging by the periodical estimates released by the Instituto Nacional de Estadística [National Statistics Institute] (INE, 2003; INE, 2004; INE, 2005; INE, 2006; INE, 2007; INE 2008 e INE 2009), prices, unitary consumptions, and efficiency have not suffered great changes (Table 5). The average unitary consumption has hardly varied (unitary packages have so far mostly depended on whether or not there is drought) whereas prices follow inflation unless a significant alteration takes place (for instance, starting to give supply to a municipality with a desalination plant).

As far as network efficiency is concerned, the values are not very reliable on the whole (Table 5), though two clarifications need to be made. Until the last report released by the INE (corresponding to 2007) no differentiation is made between apparent losses (meter sub-reading, thefts, non-measured consumptions, etc.) and leaks (INE, 2009).

The figures in Table 5 must be handled with caution, particularly those related to performance. In fact, one of the reasons which make advisable the creation of a regulatory body is the need to count on reliable data because the truth is that, at present, the administration does not have a good knowledge about how water is used in Spain. In any case, these are the most reliable global data that one can find and reality will always be near the values shown in Table 5.

German systems are also more efficient as a natural consequence of cost recovery. Figure 8 shows the annual average performance, well above that of Spain. Particularly in 2001 (the only year in which the tables attached admit a comparison and, taking into account the reliability of the data utilised, with all the advisable reservations for this case) German losses (7.3%) are less than half the Spanish losses (19.4%). This is even more striking if we bear in mind that the percentual performance data used are more favourable to systems with a higher consumption – and consumption in Spain (165 l/h day) exceeded the one in Germany for that same year (127 l/h day). This reinforces the idea that it is hard to be efficient without passing on costs.

Table 5. Evolution of the average consumption, price and efficiency in Spain during the 21st century INE).

Year	2000	2001	2002	2003	2004	2005	2006	2007
Consumption (inhabitant and day)	168	165	164	167	171	166	160	157
Average complete cycle price (€/m^3)	0.73	0.77	0.81	0.86	0.96	0.98	1.08	1.29
System efficiency (%)	73.4	80.6	80.6	81.3	82.1	82.1	83.3	76.0

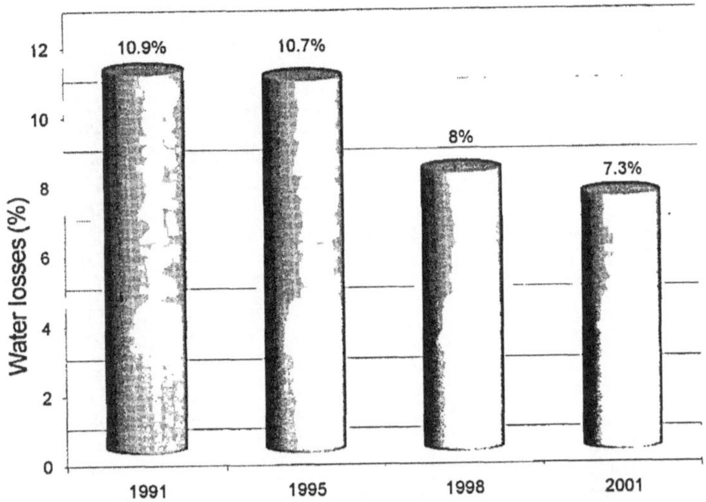

Figure 8. Evolution of (real and apparent) water losses in Germany (DVGW, 2005).

Finally, regarding urban use, Table 6 offers relatively recent data for most of the EU-25 countries. However, despite the reliability of the source (Smets, 2008) these data must be handled cautiously. The actual source warns about it too. That is why the analysis provided in Table 6 needs a number of clarifications. The first one is that, although the IWA and the OECD are two highly reliable sources, the consumptions are different (see columns b and c). The second one is a warning: the least reliable values contained in Table 6 have been marked with a star (*) or a cross (+). The third clarification focuses on the fact that Spain's prices are the cheapest ones in Europe. Water only costs less in Lithuania and Romania with Italy on a level (a higher cost for a consumption of 200 m^3 and a somewhat lower one for a consumption of 120 m^3). No data are provided about Greece, the other Mediterranean country par excellence, in a Table where the different columns provide the following information:

It is advisable to make a final clarification though. In Table 6, the Spanish consumption is situated between 119 l/inhabitant and day and 137 l/inhabitant and day, a value below that provided by INE, 160 l/inhabitant and day (Table 5, year 2006). All the same, these values are consistent, as the first ones correspond to the average for the five largest towns in each country and the second to the average for the whole country.

In short, the countries with a higher water stress are the ones in which, due to historical and cultural reasons, water is subsidised to a greater extent and, consequently, are the ones which worst manage this resource, when just the opposite would be the logical thing to expect. In fact (Table 7), nearly all the European basins that will predictably suffer from water stress problems are located in the South. In all these areas, the measure of stress, the WEI (Water Exploitation Index), the part of the total resources derived to anthropic uses presents the highest values. In addition to the WEI, Table 7 shows the percentage of water in each basin dedicated to each use (from irrigation, the percentages of which are high in the Southern countries, to energy generation).

Table 6. Drinking water consumptions and prices in Europe corresponding to 2006 (Smets, 2008).

	Occupan. Inhab. ECE a)	Consumption m³/inhab. IWA b)	m³/inhab. OECD c)	l/d/inhab. IWA d)	Price 200 m³/$/yr IWA e)	120 m³/$/yr f)
Lithuania	2.9	38	–	59–89	153	92
Belgium	2.5	56	–	78–166	575	365
Hungary	2.6	57	48	90–146	375	225
Portugal	3.0	63	65	99–144	325	222
France	2.4	67	57	70–100	681	426
Netherlands	2.3	68	61	103–149	711	395
Denmark	2.2	72	54	–	1070	680
Austria	2.4	72	47*	144–147	688	425
Finland	2.2	74	47*	140–170	. 652	403
Germany	2.1	–	45	–	564	411
Spain	2.9	76	83	119–137	301	200
Romania	2.8	–	–	83–163	146	87
Luxembourg	2.5	84	–	–	–	–
Cyprus	3.1	87	–	121–165	534	339
Sweden	2.9	88	61	180–190	602	390
Greece	2.6	100	–	–	–	–
Italy	2.6	102	84	150–229	375	170
Switzerland	2.3	121	98	119–360*	983+	721+
United Kingdom	2.3	122	–	–	–	390**
Norway	2.2	131	56*	164–200	445	282

Column a. Average dwelling occupancy in the respective countries.
Column b. Average consumption per dwelling according to the IWA (International Water Association)
Column c. Consumption per inhabitant and year according to the OECD.
Column d. Range of unitary consumptions established from the values obtained in the five largest towns of each country.
Column e. Price corresponding to a consumption of 200 m³.
Column f. Price corresponding to a consumption of 120 m³.

5 THE THREATS

With a growing population that, especially if the management modes do not change, anticipates more needs of water and, with a climate change that threatens to cause a significant reduction in the availability of resources, the challenge we are facing is a formidable one. And not only for the countries in the Mediterranean area that I will refer to later on. The threat is global. Above all in developing countries with much more serious and urgent problems to solve than those described so far. Water becomes a survival factor in poor countries and the lack of water entails a high number of deaths every year. Nearly half of the population suffers from water scarcity and, according to the UNO, the water used by man has increased twice as fast as the population.

An extreme and simultaneously paradigmatic example of the possible consequences of the water problem is found in the Sudan, more precisely in the Darfur region. The conflict which has been devastating this region since 2003 has water shortage as one of its main causes. Many people, among them Ban Ki-moon, the United Nations Secretary-General, relate it to the climate change, including the disappearance of Lake Chad (situated some 1,000 kilometres away from Darfur). The succession of images from NASA (Fig. 9) shows the regression occurred. Although it is true that rainfall has dropped by 40% during the last 20 years, there are some who blame this lack of water on agricultural development and a non-sustainable use of this resource. In any case, what seems

Table 7. European basins which are likely to suffer from water stress in 2030 (IEEP, 2008).

Member state	River basin	Total abstraction 2030 (Km/y)	Irrigation share of total (%)	Livestock share of total (%)	Domestic share of total (%)	Industry share of total (%)	Energy share of total (%)	WEI (%)
Belgium & France	Scheldt	3.7	2.6	2.1	36.8	51.4	7.2	27.2
Belgium & France & Netherlands	Meuse	13.4	14.9	2.1	28.2	27.9	26.8	26.5
Bulgaria	East Aegean	2.1	61.8	0.8	10.5	17.2	1.6	22.1
	West Aegean	1.9	87.4	0.3	9.6	2.4	0.4	63.8
Cyprus	Whole Island	0.6	82.0	0.6	17.3	0.1	0.0	69.2
Denmark	Zealand (mainly Copen-hagen, capital region)	0.2	17.0	6.8	58.1	17.7	0.4	38.2
France	Rhone Méditerranée (dry region)	0.2	64.4	0.3	19.5	11.8	4.0	26.7
	Seine Normandie Basin	11.1	10.2	0.6	23.0	15.6	50.1	42.8
Greece	Attica	0.6	35.2	0.3	61.6	2.9	0.0	108.3
	Central Macedonia	0.2	95.2	0.2	4.5	0.1	0.0	81.1
	Western Macedonia	2.5	88.8	0.1	3.5	0.1	7.5	84.6
	Thrace	1.0	91.2	0.3	7.0	1.2	0.3	33.2
	West Aegean	0.1	94.1	0.2	5.1	0.1	0.5	44.6
	Thessalia	4.4	98.0	0.0	1.9	0.1	0.0	191.8
	Eastern Sterea Elada	1.4	86.3	0.1	13.1	0.5	0.0	70.9
	Western Sterea Elada	1.3	95.6	0.2	4.1	0.1	0.1	37.5
	Eastern Peloponnese	1.0	75.9	0.2	18.5	0.8	4.7	37.9
	Northern Peloponnese	0.3	86.3	0.2	13.0	0.5	0.0	31.3
	Western Peloponnese	1.4	95.0	0.1	4.7	0.1	0.0	51.6
	Crete	1.2	92.8	0.2	6.6	0.2	0.1	53.3
	Epirus	0.9	93.2	0.3	6.4	0.2	0.0	22.5
Italy	Po	14.9	43.1	0.6	26.4	24.7	5.3	28.9
	Nortern Appennines	2.6	33.3	0.3	29.6	24.7	12.1	28.4
	Central Appennines	5.6	32.0	0.3	27.5	24.2	16.1	36.8
	Southern Appennines	12.5	50.1	0.2	22.4	20.1	6.8	72.1
	Sardinia	1.3	72.2	1.2	20.1	6.3	0.3	32.7
	Sicily	5.8	58.8	0.2	18.1	16.5	6.4	215.7
Ireland	Eastern	0.8	0.0	5.5	34.6	58.4	1.5	29.3
Malta	Whole Country	0.04	54.6	1.6	37.1	6.7	0.1	236.5
Netherlands	Rhine (NL part)	0.9	22.6	2.6	35.9	17.7	21.1	22.8
Portugal	Sado & Mira	0.4	87.7	0.6	2.4	0.0	9.4	21.1
Spain & Portugal	Guadiana	4.7	95.6	0.5	3.1	0.6	0.3	74.6
	Tajo/Tagus	7.7	75.5	0.5	15.8	5.1	3.0	44.4
	Duoro/Duero	8.2	89.7	0.5	6.9	1.2	1.7	35.7
Spain	Andalusian Mediterranean basins	2.1	84.9	0.2	10.1	4.8	0.0	147.8
	Atlantic Andalusian	0.7	80.1	0.5	13.6	5.8	0.0	40.6
	Balearic Islands	0.4	65.7	0.4	22.7	11.2	0.0	61.5
	Catalonia	1.9	42.7	0.9	36.7	19.2	0.4	49.0
	Ebro	8.0	92.3	0.6	3.9	1.4	1.9	53.7
	Guadalquivir	8.0	90.4	0.2	6.4	2.8	0.2	143.4
	Jucar Basin	4.0	84.6	0.2	9.5	4.5	1.1	120.8
	Segura Basin	3.8	88.9	0.2	7.4	3.5	0.0	630.3
UK (Engl.)	Anglian	1.5	9.9	1.0	40.4	8.0	40.1	23.2
UK (Engl.)	Humber	3.2	1.3	2.2	52.1	11.1	34.4	24.2
UK (Engl.)	Thames	3.3	0.3	0.4	70.0	15.3	14.0	55.9
Total	All River Basin Districts	148.6	57.7	0.6	19.6	12.8	9.3	

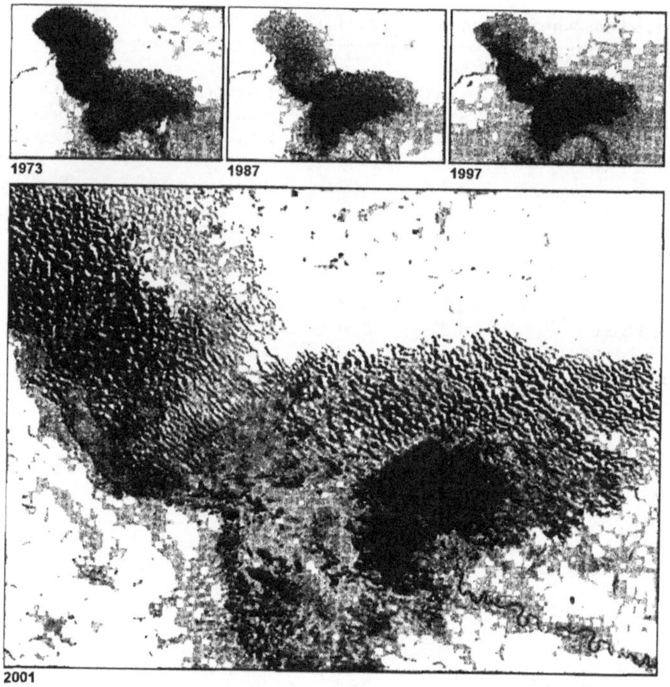

Figure 9. Evolution of Lake Chad 1977–2001 – (NASA).

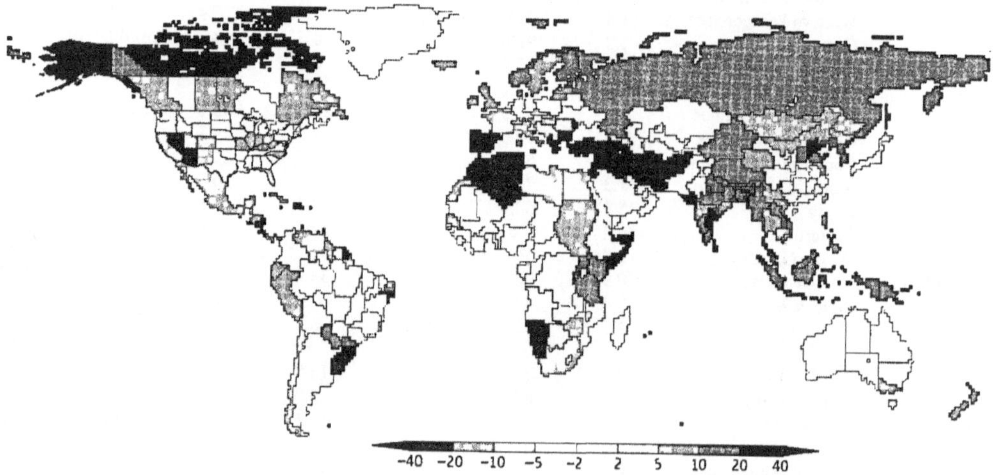

Figure 10. Estimates of variations in water resources caused by CC (Milly et al. 2008).

to be demonstrated is that water (its absence) has been one of the key factors in the conflict, something that will be increasingly frequent in the present century.

Obviously, the situation in the Mediterranean is far from being so dramatic. But it is evident that unless we manage to change the management model, one can only make out a gloomy future, above all if the predictions made using the IPCC models come true. Not in vain, according to the models elaborated, this area is the most strongly threatened in the world. This can be inferred from Figure 10 (Milly et al., 2008). It compares the average water resource availability in the

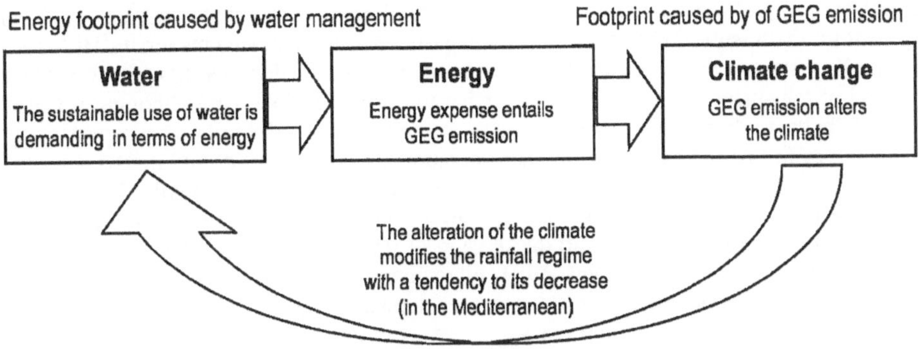

Figure 11. The water – energy – climate change loop.

period 1900–1970 with that which the world will have at its disposal halfway through the century according to the IPCC models. The conclusion is that the availability of water in the Mediterranean area will go down by 40%. The margin of error does not seem to be high, since only those areas in which at least eight of the twelve IPCC models coincide have been coloured in Figure 10.

Finally, it is worth remembering that desalination is not the remedy to water scarcity, a temptation into which these Mediterranean regions could fall. Without a doubt, it can and is indeed solving serious but punctual problems associated with urban supply. But it has an important limitation which determines its mass implementation. The energy expense that it entails is responsible for the emission of greenhouse effect gases and, consequently, for climate change. Figure 11 (Cabrera et al., 2009b) summarises the vicious circle which at least should be controlled. It thus comes as no surprise to check that the European Union regards the option of increasing the demand (desalination and transfers) as the ultimate solution to face scarcity problems (EU, 2007b). In any case, a recent report commissioned by the European Union (IEEP, 2008) has placed desalination in its right context.

6 DRIVING FORCES AND BARRIERS

It will not be easy to introduce the substantial reforms that can adapt the water policy to the current needs in the complicated board where the water policy game is played in Mediterranean countries, which are so heavily loaded with history. In this respect, it must be said that, seen from today's perspective, the water policy – as well as the culture which surrounds its use – has hardly evolved during the last one hundred years. It is practically the same policy which Costa defended at the close of the 19th century. The reasons behind this immobilism are obvious. Apart from history and culture, there are vested interests and, ultimately, there is the brightness and short-term approach which characterises the management of the supply (providing more water) as opposed to the discretion in the management of the demand, the results of which can only be appreciated in the medium-long term.

This explains why the different political parties, which take turns in power in Spain, despite strongly criticising the policy developed by the 'rival' government of each moment, when they are ruling the country and have to make decisions, they invariably opt for the supply, either through transfers or with desalination plants. But only the weight of history can be felt in the culture of citizens. As all these countries have distinct geographical regions and different regional governments, the conflicting interests have grown, and the difficulty to tidy up such a big mess has accordingly grown too. Regional politicians have found in water policy a 'gold-mine' of votes that is very easy to exploit. They only need to resort to populist and demagogic messages which never include the necessary recovery of water costs. In other words, they keep the facts from the past which can be convenient for them (subsidies) while they enthusiastically brandish the new territorial flags.

All in all, just as it happened in the Northern countries two decades ago, the symptoms of exhaustion of the current water policy are starting to become visible. And although, for the time being, right in the middle of the economic crisis, the governments of Southern European countries are reluctant to follow the examples of their neighbours from the North and implement the cost recovery principle, the everyday realities perceived by society will eventually force the public opinion to demand the introduction of the necessary changes. Many rivers and aquifers are already exhausted and contaminated today and the passing of time only increases the concern about the preservation of the natural environment. What still remains to be discovered is whether governments will be ahead of their time and adopt the measures demanded by the new times or whether, as is nearly always the case with unpopular measures (we are going to refer to them in the next section), the solution will arrive with a crisis.

7 THE GREAT CHALLENGES

With such a clear diagnosis in our hands, there are few doubts about what needs to be done. The problem lies in how it should be done. That is why this section is only going to provide an overview of the challenges leaving for the next one the intricate way that will make it possible to face them. It all with common sense, of course, because we should not try to reinvent the wheel. Therefore, the best possible option is a review of the documents that the European Union itself and its Environmental Agency have elaborated in relation to this. Among the numerous existing documents, we are going to focus on three of them, selected according to two criteria, the date of publication (it must be recent) and the condition that they deal directly with the issue we are interested in. In chronological order, they are the following:

- Facing the challenge of scarcity and drought in the European Union (UE, 2007b)
- Water resources across Europe. Confronting water scarcity and drought (EEA, 2009)
- Incorporation of sustainable development into EU policies: 2009 Report on the European Union strategy for sustainable development (UE, 2009a).

The conclusions drawn from these reports confirm what has been said so far. Thus, the first one lists the five challenges identified on the path that should lead to find solutions to the water problem:

- Making progress in the application of the WFD. And the assessment of the difficulties to achieve that progress allows us to highlight the biggest one, namely the fact that ineffective water rating policies are currently in force.
- The planning of soil uses, an extremely relevant issue which has only been treated collaterally in this paper for reasons of space.
- Encouraging saving because the present-day situation has an enormous margin for improvement.
- Going ahead in the coordination of water saving policies on a European, national, regional and local scale. Or to put in another way, to adapt the administration to the current world.
- Improving the existing information about availabilities and consumptions. We have verified throughout this analysis that the data used are very often not consistent enough, which fully justifies a challenge that, in the case of Spain, implies reforming the administration.

All things considered, we would like to highlight the third of these five challenges. After all, the only 'advantage' derived from not having adapted water policies to the current context is the impressive margin for improvement that still exists. So far, arguments such as the mere assignation of a price for water (even if it is far away from the complete cost recovery) and the growth of environmental sensitivity have sufficed to uncouple water expense from economic growth in the US context. It is demonstrated by Figure 12 (Gleick, 2003), which additionally confirms what

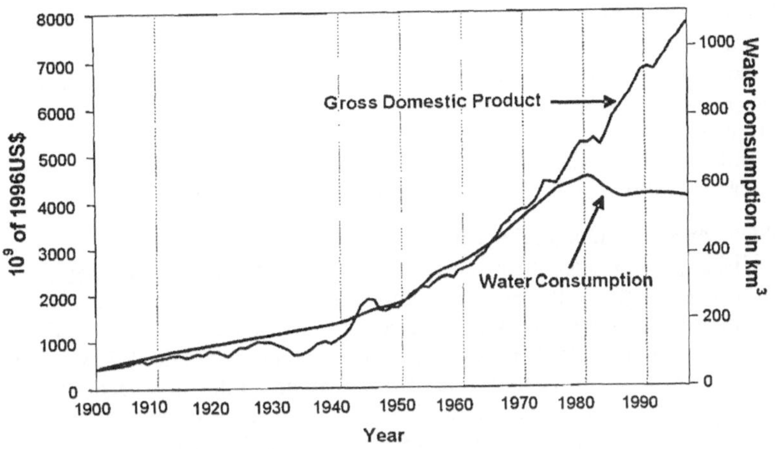

Figure 12. Evolution of the US gross domestic product and of water consumption (Gleick, 2003).

happened in Germany with unitary consumptions and with the evolution of its networks (Figs. 7 and 8). It can be seen that there is coincidence in time too, since the uncoupling started in the 1980s. In the Mediterranean countries, the possibilities for improvement in the area of management, combined with the reutilisation and the use of rain water have hardly been explored, which means that the margin for improvement is enormous.

As for the document mentioned in second place, it reminds in its executive summary that the European water policy has traditionally been based on supply management (increasing the resources available) and suggests managing the demand for the future, which means:

- Implementing a suitable price policy in all the uses based on the universal measurement of consumptions and on a rating system which favours efficient use.
- Along the same lines, making sure that agricultural water subsidies promote its efficient use.
- Investing in new technologies which can improve water use efficiency and renovate the existing infrastructures.
- Only when all the possibilities to reduce consumptions and improve efficiency have not only been explored but also implemented, increasing the supply choosing the most sustainable options.

The last and most recent of these three documents is less specific, as it deals with sustainable development in general and not so much with water policy in particular. In any case, it pays a lot of attention to our study object. Among the recommendations contained in the document we are going to extract what we think can be of interest to us. It is textually said that: The key action in water management includes continuity in the application of the Framework Directive on water and the EC policy on water scarcity and drought in the European Union. In other words, it insists on the need to implement the WFD (and therefore cost recovery) while simultaneously referring the reader back to the first document examined in this section. It also refers to seawater policy – which is beyond the scope of this paper – referring the reader back to the corresponding Framework Directive (EU, 2008).

Therefore, the diagnosis (based on which the treatment has to be specified) leaves no room for discussion. We must use water efficiently. These are the four facts that explain the need:

- In countries with a high water stress, like those in the South of Europe, the need to promote efficient water use becomes a critical issue, a need that is particularly felt in dry years. However, as has been said above, and due to cultural tradition, these countries have so far faced these problems trying to increase the resource volume. It is not easy at all to change this mentality.

- In highly developed countries, efficient water use has become the most effective measure to control contamination.
- In less developed countries with high birth rates and where the population concentrates in large cities, the efficient use of water is vital to prevent the capacity of hydraulic infrastructures (water treatment plants, pumping systems, pipes, etc.) from being collapsed in few years. Efficiency delays investments and makes it possible to be economically sustainable.
- Finally, it is the best remedy to fight the looming climate change.

The problem, above all in countries with so much history, is how to do it. Because even though the EU establishes goals and identifies guidelines, it says nothing about the way to reach them, as that would mean interfering with the sovereignty of the Union member states. Furthermore, the specific circumstances in each case may suggest the convenience of adopting different strategies. It has already been explained that the conditioning factors in the South of Europe have nothing to do with the factors that are influential in the North. Taking all this into account, the next section provides an action proposal that can prove useful for Spain and also for the Mediterranean areas, as it contemplates history along with the circumstances described above.

8 THE PATH TO FOLLOW

All goals converge into the most important one, using the natural resources in a sustainable way. And until very recently, those who defended the supply-based policy as opposed to the demand-control approach claimed that not doing it meant limiting economic growth, an argument which is not very solid at present (Figure 12). However, the problem, especially in the Mediterranean, lies in the fact that the measures which need to be adopted seeking to become more efficient entail sacrifice and are unpopular (cost recovery) and control (monitoring the degree of compliance with the objectives or goals). This is why the steps to be taken are the following:

1. Taking water management far from the political arena.
2. Educating the citizens.
3. Reforming the Administration.
4. Implementing the kind of legal reforms that can encourage efficient use.
5. Setting up monitoring and control mechanisms.

Below is a brief discussion of the relevance and scope of each one of them.

8.1 *Taking water management far from the political arena*

The solutions which are convenient for the future clash with the short-term interests of politicians. It is always brighter and more colourful to transfer or desalinate water (very often the execution is completed within one term of office) than to renovate the water distribution networks, above all if the waters are subsidised and their costs are not directly passed on users. This fact explains why the two main parties have changed their positions regarding this issue during the last two terms of office, defending supply-based or demand-based options depending on whether they were in the government or in the opposition. The defence of territorial interests ignoring the foundations of sustainable water management is also explained by the time periods between terms of office which differ so much from those required for water sustainable management. These problems have already been presented in great detail within a previous study (Cabezas et al., 2008)

8.2 *Educating the citizens*

The undoubtedly noble vocation of a politician is to exercise power and that is why he tries to be in tune in with the majority's opinion. It is an unquestionable fact which has become more and more evident in recent years, especially after the collapse of the communist bloc, that the traditional labels (left or right) matter less now, and the voter bets on efficiency, honesty or, ultimately, on the sensitivity shown to attend to the most unprivileged social groups.

Taking the above into account and being aware of the fact that a large part of the measures to be adopted are unpopular in the short term (starting with cost recovery and the later rise in the price of water) we need to educate a collective of citizens who still do not notice the risk of collapse which already exists if we maintain the same patterns. Therefore we have to replace the populist messages, which are so common in the context of water policy, with environmental education. In that way, the citizens will understand the reasons for the sacrifice that is requested from them and will support the political actions.

8.3 *Reforming the administration*

It is the first of the actions directly related to water policy. We have already said that the management structures have not been adapted to the current needs. The need for this adaptation was underlined in its time by the Brundtland report. In Spain, the water Administration not only has not been adapted but has additionally become fragmented – and with negative consequences. On the one hand, the organisation chart is more confusing and, on the other hand, the map of competences is more complex, which largely hinders the adoption of relevant decisions like the one linked to the possibility of carrying out global analyses. Since most of the current solutions have advantages and drawbacks (it has already been said, for example, that drip irrigation increases efficiency at the expense of increasing the energy bill), the need to perform global analyses is increasingly high.

Only an administration with a clear organisation chart, a strong leadership and well defined competences can confront change. This issue has not only been emphasised in the Brundtland report but also appears time and time again in the conclusions of international forums. But again, the difficulty involved in introducing changes in countries with a long history is much greater than in younger countries, such as Australia or Israel. The history of water administration in Israel started in 1959 with the nationalisation of water and a very clearly defined competence structure. It is the Israel Water Law (Arlosoroff, 1974). However, seeking to keep pressure groups away from the decisions about water policy, the administration was deeply reformed in 1999 (Shuval, 1999). Due to its peculiar characteristics, Israel's experience cannot be extrapolated, but it is advisable to take account of the goals around which its reforms have been structured.

The case of Australia is different but equally interesting. The planet's largest island counts on a huge central desert and big water availability problems along the coast, where droughts are severe and recurrent. The same as Israel, Australia has become a referent in water management. They have taken advantage of the country's youth to restructure the administrations responsible for water management on a national scale. Ever since the reform in the sector started in 1994, Australia has been shaping one of the first legislations that include the new water management (economic, social and environmental) dimensions. The last of these steps was taken recently and it is a mirror in which Spain, and probably other Mediterranean countries, should look at themselves. Despite having a decentralised government in which the different states have their competences delegated, the recent 'Australian Water Act' (2007) has entered the difficult debate on the distribution of water between interstate basins, takes into account the sustainability of the different uses and places a strong emphasis on the collection of information about the resource (OLDP, 2009).

Other countries have opted for smaller but equally interesting reforms. It is the case of Portugal which, seeking to organise the urban water cycle, has created the Instituto Regulador de Águas e Resíduos [Water and Waste Regulatory Institute]. It is a very partial but also necessary reform (IRAR, 2008). Italy had previously moved in the same direction with the Galli Act. Regulating water supply and creating the ATO (Ambito Territorial Ottimale [Optimum Territorial Context]) is an experience that has gone on for more than ten years, which allows making critical assessments about its results (Citroni et al., 2007).

Despite the awareness of the difficulties involved in reforming the administration, we must realise that those difficulties do nothing but grow as time goes by. For this reason, the first essential step required to undertake the reforms is to reach a previous political agreement. There are no doubts about the fact that the execution and development of the new water policies suited to the challenges posed by the 21st century demand an agile instrument with capabilities and attitudes that are currently non-existent in Spain in particular and in most of the water administrations of Mediterranean countries in general.

8.4 *Implementing the kind of legal reforms which can favour efficient use*

Many of the laws in force which used to rule and still rule the world of water were enacted within a context and under circumstances which have nothing to do with the present-day situation. The environmental problems which raise so much concern today (for example, the determination of minimum environmental volumes of flow during drought periods) and are present throughout the WFD did not exist when the essential part of the Spanish legislation was enacted. The WFD, along with the environmental aspects that it entails, was awkwardly inserted in the Spanish legislation in 2003 (Embid, 2007). But some historical rights continue to be in force, and that is only one example.

All these reforms are harder to introduce in countries with a considerable political decentralisation like Spain A recent publication which gathers the lectures of the seminar Gestión del Agua y Descentralización Política [Water Management and Political Decentralisation] held within the framework of the EXPO-Zaragoza 2008, refers both to the need and to the difficulty involved in introducing the changes required to adapt the legislation to the WFD (Embid and Höllingt, 2009). But we are not going to deepen any more in this issue because we are not legal experts and also because that is beyond the scope of this paper. But, despite the complexity which is associated with it, it is indeed necessary to underline the need to adapt the water legislation to the present-day problems.

8.5 *Setting up monitoring and control mechanisms*

If we do not check the degree of compliance with the objectives sought, it is impossible to assess the effectiveness of the measures applied. It is therefore necessary to set up mechanisms which can verify that level of compliance – and, of course, these mechanisms do not exist in many countries. For instance, the protection of overexploited aquifers must necessarily include monitoring their extractions. If we do not have available an effective flow measuring system and we cannot calculate the volumes detracted, it is very difficult to protect the aquifer. Another example: if a decision is made to extend the subsidies for agricultural water, this must be done in a way that encourages efficiency. And that once again requires another control mechanism. How to implement it (with regulatory offices for each one of the different uses or, like in the case of Israel, by the Administration itself) is something beyond the scope of this paper, which confines itself to highlighting the inescapable need to undertake these initiatives.

To finish with, it is worth saying that putting in practice these five action lines demands political will, good judgment and a lot of time. Furthermore, it is essential to count on citizens' involvement. The reforms can only be stable in that way. When we no longer talk about government, but about governance, the good practices that lead to an efficient sustainable management require a broad consensus that can only be achieved with a deep, thorough environmental education.

9 CONCLUSION

The objectives at which the 21st-century water policy must aim are perfectly identified and so is the path that needs to be followed to come gradually closer to them. However, the complexity of the challenge, especially in countries with a water tradition and culture that has consolidated and become deep-rooted through several millennia, leads to political indecision when the time comes to adopt the decisions which are convenient for the future. And it can be understood. In everything that refers to water use, the changes that mankind has witnessed during the last one hundred years, firstly with the hydraulic overdevelopment and to go to the other extreme, with the recent environmental concern, largely exceed the changes that mankind had lived through from its origins to the end of the 19th century.

Water management has evolved slowly throughout history and, consequently, its government has adapted to the needs of each moment without great surprises. But the changes along with the problems generated in recent years have been so fast that the countries with a longer tradition, and therefore, with more consolidated institutions and legislations have been unable to articulate changes at the speed required by the problems which have gradually but also quickly emerged. That is why the trinomial water-history-sustainability is so hard to harmonise today.

Nobody argues that water is scarce, and consequently precious, in Southern European countries; but its management is not intrinsic to its culture, above all because the costs associated with it have always been subsidised so far. And what used to be logical is no longer logical now. Spain provides a paradigmatic example. As said above, cultural reasons, vested interests, inertias and a great deal of respect, if not fear, among politicians, to introduce the necessary changes, explain the delay. Nevertheless, there are factors and conditions that will sooner or later drive this badly-needed reform. Among these stand out the growing environmental awareness and the need to confront the increasingly frequent droughts successfully, as well as the formidable challenge posed by climate change. The path to be followed must take into account both the guidelines identified in all the international forums and the experience acquired in this respect in Northern European countries, all of it within the new scenario shaped by a European Union Water Framework Directive which, despite not having led to many practical results yet, will soon celebrate its tenth birthday.

REFERENCES

Agustí, M. (2000) *Citricultura*. Madrid, Ediciones Mundi-Prensa.

Arlosoroff, S. (1974) Legal, Administrative and Economical Means for the Preservation and Efficient Use of Water in Israel. *Water in Israel*. Hakirya, Tel Aviv, Ministerio de Agricultura. Comisión del Agua. pp. 1–29.

Blackman, D.R. & Hodge, A.T. (2004) *Frontinus' Legacy. Essays on Frontinus' de aquis urbis Romae*. Ann Arbor, Michigan, USA.

BOE (Boletín Oficial del Estado) [Official Gazette], 2003 RD 140/2003, de 7 de febrero, por el que se establecen los criterios sanitarios de la calidad del agua de consumo humano. BOE No. 45 of February 21 2003, pp. 7228–7245.

Bonnin, J. (1984) L'eau dans l'antiqueté. L'hydraulique avant notre ère. Paris, Editions Eyrolles.

Bruins, H. (1993) Drought Risk and Water Management in Israel: Planning for the future. In: Wilhite, A.D. (ed.) *Drought Assessment, Management & Planning: Theory and Cases Studies*. Dordrecht, Holland, Kluwer Academic Publishers.

Brundtland, B.H. (1987) *Our Common Future, Brundtland Report*. Oxford, Oxford University Press.

BUNR (Bundesminiterium für Umwelt, Naturschutz und Reaktorsicherheit) (2001) *Water Resources Management in Germany*. Bonn, Germany, Federal Ministry for the Environment.

Burgi, P.H. (1998) Change in emphasis for hydraulic research at the Bureau of Reclamation. *Journal of Hydraulic Engineering*. ASCE, July.

Cabezas, F., Cabrera, E. & Morell, I. (2008) El agua, cuestión de Estado. Perspectiva desde la Comunidad Valenciana. Valencia, AVE (Asociación Valenciana de Empresarios).

Cabrera E., Cabrera, E. Jr., Cobacho, R. (2009a) Water supply in urban areas. In: Llamas, M.R. & Garrido, A. *Water Policy in Spain*. Balkema, CRC Press. ISBN 978-0-415-55411-4, pp. 77–84.

Cabrera, E., Pardo, M.A., Cabrera, E. Jr., Cobacho, R. (2009b) *Agua y Energía en España, un reto complejo y fascinante. Primeras Jornadas de Ingeniería del Agua*. Madrid, Fundación para el Fomento de la Ingeniería del Agua y Centro de Estudios Hidrográficos.

Citroni, G., Gianelli, N., Lippi, A. & Profeti, S. (2007) Chi governa l'aqua?: Regolazione, potere locale e arene della rappresentanza nella governance del servizio hídrico integrato.

Cobacho, R. (2000) La gestión de la demanda en el contexto de una nueva política integral del agua. Doctoral Thesis. Valencia, Spain, Polytechnic University.

Convegno SISP (2007) *Catania*, Sicily, September 20–22.

Corominas, J. (2000) El papel económico de las aguas subterráneas en Andalucía. Madrid, Fundación Marcelino Botín.

Corominas J. (2009) Agua y energía en el riego en la época de sostenibilidad. Primeras Jornadas de Ingeniería del Agua. Madrid, Fundación para el Fomento de la Ingeniería del Agua y Centro de Estudios Hidrográficos.

Costa, J. (1911) Política Hidráulica. Misión social de los riegos en España. Madrid, Colegio de Ingenieros de Caminos Canales y Puertos.

Domínguez-Castro, F., Santisteban, J.I., Barriendos, M. & Mediavilla, R. (2008) Reconstruction of drought episodes for central Spain from rogation ceremonies recorded at the Toledo Cathedral from 1506 to 1900: a methodological approach. *Global and Planetary Change* 63, 230–242.

DVGW (The German Technical and Scientific Association for Gas and Water) (2005) Profile of the German Water Industry 2005. Wirtschafts und Verlagsgesellschaft Gas und Wasser GmbH, Bonn, Germany.

EEA (European Environment Agency) (2009) Water resources across Europe. *Confronting water scarcity and drought.* EEA Report No. 2/2009, Copenhagen, Denmark.

Embid, A. (2007) La Directiva Marco del Agua y algunos de los problemas de su proceso de implantación en España y otros países europeos. *Revista Ingeniería y Territorio, CICCP No. 80, 2007.*

Embid, A. & Hölling, M. (2009) *Gestión del Agua y Descentralización Política.* Conferencia Internacional de gestión del agua en países federales y semejantes a los federales. Zaragoza, Julio de 2008. Editorial Aranzadi. Navarre, Cizur Menor.

Escudero, G. (1986) *Actividad, ocupación y productividad agrarian en España: un análisis de la población y del empleo.* Revista de Estudios Agro-Sociales No. 137, Extra issue September 1986, pp. 380–417.

EU (European Union) (2000) Directive 2000/60/EC of the European Parliament and of the Council establishing a framework for the community action in the field of water policy. *Official Journal of European Union*, L 327, P. 0001–0073.

EU (European Union) (2007a). European Commission Agriculture and Rural Development. The Common Agricultural Policy Explained. European Commission Directorate-General for Agriculture and Rural Development. Publications Office -KF-81-08-237-EN-C.

EU (European Union) (2007b) Commission of the European Communities addressing the challenge of water scarcity and droughts in the European Union. Communication from the Commission to the European Parliament and the Council. Brussels, COM 2007) 414 final.

EU (European Union) (2008) Directive 2008/56/EC of the European Parliament and of the council establishing a framework for community action in the field of marine environmental policy (Marine Strategy Framework Directive) of 17 June 2008. *Official Journal European Union*, L 164, 00019–0048.

EU (European Union) (2009a) Mainstreaming sustainable development into EU policies: 2009 review of the European Union Strategy for Sustainable Development. Communication from the commission to the European Parliament, The Council, The European Economic and social committee and the Committee of the Regions. Brussels, COM(2009) 400 final.

EU (European Union) (2009b) Directive 2009/128/EC of the European Parliament and of the council of 21 October 2009 establishing a framework for community action to achieve the sustainable use of pesticides. *Official Journal European Union*, L 309/71, 00071–0086.

Evans, H.B. (2000) Water Distribution in Ancient Rome: The Evidence of Frontinus. USA, The University of Michigan Press.

Fewkes, A. (2006) The technology, design and utility of rainwater catchment systems. In: Butler, D. & Fayaz, F.A. (eds.) *Water Demand Management.* IWA Publishing.

Giner Boira, V. (1997) *El Tribunal de las Aguas de Valencia.* Valencia, Spain, Fundación Valencia III Milenio.

Gleick, P.H. (2003) *Water Use. Annual Review of Environmental Resources.* Palo Alto CA, USA, Annual Reviews Publisher. pp. 275–314.

IAHR, (International Association of Hydraulic Research) (1987) *Hydraulics and Hydraulic Research.* In: Garbrecht, G. (ed.) An Historical Review. Rotterdam, The Netherlands, Balkema.

IEEP (Institute for European Environmental Policy) (2009) Potential impacts of desalination development on energy consumption. DG Environment Study Contract 07037/2007/486641/EUT/D2, 7 April 2008, Brussels.

INE (Instituto Nacional de Estadística) [National Statistics Institute] (2003) *Encuestas del agua 2001. INE, Press Release of July 1 2003.*

INE (Instituto Nacional de Estadística) (2004) Encuestas del agua 2002. INE, Press Release of July 1 2004.

INE (Instituto Nacional de Estadística) (2005) Encuestas del agua 2003. INE, Press Release of August 3 2005.

INE (Instituto Nacional de Estadística) (2006) Encuestas del agua 2004. INE, Press Release of August 17 2006.

INE (Instituto Nacional de Estadística) (2007) Encuestas del agua 2005. INE, Press Release of October 17 2007.
INE (Instituto Nacional de Estadística) (2008) Encuestas sobre el suministro y el tratamiento del agua 2006. INE, Press Release of July 17 2008.
INE (Instituto Nacional de Estadística) (2009) Encuesta sobre el suministro y saneamiento del agua. Año 2007. INE, Press Release of July 17 2009.
IRAR (Instituto Regulador de Aguas y Residuos) [Water and Waste Regulatory Institute] (2008) *Nota sobre a nova legislação relativa a serviços públicos essenciais.* Lisbon, Portugal, IRAR.
Kerisel, J. (1999) *Le Nil: l'espoir et la colère. De la sagesse à la démesure.* Paris, Presses de l'école nationale des Ponts et Chaussées (ENPC).
Levi, E. (1995) *The Science of Water. The Foundation of Modern Hydraulics.* New York, American Society of Civil Engineers.
Mehan, G.T. (2003) *Water Challenges in the 21st century.* Philadelphia, EWRI 2003, World Water and Environmental Congress, ASCE.
Merkel, W. (2003) El Futuro de la Industria de Agua en el mundo. *Ingeniería del Agua*, 10(3).
Milly, P.C.D., Betancourt, J., Falkenmark, M., Hirsch, R.M., Kundzewicz, Z.W., Lettenmaier, D.P. & Stouffer, R.J. (2008) *Stationarity is dead: whither water management? Science*, 319, 573–574.
MIMAM (Ministerio de Medio Ambiente) [Environment Ministry] (1998) *El Libro Blanco del Agua en España.* Madrid, MIMAM.
NAPSTJERT, L. (2002) *Water Savings in Copenhagen. Watersave Network Fourth Meeting.* UK, Loughborough University.
NAS (National Academy of Sciences) (2004) *Confronting the Nation's Water Problems.* Washington. USA, The National Academies Press.
OECD, Organisation for Economic Cooperation and Development (1999) *Agricultural Water Pricing in OECD Countries.*
OECD, Paris.
OLDP (Office of Legislative Drafting and Publishing) (2009) *Water Act 2007.* Canberra, Office of Legislative Drafting and Publishing. Attorney-General's Department.
Schnitter, J. (1994) *A History of Dams. The useful pyramids.* Balkema, Rotterdam.
Shuval, H. (1999) Sustainable water development under conditions of scarcity: Israel as a case study. In: Sergio Marchisio, Gianfranco Tamburelli & Liana Pecoraro (eds.) Sustainable Development and Management of Water Resources: A Legal Framework for the Mediterranean. Rome, Institute for Legal Studies on the International Community. pp. 196–223.
Smet, H. (2008) *De l'eau potable a un prix abordable. La practique des États. Académie de l'Eau.* Available from: www.academie-eau.org.
Tritti,n J. (2001) *A First Framework for Action Has Been Set Up. Water Supply and Development Go Hand in Hand.* Available from: http://www.bmu.de/english/water_management/doc/3470.php.
UN (United Nations) (2008) *World Urbanization Prospects. The 2007 Revision.* New York, United Nations.
UNFCCC (United Nations Framework Convention on Climate Change) (2009). *Greenhouse Gas Inventory Data.* Available from: http://unfccc.int/ghg_data/items/3800.php.
Viollet, P.L. (2000) *Hydraulique dans les civilisations anciennes.* Paris, Presses de l'Ecole National des Ponts et Chaussés.

CHAPTER 8

Water and agriculture. Current situation and future trends

Martín Sevilla Jiménez
Senior Professor of Applied Economics, University of Alicante, Spain

ABSTRACT: Humankind's goal of ensuring a sufficient food production for the planet's growing population has led to a constant search for elements and procedures that would help to attain that goal: An expansion of farmed lands, fertilization, plague eradication, prevention against natural disasters, mechanization, computerization, and … finding water resources. Water has always been a central issue.

In this paper, the author highlights the important role played by irrigation systems in meeting new food demands and their foreseeable evolution in the coming years. In addition, the author links that need (meeting growing food demands) with the conception of water as a natural, social, and economic, renewable resource which, in the European Union, with the passing of the European Water Directive, calls for the inclusion of the "cost recovery" principle as an element that will be essential for the environmental protection of water resources in the future. Finally, the author analyzes the changes that are occurring in the Common Agricultural Policy with regard to irrigated crops and also the impact of Spain's National Irrigation Plan on irrigated crops.

1 INTRODUCTION: AGRICULTURE & NUTRITION

Talking about the connection that exists between agriculture and water in the early 21st century calls for a great effort to try and identify the essential elements that determine how agricultural systems function today. More precisely, how these function in terms of the production aspect, the trade aspect and the destination of produce in the planet, and the current situation of the water resources that are under exploitation.

At present we are witnessing a sharp contrast between, on the one hand, the affluence and food abundance of developed countries and, on the other hand, the food scarcity and frequent famines suffered by other countries. And this reality is a sheer insult to those of us who believe that our current level of scientific knowledge should not allow this to happen.

It could be argued that there are a number of general factors that do influence our chances of fixing the above problem: Political imbalances, dominated areas, bilateral/multilateral agreements, and so forth. Irrespective of such conditions, in my opinion, there are three key factors to understand the causes of the current reality: 1) The production and supply of agricultural produce; 2) the political systems in place in each country; and 3) the changes in food demand and consumption.

Technical advances and productivity increases were spectacular throughout the 20th century in the agricultural sector. Thus, the final agricultural production levels increased despite the gradual drop in the percentage of the population that worked in agriculture (see Table 1). The more developed countries generated policies that attempted to preserve their agricultural populations by subsidizing the prices of their products or supplementing their incomes. As a matter of fact, however, such policies did little to curb the decrease in the agricultural population. Instead, they caused significant problems in terms of surplus generation and (due to protectionist policies in developed countries) hindering the trade of agricultural produce from less developed countries.

Table 1. Population working in agriculture (Millions and %).

	1970	1980	1990	2000	2010
All developing countries	790	923	1051	1130	1190
(% of total econ. active popul.)	(71)	(66)	(60)	(53)	(47)
93 Study countries	780	912	1039	1120	1180
(% of total)	(71)	(66)	(60)	(53)	(47)
Africa (sub-Sahara)	98	118	140	170	205
(% of total)	(81)	(76)	(71)	(66)	(60)
Middle East/North Africa	31	32	35	38	39
(% of total)	(57)	(46)	(37)	(30)	(24)
East Asia	411	488	549	550	530
(% of total)	(76)	(71)	(63)	(55)	(47)
South Asia	203	235	275	320	365
(% of total)	(71)	(68)	(65)	(61)	(57)
Latin America/Caribbean	37	39	41	41	40
(% of total)	(41)	(32)	(26)	(21)	(17)

*The data, and in particular the projections, should be understood as indicative of broad orders of magnitude. They are, as far as possible, standardized for comparability among countries and regions. They may differ from those obtained from the routine labor force survey statistics. For discussion, see FAO (1986). Data by country are given in Appendix 3. The basis of these estimates is the historical data up to the early 1980s from ILO's work providing internationally comparable statistics. ILO is in the process of updating these data.
Source: FAO (2002) World Agriculture: Towards 2010. An FAO Study...

In spite of that, the good news is that the increases in productivity have helped some less developed countries to reduce or even solve their food supply problems.

The advances in terms of productivity have not yet reached their ceiling. Rather, the forecasts for the coming years suggest that there will be an across-the-board crop productivity increase. It is estimated that by 2030 productivity will have doubled or even tripled the productivity levels reached back in the 1960s (see Table 2).

Even though agricultural techniques have gradually spread to all countries, putting them in practice in less developed countries has been and remains a quite difficult task. The availability of individuals with a certain basic knowledge is essential for implementing and using innovative agricultural techniques such as new soil management systems, seed selection and use, introduction of intensive and localized irrigation systems, manipulation of production cycles, etc. This delay in the implementation of technical innovations, due to the absence of specialized manpower, is not just a temporarily current phenomenon – the most capable individuals keep migrating from rural areas to cities, and from less developed areas to more developed ones. Therefore that delay will tend to persist in the future.

Based on the latest estimates and projections, the majority of the world's population will live in urban centers by 2007. The urban population will grow from 3 billion in 2003 (48% of the total population) to 5 billion by 2030 (60%). A larger part of that increase in urban population will be due to vegetative growth, not migrations. In that same period, the rural population will decrease slightly from 3.3 billion to 3.2 billion. Projections suggest that, between 2000 and 2030, the urban population will increase at a yearly rate of 1.8%, i.e. almost twice the growth rate of the world population. The percentage of urban population in the least developed regions will grow at a rate of 2.3% and by 2017 more than half of their population will be concentrated in urban areas. By 2030, there will be a majority of urban population in all regions of the world (Africa: 54%; Asia: 55%). In the same period, almost the entire growth of the world population will occur in urban areas of developing countries (UNFPA, 2004).

It seems reasonable to affirm that there are not any problems of a technological nature for producing enough food to meet the expected growing demand for food of the urban population, based

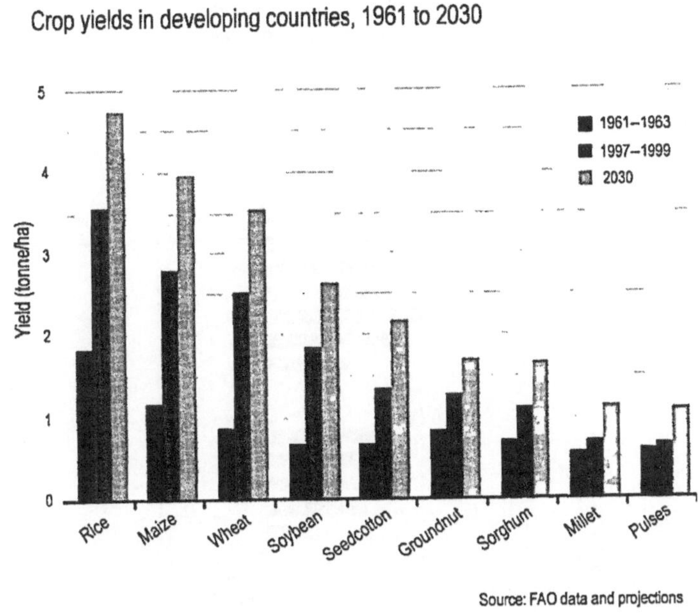

Figure 1. Evolution in the productivity of some crops.

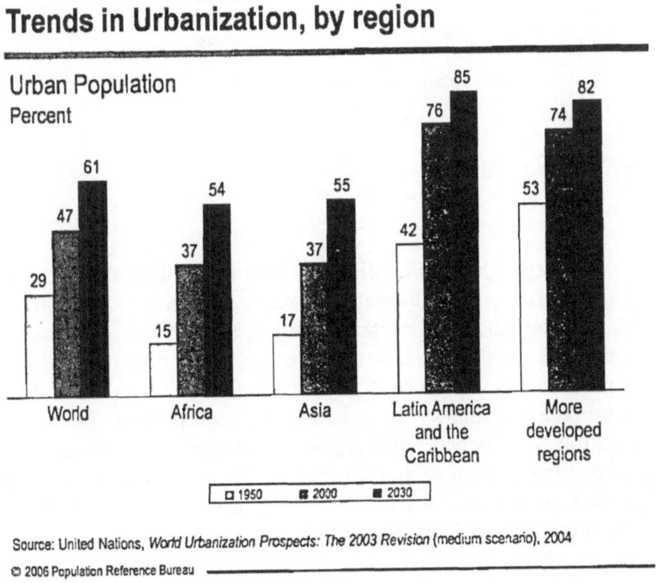

Figure 2. Urban concentration of the population 1950–2030. http://www.prb.org/presentations/283,1,trends in Urbanization, by region.
Source: Population Reference Bureau (2006)

on FAO forecasts. Instead, it seems that the main challenges for the less developed countries are originated by their social organization when it comes to applying new techniques in their production processes, techniques which will be at the root of future increases in agricultural productivity (along with adequate and sufficient investments).

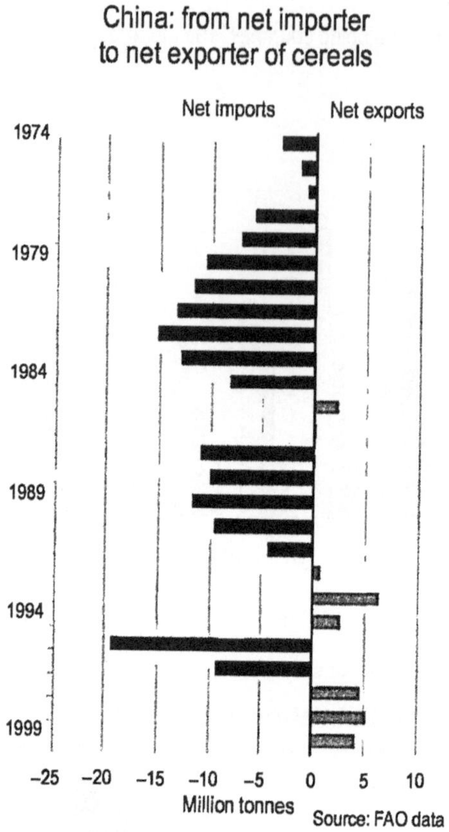

Figure 3. China, cereal net trade balances.

In addition, let us not forget that the application of some technical solutions should take into account specific local conditions (soil, climate, previous experiences, traditional knowledge, the social role of women, etc.) and also the availability of adequate financial resources.

The political organization and the social structure of a country do have an impact on the chances of solving all these problems, too. Successfully establishing the right conditions to encourage production and develop the production units is a basic requirement. Less developed countries should encourage the initiative of private entrepreneurs and remove bureaucratic and administrative obstacles that may block the ways out of their problems. There are numerous other issues that need to be considered before implementing any solution, such as land ownership, the capacities of trade organizations, and the regulation of labor relations.

The best examples of reforms that were boosted following this approach are found in China and India. The most populous countries in the planet have been gradually changing their agricultural policies in ways that have not only enabled them to cover their needs for staple foods, but have also generated positive net balances in their external trade exchanges.

Even though their organizational and political characteristics cannot be reproduced completely in other countries, China and India did demonstrate that it is possible for other countries to modify their internal policies in such a way as to escape their current imbalances and put an end to the malnutrition of their populations. Having said that, it should be admitted that some countries (especially those in Sub-Saharan Africa) will have a hard time unless the international community intervenes and political changes take place to ensure more stability. There are several initiatives,

Table 2. Food production and consumption by country. A2 growth rates of aggregate demand and production (percent per annum).

	1969 to 1999	1979 to 1999	1989 to 1999	1997–99 to 2015	2015 to 2030	1997–99 to 2030
Demand						
World	2.2	2.1	2.0	1.6	1.4	1.5
Developing countries	3.7	3.7	4.0	2.2	1.7	2.0
Idem, excluding China	3.2	3.0	3.0	2.4	2.0	2.2
Sub-Saharan Africa	2.8	3.1	3.2	2.9	2.8	2.9
Idem, excluding Nigeria	2.5	2.4	2.5	3.1	2.9	3.0
Near East and North Africa	3.8	3.0	2.7	2.4	2.0	2.2
Latin America and Caribbean	2.9	2.7	3.0	2.1	1.7	1.9
Idem, excluding Brazil	2.4	2.1	2.8	2.2	1.8	2.0
South Asia	3.2	3.3	3.0	2.6	2.0	2.3
East Asia	4.5	4.7	5.2	1.8	1.3	1.6
Idem, excluding China	3.5	3.2	2.8	2.0	1.7	1.9
Industrial countries	1.1	1.0	1.0	0.7	0.6	0.7
Transition countries	−0.2	−1.7	−4.4	0.5	0.4	0.5
Production						
World	2.2	2.1	2.0	1.6	1.3	1.5
Developing countries	3.5	3.7	3.9	2.0	1.7	1.9
Idem, excluding China	3.0	3.0	2.9	2.3	2.0	2.1
Sub-Saharan Africa	2.3	3.0	3.0	2.8	2.7	2.7
Idem, excluding Nigeria	2.0	2.2	2.4	2.9	2.7	2.8
Near East and North Africa	3.1	3.0	2.9	2.1	1.9	2.0
Latin America and Caribbean	2.8	2.6	3.1	2.1	1.7	1.9
Idem, excluding Brazil	2.3	2.1	2.8	2.1	1.8	2.0
South Asia	3.1	3.4	2.9	2.5	1.9	2.2
East Asia	4.4	4.6	5.0	1.7	1.3	1.5
Idem, excluding China	3.3	2.9	2.4	1.9	1.8	1.9
Industrial countries	1.3	1.0	1.4	0.8	0.6	0.7
Transition countries	−0.4	−1.7	−4.7	0.6	0.6	0.6
Population						
World	1.7	1.6	1.5	1.2	0.9	1.1
Developing countries	2.0	1.9	1.7	1.4	1.1	1.3
Idem, excluding China	2.3	2.2	2.0	1.7	1.3	1.5
Sub-Saharan Africa	2.9	2.9	2.7	2.6	2.2	2.4
Idem, excluding Nigeria	2.9	2.9	2.7	2.6	2.3	2.4
Near East and North Africa	2.7	2.6	2.4	1.9	1.5	1.7
Latin America and Caribbean	2.1	1.9	1.7	1.3	0.9	1.1
Idem, excluding Brazil	2.1	1.9	1.8	1.4	1.0	1.2
South Asia	2.2	2.1	1.9	1.6	1.1	1.3
East Asia	1.6	1.5	1.2	0.9	0.5	0.7
Idem, excluding China	2.0	1.8	1.6	1.2	0.9	1.0
Industrial countries	0.7	0.7	0.7	0.4	0.2	0.3
Transition countries	0.6	0.5	0.1	−0.2	−0.3	−0.2

Source: FAO (2002)

such as the Africa Plans (under the auspices of the United Kingdom and Spain), which should become involved very actively in this task, if they are really to put an end to poverty.

Food demands and food consumption are changing, too. The fast demographic growth in the less developed countries, and the changes caused by migrations from rural to urban areas, as well as the

Table 3. Changes in commodity composition of food, 1980–2030.

	Cereals	Roots and tubers	Sugar (raw eq.)	Pulses (dry)	Veg. oils, oil seeds (oil eq.)	Meat (carcass weight)	Milk and dairy[1]
	Kg/capita/year						
World							
1979–81	160	74	23.5	6.5	8.4	29.5	77
1997–99	171	69	24.0	5.9	11.4	36.4	78
2015	171	71	25.1	5.9	13.7	41.3	83
2030	171	74	26.3	6.1	15.8	45.3	90
Industrial countries							
1979–81	139	67	36.8	2.8	15.7	78.5	202
1997–99	159	66	33.1	3.8	20.2	88.2	212
2015	158	63	32.4	4.0	21.6	95.7	217
2030	159	61	32.0	4.1	22.9	100.1	221
Transition countries							
1979–81	189	119	45.9	3.1	9.2	62.9	181
1997–99	173	104	34.0	1.2	9.3	46.2	159
2015	176	102	35.0	1.2	11.5	53.8	169
2030	173	100	36.0	1.1	14.2	60.7	179
Developing countries							
1979–81	162	70	17.6	7.8	6.5	13.7	34
1997–99	173	67	21.3	6.8	9.9	25.5	45
2015	173	71	23.2	6.6	12.6	31.6	55
2030	172	75	25.0	6.6	14.9	36.7	66
Sub-Saharan Africa							
1979–81	115	172	9.9	9.8	8.5	10.6	34
1997–99	123	194	9.5	8.8	9.2	9.4	29
2015	131	199	11.3	9.8	10.7	10.9	31
2030	141	202	13.0	10.5	12.3	13.4	34
Near East and North Africa							
1979–81	199	26	28.2	6.4	11.1	17.4	85
1997–99	209	34	27.6	6.7	12.8	21.2	72
2015	206	33	28.7	6.9	14.4	28.6	81
2030	201	33	29.9	6.9	15.7	35.0	90
Latin America and Caribbean							
1979–81	130	74	48.5	12.6	10.2	40.6	97
1997–99	132	62	48.9	11.1	12.5	53.8	110
2015	136	61	48.2	10.7	14.5	65.3	125
2030	139	61	47.9	10.6	16.3	76.6	140
South Asia							
1979–81	151	20	20.7	11.2	5.8	4.0	42
1997–99	163	22	26.7	10.9	8.4	5.3	68
2015	177	27	29.5	9.1	11.6	7.6	88
2030	183	30	32.2	7.9	14.0	11.7	107
East Asia							
1979–81	181	83	8.1	4.3	4.7	13.0	5
1997–99	199	66	12.4	2.1	9.7	37.7	10
2015	190	64	14.6	2.0	13.1	50.0	14
2030	183	61	16.6	2.1	16.3	58.5	18

[1] Fresh milk equivalent
Source: FAO (2002)

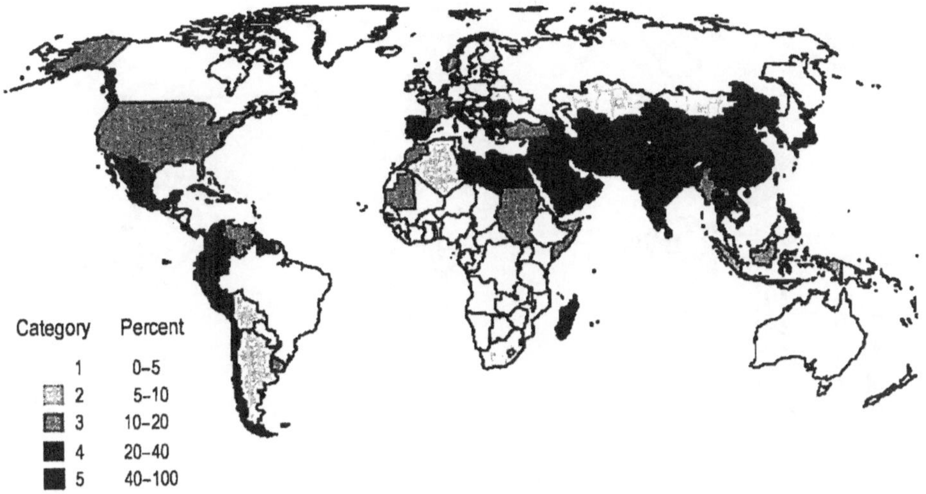

Figure 4. Area equipped for irrigation as a percentage of cultivated land in 1998.
Source: FAO (2000)

shift from production for own consumption to market consumption have exacerbated (and continue to exacerbate) the imbalances between the production levels in the less developed countries and food demands.

Diets are another decisive factor in the balance of agricultural produce consumption. We are witnessing a shift from rice consumption to cereal consumption in Asian countries which is changing (and may change even more significantly) the world's diet and trade balance. Such diet changes may alter the overall demands for food of vast masses of population in the future, which would have an impact on farmed lands and prices of produce.

It is not the aim of this paper to cover all those topics. However, it is important to remember them all, considering their impact on the nutritional situation of the world's poorer population and the gloomy prospects for the future which have been repeatedly announced by the so-called Millenium Summit, among others.

We will now focus on analyzing the past and present role of water in all the changes occurred in agriculture and on examining to what extent an adequate use of this resource might contribute to solving the nutritional problems that are expected for the coming years.

2 IRRIGATION ISSUES RELATING TO THE USE OF WATER IN AGRICULTURE. DEVELOPED COUNTRIES AND DEVELOPMENT

Water use in agriculture has been and remains to be the key to understanding the progressive enhancements in agricultural production and productivity. From the "passive" use of water (in the form of rain) in farmlands as a natural resource that contributes to the improvement of agricultural production (the Andes High Plain is a good example) to the sophisticated use of localized irrigation systems in the modern "agricultural production factories", water has been and will be an essential factor in agricultural production.

Today, a significant share of the world's crops is irrigated. Around 1997–1999, irrigated lands represented only a fifth of the total arable land in developing countries. However, as a result of increased yields and more frequent crops, irrigated lands generated two fifths of the entire production, and about three fifths of the total cereal production.

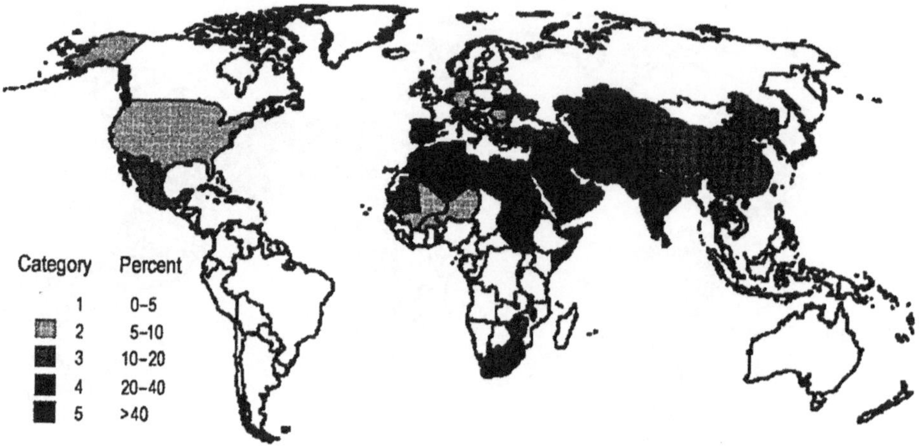

Figure 5. Water Withdrawals by continent and by sector.
Source: FAO (2000)

That ratio is expected to increase even more in the next three decades. Based on several factors, such as irrigation potential, specific national plans, and water dependency of the crops, it is expected that the irrigated lands in developing countries as a whole will increase from 202 million hectares in 1997–1999 to 242 million hectares in 2030. That is a net forecast, i.e. it is based on the assumption that all those arable lands that will be lost due to water salinity and scarcity (or other causes) will be recovered through land remediation or substituted by new arable lands.

The bulk of such expansion will occur in areas where arable land is limited and irrigation systems are already crucial. For instance, East Asian and Southern Asia will add 14 million hectares each.

There will be a significant expansion in the Near East and Northern Africa as well. As for Sub-Saharan Africa and Latin America, where lands are abundant, the need for irrigation is smaller, and the potential is smaller, too, the increase is expected to be more moderate (2 million and 4 million hectares respectively). The expected expansion is ambitious, but much less impressive than the expansion achieved in the past.

Since the early 1960s, no less than 100 million hectares of new irrigated land were created. The expected net increase in the next three decades is just about 40 percent of that figure. In a like manner, the expected annual growth rate of 0.6 percent is merely a third of the annual growth rate achieved in the past 30 years.

The FAO forecast does not mention the potential expansion of irrigated lands in developed countries, which currently represent about a fourth of the world's irrigated areas. Whereas irrigation expanded very quickly in developed countries in the 1970s, the annual growth rate fell to only 0.3 percent in the 1990s.

The two main issues under discussion are the availability of arable land (or lack thereof) and the availability of water resources to exploit the land. With regard to the first issue, some believe that soon there will be a scarcity of land that is adequate for irrigation (and a scarcity of land in general, for that matter). Also there are increasing concerns about the possibility that vast irrigated areas may suffer serious damage due to salinity and pollution.

Those concerns might be exaggerated on the global scale, but indeed some specific areas might face serious trouble. FAO studies suggest that there is still room for expanding irrigation and thus meeting future demands. However, it is hard to estimate accurately the irrigation potential, since it depends on complex variables such as soil, rainfall and terrains. Therefore, all the figures should be taken merely as rough estimations.

The total irrigation potential of developing countries is estimated at 402 million hectares, roughly. Half of that area was under exploitation in 1997–1999, so the remaining unused potential should

Table 4. Land use by country in the period 1997–2030.

	Arable land (million ha)			Harvested land (million ha)			Cropping intensity (%)		
	Total	Rainfed	Irrigated	Total	Rainfed	Irrigated	Total	Rainfed	Irrigated
Developing countries									
1997–99	956	754	202	885	628	257	93	83	127
2015	1017	796	221	977	671	306	96	84	138
2030	1076	834	242	1063	722	341	99	87	141
Sub-Saharan Africa									
1997–99	228	223	5.3	154	150	4.5	68	67	86
2015	262	256	6.0	185	179	5.7	71	70	95
2030	288	281	6.8	217	210	7.0	76	75	102
Near East and Northern Africa									
1997–99	86	60	26	70	43	27	81	72	102
2015	89	60	29	77	45	32	86	75	110
2030	93	60	33	84	46	37	90	78	112
Latin American and Caribbean									
1997–99	203	185	18	127	112	16	63	60	86
2015	223	203	20	150	131	19	67	64	95
2030	244	222	22	172	150	22	71	68	100
Southern Asia									
1997–99	207	126	81	230	131	100	111	103	124
2015	210	123	87	248	131	117	118	106	134
2030	216	121	95	262	131	131	121	109	137
East Asia									
1997–99	232	161	71	303	193	110	130	120	154
2015	233	155	78	317	186	131	136	120	168
2030	237	151	85	328	184	144	139	122	169

Source: FAO (2002)

be 200 million hectares. In other words, the increase that is expected by 2030 would amount to only 20% of such unused potential. However, some regions will be much closer to exploiting their full irrigation potential: By 2030, in the Near East, Northern Africa, and East Asia, three fourths of the area that is available for irrigation will be in use, whereas Southern Asia (not including India) will be using almost 90 percent.

Another source of concern is that most of our planet is moving toward a situation of water scarcity. Since agriculture consumes about 70 percent of all the water for human use, there is fear that the above scenario might impact the future food production. Even though, apparently, there are no reasons to become alarmed at the global level, it is very likely that some countries and regions will face a severe water scarcity.

With regard to this, it should be noted that the FAO assessment of irrigation potential (2002) did take into account water constraints. The renewable water resources available in a given area are its rainfall recharge plus its river inflows minus the volume of water lost through evapotranspiration. These phenomena can vary considerably from one region to another. For instance, in an arid region such as the Near East and Northern Africa, only 18 percent of rainfall and river inflows remain after the process of evapotranspiration, whereas in the humid area of East Asia that figure can be over 50 percent.

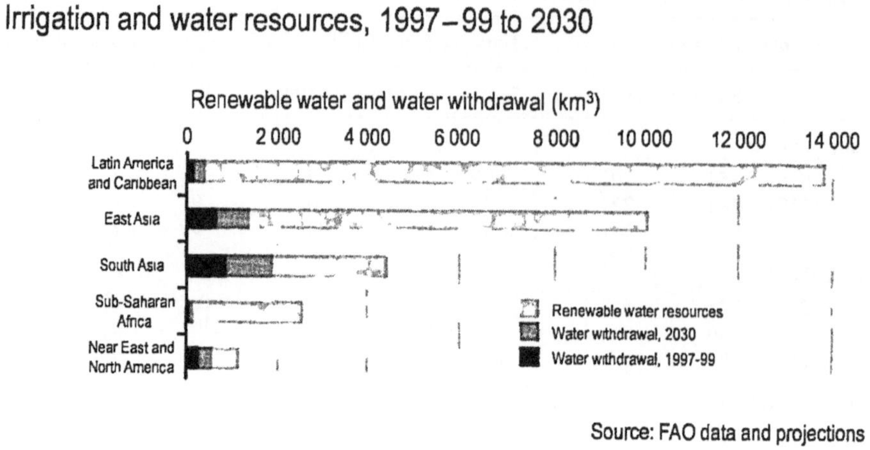

Figure 6. Irrigation vs. water resources. 1997–2030.

The amount of water used in irrigated lands is the amount of water absorbed by the cultivated crop plus all the water poured on the fields, which can be considerable in the case of aquatic crops such as rice. Water losses occur through leaks and evaporation before water reaches the fields and also through drainage of water in the fields that is not absorbed by the crops. The ratio between the amount of water that is really used for growing the crops and the amount of water withdrawn from the existing hydric resources is what we call water use yields. The yields are very different from one region to another. As a general rule, the yields are higher wherever water resources are limited. For illustrative purposes, here are some yield figures: Latin America, 25%; Near East and Northern Africa, 40%; Southern Asia, 44%. In 1997–1999, the developing countries withdrew only about 7% of their renewable water resources for irrigation purposes.

Due to differences in terms of yields and water availability, some regions are using much higher ratios than others. In Sub-Saharan Africa, where irrigation systems are less spread, only about 2 percent was used, whereas in Latin America (rich in water resources) merely 1 percent was used. In sharp contrast, the figures for Southern Asia and Near East/Northern African were 36 and 53 percent, respectively. According to projections on developing countries, by 2030 there will be an increase of 14 percent in their water extraction levels for irrigation purposes. Should that scenario become true, such countries would be using only 8 percent of their renewable water resources for irrigation. The ratios will remain very low in Sub-Saharan Africa and Latin America.

Water availability is considered a critical issue when over 40 percent of the existing renewable water resources is being used for irrigation purposes. It is then that a country is forced to make tough choices between water supply for agricultural purposes and water supply for urban use. By 2030, Southern Asia will have reached that figure, whereas the Near East and Northern Africa will be at no less than 58 percent. Among 93 developing countries studied by the FAO in 2002, 10 countries were using more than 40 percent in 1997–1999, and 8 countries were using over 20 percent, which is the threshold that is taken as the indicator of imminent water scarcity. By 2030, two other countries will have crossed that threshold, whereas one in five developing countries will be facing water scarcity or just about to face it.

Groundwaters will be impoverished at local level in some countries of the Near East, Northern Africa, Southern Asia and East Asia. In large areas of India and China, groundwater levels are decreasing at 1 to 3 m per year, which is causing buildings to collapse, seawater infiltration in aquifers, salinization, and rising pumping costs. In those countries and regions, policy changes and investments will be necessary in order to improve water yields and to adjust consumption patterns to the total costs of the resource, along with innovations for better water catchment and infiltration.

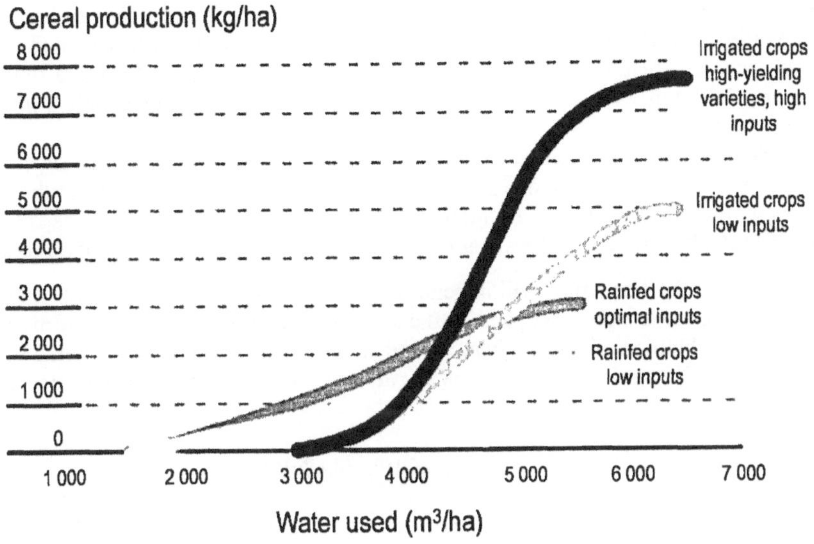

Figure 7. Productivity of irrigated vs. rainfed agriculture.

The bigger part of the future growth in crop production will be achieved through better yields (more efficiency), and water will play a central role. Yield advances have been inconsistent in the past three decades. The yields in cereal production rose quickly between 1961 and 1999 at the world level, and reached an average annual yield of 2.1 percent. Thanks to the green revolution, they increased even faster in developing countries, at an annual average rate of 2.5 percent. The fastest growth rates were achieved for wheat, rice and corn production. Those are considered the most basic foods in the world and therefore all international efforts concentrated on improving their efficiency.

The yields in the main commercial crops (soybean and cotton) increased quickly, too. At the opposite end of the scale, the yields in millet, sorghum and leguminous crops increased very slowly. These crops are exploited mainly in semiarid regions by poor farmers with limited resources. International researchers have failed to provide ways to achieve great yield increases in these farming exploitations.

Rice yields increased at an annual average of 2.3 percent between 1961 and 1989, but the rate dropped to less than a half between 1989 and 1999 (1.1 percent). However, valuable improvements were achieved and farmers' yields have become more stable than they used to be, thanks to the introduction of new features such as early ripening. The growth of cereal yields slowed down in the 1990s globally. Corn yields in developing countries maintained their upward trend, but the improvements in wheat and rice production were much slower.

Is the anticipated yield growth a realistic one?

The slower production growth expected for the coming 30 years involves that the yields will not need to grow as quickly as in the past. It is expected that the increase in wheat yields will be around 1.1 percent annually in the next 30 years. Rice yields will increase at a rate of 0.9 percent annually.

However, better yields will be necessary, so the real question is: Is the expected growth feasible? One way of assessing that question is to analyze any behavior differences from one group of countries to another. Some developing countries have attained very high crop yields. For instance, in 1997–1999, the top 10 percent attained average wheat yields six times higher than the bottom 10 percent and double the average yields attained by the main producers (China, India and Turkey). With regard to rice, the differences were about the same. Such differences in terms of yield performance from one country to another are due to two main types of causes:

- Different conditions in terms of soils, climate, and geography. For instance, most of Mexico is arid or semiarid and less than a fifth of the cultivated land dedicated to corn is good for enhanced hybrid varieties. As a result, Mexico's corn yields (2.4 tonnes/ha) are just about a fourth of the USA average. Differences in yields such as this, which are due to agroecological differences, cannot be reduced.
- There are other yield differences that are caused by differing crop management procedures, e.g. the amount of fertilizer being used. These differences can be reduced as long as the solutions are financially feasible for farmers. In order to estimate the potential progress in terms of yields one needs to calculate the differences that can be reduced and those that cannot be reduced.

In short, the future food demands will depend on the expansion of irrigated lands (particularly in areas with high conversion potential) and on the increase in crop yields, which in turn relies both on technological and biotechnological advances and social organization enhancements in order to face the potential conflicts relating to resource use and scarcity in the most critical areas.

3 IRRIGATED AGRICULTURE IN THE EUROPEAN UNION

In most countries (and particularly in developed countries), agriculture is relatively less efficient than the industrial and services sectors in terms of international prices, in spite of the considerable technological component which has been added in the past decades and which has pushed productivity.

Table 5. Agricultural production protection in OECD countries 2000–2002.

	ESP/PPF (%)		ESP/UTA (000 $/UTA)		ESP/SAU ($/Ha)	
	2000–02	2002	2000–02	2002	2000–02	2002
Australia	4	5	2	3	2	2
Canadá	19	20	10	11	57	62
R. Checa	23	28	5	7	196	254
UE	35	36	15	17	670	730
Hungría	24	29	5	6	205	265
Islandia	63	63	27	27	65	64
Japón	59	59	23	21	9.828	9.028
Corea	66	66	23	23	9.307	9.341
Méjico	22	22	1	1	71	75
N. Zelanda	1	1	1	1	5	7
Noruega	68	71	38	45	2.254	2.526
Polonia	15	14	1	1	114	114
R. Eslovaca	21	21	3	3	127	137
Suiza	73	75	30	32	2.958	3.197
Turquía	18	23	n.d.	n.d.	125	151
USA	21	18	19	16	112	94
OCDE 24	31	31	11	11	182	182

Table guide:
ESP: Production Subsidy Equivalent
PPF: Partial Factor Production
UTA: Annual Work Unit
SAU: Farmland Area
UE: European Union
OCDE: OECD
Source: Agricultural Policies of the OECD Countries: Monitoring and Assessment, 2003.
Spanish Ministry of Agriculture, Fisheries and Food (2002)

There are several causes, from the income level of farmers (much higher than that of farmers in developing countries or chiefly agrarian countries), to the difficulties to technify, automatize and profit from economies of scale in a biological process that is greatly climate-dependent – from the fact that the value of land as a limited asset is higher than merely its productive value (due to urban development processes), to the inelasticity of the prices of the main agricultural produce in the world's markets.

Despite such relative inefficiency, developed countries have traditionally protected their agriculture and are likely to continue to do so in the future. This is mainly due to issues of food self-sufficiency linked to the idea of sovereignty, but there are other reasons such as the influence and political power that are involved in the export of basic agricultural produce (commodities), the role of agriculture in environmental and nature conservation, land use and planning, and finally, rural development and the political pressure exerted by powerful farmers' lobbies.

The OECD has come up with a series of indicators that measure the support given to the agricultural sector by its Member States. The most widely used indicator is the Production Subsidy Equivalent (PSE), which reflects not just direct subsidies, but the entire protection to farmers (price support including customs tariffs, all kinds of direct coupled and decoupled subsidies, tax rebates, etc.)[1]

For the period 2000–2002, the PSE of the final agricultural production in the European Union was 35%; in the USA, 21%; and the average in OECD States was 31%. If the PSE is related to the available farmland, the figures are \$670/Ha in the EU; \$112/Ha in the USA, and \$182/Ha for the whole OECD. If we refer the PSE to real labor (annual work unit or AWU), the resulting figures are \$19,000 AWU in the USA; \$15,000 per AWU in the EU; and \$11,000 for the global OECD. (See Table 5).

These figures show that the EU, the USA and other developed countries protect their agriculture significantly, both through direct subsidies and different barriers (customs tariffs among others) that hinder imports from less developed countries.

These data do not discriminate the different agrarian productions. However, one should remember that, even if the agricultural sector of developed countries does not receive direct subsidies, the protectionist regimes do benefit their agrarian productions (i.e. irrigated crops of fruits and vegetables in developed countries).

Table 6. Distribution of EAGGF guarantee's direct subsidies in Spain by strata of beneficiaries (financial year 2001).

	Total amounts granted (in thousands of € and %)			Beneficiaries (in thousands and %)		
Subsidy Strata	Amounts	By stratum (%)	Cumulative (%)	Number of beneficiaries	By stratum (%)	Cumulative (%)
0 to <5,000	845,479	21.21%	21.21%	737,990	79.45%	79.45%
>5,000 to <10,000	678,213	17.01%	38.22%	96.18	10.35%	89.80%
>10,000 to <20,000	811,692	20.36%	58.58%	58.31	6.28%	96.08%
>20,000 to <50,000	828,640	20.78%	79.36%	28.22	3.04%	99.12%
>50,000 to <100,000	394,087	9.88%	89.25%	5.82	0.63%	99.74%
>100,000 to <200,000	250,349	6.28%	95.52%	1.88	0.20%	99.94%
>200,000 to <300,000	78,778	1.98%	97.50%	0.33	0.04%	99.98%
>300,000	99,646	2.50%	100.0%	0.19	0.02%	100.0%
Total for Spain	3,986,884	100.00	–	928.92	100.00	–

Source: COMMISSION EC (2003)

[1] http://www.libroblancoagricultura.com/libroblanco/jtematica/pac/ponencias.asp

Table 7. The mid-term review of the CAP (2003).

Main elements of the mid-term review of the cap	
De-coupling	– The subsidies are not production-related. A single payment is established per production unit based on subsidies received in the 2000–2002 period. – This de-coupling can be a partial one in some crops, in a limited percentage.
Cross-compliance	– Receiving the subsidy is dependent upon compliance with specific common norms in the areas of environment, public health, vegetable and animal health, animal wellbeing, and preservation of lands in good agronomic conditions.
Modulation	– Direct payments over €5,000 must be reduced by 3% in 2005, by 4% in 2006, and by 5% from 2007 to 2012. – The withheld amounts will be devoted to rural development programs of EAGGF-Guarantee. – At least 80% of such withheld amounts will be invested in the Member State which generated those moneys. The remainder will be distributed among Member States on the basis of farmland area, agrarian employment rate, and GDP per capita in pps.
Financial limitation	– Direct payments will be adjusted when the envisaged amounts exceed in more than 300 million euros the annual allocation of funds set in the financial prospects for heading 1A of the budget for agriculture. – Any new agrarian expenditure should be financed by transferring funds from one sector to another.
Rural development	– Subsidies increased to support new, young farmers and the structural adjustment of their exploitations. – Subsidies to promote compliance with environmental, animal and vegetable health, public health, animal wellbeing and work safety rules. – Subsidies to cover expert advice costs. – Incentives to adhere to food quality encouragement schemes and for producer associations to launch consumer information and product promotion campaigns. – Incentives for innovative food processing.

Source: EESC (2005)

Being above or below the indices does reflect the comparative advantages or structural problems of a country's agricultural sector. This confirms the fact that agriculture is something more than its purely economic value.

Through the CAP, the EU shows that its agrarian policy is a top priority. The CAP resulted from the competition problems that existed for the free exchange of agrarian produce in the new common market, since all the founding States had quite protectionist national agrarian policies. The solution chosen by the EU institutions in order to ensure the exchange of agrarian produce in free competition conditions was to eliminate the national agrarian policies, which were replaced by a common agricultural policy.

The goals of the CAP (envisaged in the Treaty of Rome, which laid the foundations of the European Community) were further developed mainly on the basis of the conclusions of the Stresa Conference and through the creation of two special sections (Guarantee and Guidance) within the European Agriculture Guarantee and Guidance Fund (EAGGF) that are dedicated to the market policy and the structural policy (respectively). These sections continue to be the most important ones (from a financial point of view) in the Expenditure Budgets of the EU.

Pressure by the less developed countries and the acceptance by EU countries that the CAP would be unsustainable if based on production support through the EAGGF (Guidance) have led to gradual reforms of the CAP aimed at correcting its most negative effects, namely: Damaging trade relations with less developed countries, inefficiency (due to market distorsion), damaging the Union's coffers, and criticisms about the sharp disparity in contributions and subsidies (which were granted mainly to large land owners in the European Union). See Garcia Delgado, J.L. 2006.

Table 8. Land use in the agriculture of the EU.

Country	Total area (km^2)	Farmland area in 1999 (1.000 ha)	Farmland vs total area (%)	Irrigated area in 1999 (1.000 ha)	Irrigated v total area (%)
Germany	357.028	17.152	48	485	3
Austria	83.878	3.381	40	46*	1
Belgium- Luxemburg	33.114	1.521	46	40	3
Denmark	43.096	2.668	62	447	17
Spain	506.470	25.729	51	3.640	14
Finland	338.145	2.192	6	64	3
France	543.965	29.899	55	2.100	7
Greece	131.957	5.109	39	1.441	28
Irland	70.224	4.418	63	0	0
Italy	301.277	15.365	51	2.698	18
Netherlands	41.864	1.954	47	565	29
Portugal	92.389	3.942	43	650	16
UK	244.100	16.451	67	108	1
Sweden	449.960	3.157	7	115	4
EU	3.237.467	132.938	41	12.357	9

Notes: * Taken from Döll y Siebert, 2000 y Siebert y Döll, 2001
Farmland area correspond to: 1996 for Greece; 1997 for The Netherlands, Portugal and Spain; and 1998 for Irland, Italy and Sweden.
Sorce: FAOSTAT, 2001; EEA, 2001c., Spanish Ministry of the Environment (2004), page 79.

The latest boost to these reforms was the so-called "Mid-Term Review" passed in 2003 and which is shown in Table 7.

Many authors have emphasized how profound these reforms are with regard to the previous situation, and they particularly highlight the principle of separating (de-coupling) production and subsidy payment, thus directing agriculture to those production levels that are demanded by the market, while requiring farmers to preserve their lands in adequate environmental conditions ("cross-compliance").

Does this reform have any impact on the irrigated crops of the European agriculture? It is still too soon to speculate on the impact of these reforms. However, it is certain that they will have an impact, particularly when this process will coincide with the implementation of the European Water Directive, which envisages higher prices for the use of water for farmland irrigation purposes ("full recovery cost principle").

It is common knowledge that the CAP was meant mainly for continental crops (cereal, oilseeds, protein plants, beet, vine) that are usually rainfed. But, in view of the higher productivities attained with irrigated crops, irrigation methods have been introduced in vast areas of Europe in the past few years in order to increase the productivity and income levels. Such de-coupling of production and subsidy payment through a single payment might curb or even decrease that process.

Table 8 shows the great diversity of realities across the European Union. While the southern countries consume great amounts of water for agricultural purposes, the consumption of water in northern countries is much smaller, even negligible in some cases, so one can expect a considerably minor impact of the reforms in northern countries. Naturally these figures are dependent on the water withdrawals that are required for irrigating these expanses of land, and those figures are offered rather inconsistently by the EU's statistics agency.

In the past decades, the marked difference between the yields from irrigated crops and the yields from rainfed crops has encouraged those countries that had more opportunities to increase their irrigated lands to do so indeed in order to receive more EU subsidies. The most conspicuous case is that of France, where irrigated farmlands have grown dramatically. In the case of Spain (where the

Table 9. Water withdrawals for agriculture in the european union (Hm^3).

	1992	1993	1994	1995	1996	1997	1998	1999	2000	2001	2002	2003
Belgium	:	:	10.0	:	:	:	:	:	:	:	:	:
Czech Rep	:	:	:	37.0	31.0	20.0	10.3	13.4	14.5	12.0	18.7	:
Denmark	:	:	:	295.0	360.0	:	:	:	:	:	165.2	:
Germany	:	:	:	:	:	:	:	:	:	:	:	:
Estonia	:	:	:	:	:	:	29.6	36.7	36.4	40.3	30.0	:
Greece	7900.0	7600.0	:	:	:	:	:	:	:	:	:	:
Spain	:	17000	:	24116.0	:	23413.5	25011	26325	24070	24568	24160	:
France	4919.0	4949.0	4971.0	:	:	:	:	:	4871.9	4767.7	4535	:
Ireland	:	:	179.0	:	:	:	:	:	:	:	:	:
Italy	:	:	:	:	:	:	:	:	:	:	:	:
Cyprus	:	:	:	:	:	:	132.3	145.6	135.5	143.2	158.4	165.8
Latvia	135.5	74.6	62.8	:	:	58.6	53.4	50.0	48.4	46.6	53.2	52.4
Lithuania	:	:	:	:	:	:	:	:	:	53.0	59.7	84.2
Luxembourg	:	:	:	0.2	:	:	:	0.2	:	:	:	:
Hungary	949.0	945.0	711.0	662.1	455.6	407.5	407.2	441.5	720.7	716.3	679.6	:
Malta	:	:	:	:	:	:	:	:	:	:	:	:
Netherlands	:	:	:	260.0	230.0	90.0	53.0	76.0	:	:	:	:
Austria	100.0	100.0	100.0	100.0	100.0	100.0	100.0	100.0	100.0	100.0	100.0	:
Poland	1369.3	1392.8	1237.8	1176.8	1057.5	1082.9	999.2	1045.4	1060.6	1033.3	1108	1014
Portugal	:	:	10000	10000	:	:	8754.6	:	:	:	:	:
Slovenia	3.7	3.5	5.5	4.8	0.2	:	:	:	:	:	6.6	:
Slovakia	152.0	192.0	138.0	96.0	75.0	67.0	60.4	37.6	91.3	69.6	56.0	89.2
Finland	:	:	50.0	50.0[e]	50.0[e]	50.0[e]	50.0[e]	50.0[e]	:	:	:	:
Sweden	174.0	174.0	176.0	150.0	150.0	150.0	150.0	150.0	150.0	135.0	135.0	:
United Kingdom	:	:	:	:	:	:	:	:	:	:	:	:
Bulgaria	2243.3	1078.2	1249.5	579.3	1006.8	831.6	800.9	759.5	1184.6	865.2	743.0	1097
Romania	2794.0	2980.0	2520.0	1910.0	2320.0	1030.0	1300.0	1027.0	940.0	1018.0	1192	1283
Iceland	70.0	70.0	70.0	70.0	70.0	70.0	70.0	70.0	70.0	70.0	70.0	70.0
Norway	:	:	:	271.1	228.1	762.7	696.4	769.3	769.8	877.8	903.5	:

Source: http://epp.eurostat.cec.eu.int/portal/page?_pageid=1996,39140985&_dad=portal&_schema=PORTAL&screen=detailref&product=Yearlies_new_environment_energy&language=en&root=Yearlies_new_environment_energy/H/H1/H11/dda13072

figures show a global decrease), there has been a shift from traditional irrigated crops to new types of irrigated crops (e.g. in the region of La Mancha) in order to qualify for the European subsidies.

That phenomenon is once again shown by Figure 8. Even though the increase has been more moderate in the past few years, the irrigated area in Europe has grown at the rate of 120,000 hectares annually. The problem may become more serious should that trend continue.

The new CAP passed in 2003 may mark a trend change. If the top priority will be to support farmers' incomes rather than production, then logically there will be change in the incentives to crop productivity through irrigation. As stated earlier, it may be too soon for analyzing the possible changes. However, based on the logics of economics, there are hardly any doubts that incurring large investments or energy expenses in order to withdraw water and increase productivity is not cost-effective, considering that the subsidies will be the same. It is true that those production units which are not subsidized but are still profitable enough will continue to put pressure on water demands. However, one should remember that those lands that have joined this process cannot compete with other irrigated lands whose climate blesses them with a much higher yield potential.

In these cases one would expect a shift toward crops that are less resource-demanding and which contribute to meeting the principle of cross-compliance, thus preserving the environmental

Table 10. Evolution of the Irrigated area in the European Union 1961–1996.

	Trend (1.000 ha)			Differences (1.000 ha)	
	1961–80 (a)	1980–96 (b)	1990–96 (c)	(b)–(a)	(c)–(b)
EU 15	152	146	123	−6	−23.6
F	25	48	59	23	10.8
EL	28	27	28	−0.8	1.7
E	58	34	NS	−24.3	NP
I	NS	25	NS	NP	NP
P	0.6	0.1	0.4	−0.5	0.3
Others	41	12	NS	−28.8	NP

Source : FAO
NS : statistically non significant (p>0,01 determined by t-student)
NP : non pertinent

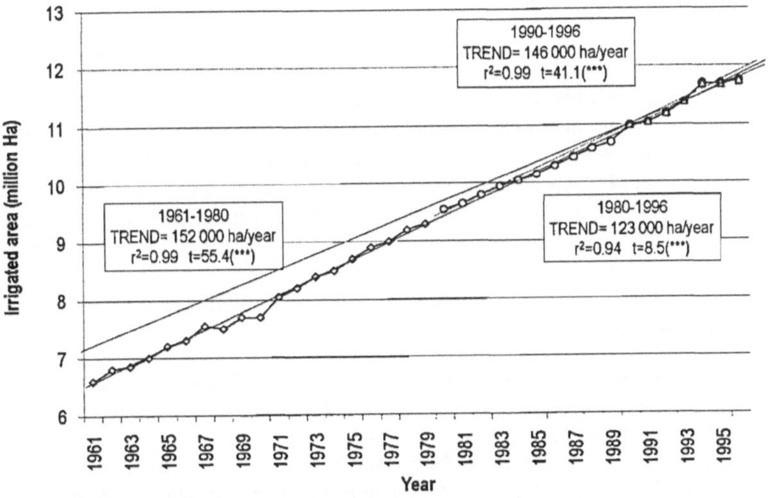

Figure 8. Evolution of Irrigated area in the EU-15 (million Ha).
Source: FAO *** highly significant

conditions, not incurring any additional costs, and adding efficiency in terms of agrarian production and water consumption.

4 THE IMPLICATIONS OF THE WATER FRAMEWORK DIRECTIVE FOR IRRIGATED AGRICULTURE COST RECOVERY

The implementation of the new CAP from 2003 has happened to coincide in time with the implementation of the brand new European Water Framework Directive, which was passed in 2000. Since that year, all European governments must transpose said directive into their national legal systems, and not simply the directive's formal aspects, but also to carry out institutional and economic changes. This shook the balances in place until recently, i.e. it altered the diversity that existed across the European Union when it came to the role of water in the economic and social sectors of each Member State. Table 11 shows the deadlines that European governments must meet in the process of implementing and complying with this directive.

Table 11. Implementation of the European water directive.

Date	Subject	Reference
22/12/2000	Directive comes into force	Art. 25
22/12/2003	Water Directive transposed into Member States' domestic legislation. Hydrographic basins and competent authorities identified.	Art. 24 and Art. 3
22/12/2004	Assessment of water-related pressures and impacts, plus economic assessment, completed	Art. 5
22/12/2006	Monitoring programs must be operational to serve as basis for water management Public information and consultation campaign begun by this date.	Art. 8 and Art. 14
22/12/2008	Public presentation of tentative Hydrographic Basin Management Schemes	Art. 13
22/12/2009	Final Hydrographic Basin Management Schemes, and action programs, published	Art.11 and Art.13
2010	Prices policy established.	Art. 9
22/12/2012	Action programs become operational.	Art. 11
22/12/2015	The environmental goals should be met as of this date.	Art. 4

The impact of implementing the Water Framework Directive on the Spanish and European agricultures has been the subject of extensive discussion. In the coming years, this subject is bound to bring about even more intense discussions and also social and political clashes.[2]

In Spain, from the legal point of view, this issue is regulated under article 129 of Act 62/2003 (dated 30 December), which describes a scheme of fiscal, administrative and social measures, and which amends all those aspects of the Spanish Water Act which needed to change as a result of transposing the Water Framework Directive.

More specifically, a new article (111 bis) is included in the Water Act, in section 36. Here it is:
Article 111 bis – General Principles

1. The competent public authorities will take into account the principle of cost recovery in all services relating to water management, including environmental and resource costs, based on long-term supply and demand projections.

2. The principle of cost recovery should be applied in such a way as to stimulate the efficient use of water, thus contributing to the compliance with the established environmental goals.

In a like manner, the application of such principle should be done with an adequate contribution from the different uses, based on the principle of "Polluters pay", and by considering at least these main uses: public water supply, agriculture, and industry. All of the above should be done with transparency and open accountability.

3. When applying the cost recovery principle, the authorities will consider the social, environmental and economic consequences, as well as the local geographic and climatic conditions, provided that this will not compromise the established purposes and environmental goals. All exceptions made should be justified by the hydrological basin plans.

In other words, the idea is to reproduce almost literally article 9 of the Water Framework Directive when it deals with the cost recovery principle.

How will this article be implemented in practice? There are great concerns about the implications of this article for irrigated agriculture in countries where irrigation implies high water consumption, as is the case in Spain. Statements by politicians have tried to reassure irrigators about the upcoming changes by emphasizing the flexible nature of the regulations (e.g. by emphasizing the phrase "will take into account") and their adaptive (case-by-case analysis) attitude, especially as regards the cost recovery principle.

[2] See Sumpsi Viñas J.M. *et al.* (1998) and Garrido Colmenero, A. & Martínez Valderrama, J. (2003)

The Spanish Ministries of the Environment and Agriculture felt the need to make public statements (EFE, 2005) concerning this: The General Secretary of Agriculture affirmed that a small fluctuation in the water price could have very large financial repercussions in a specific crop, and a really negligible impact in other crops. Consequently, he defended taking "heterogeneous" approaches. It is in this manner that the Government expects to prevent the social, environmental, and economic impact of implementing the cost-recovery principle and will present reasons that justify not applying that principle in specific geographic areas.

Antonio Serrano, from the Ministry of the Environment, affirmed that "we are not shelving" any processes (referring to the water policy imposed by the European Union), but rather "we are trying to rationalize them" and to begin a process of gradual implementation, while making a "basin by basin" impact assessment.

Mr. Moraleda (from the Ministry of Agriculture) insisted that the implementation of this measure will not translate into an additional increase in the price of water for irrigators, and both general secretaries pointed out that the price might well drop in some geographic areas. Both Mr. Serrano and Mr. Moraleda said that the implementation of the European water directive will have a different impact in each basin, and that there will be different price policies for each basin. Having said that, they reminded that things are not so different today, since irrigators and urban consumers are paying very different prices for the water they use.

Undoubtedly this approach will lead to a profound and thorny debate about the following issue: Which are the regions where special (reduced) water prices will be charged for cost recovery purposes? Let us remember that the Spanish regions have regional governments, that these are controlled by different political parties, and that important election campaigns are coming up.

The Spanish "Hydrographic Confederations" (Water Basin Authorities) undertook to present the assessments mentioned in articles 5 and 6 by the year 2005. On the sole basis of such assessments it is hard to predict the consequences for water prices. (Sevilla, M., 2006).

This debate is not exclusive to Spain – similar cases are found across the European Union. This was the situation in 2004: "The water prices that farmers pay do not cover the capital cost of supplying water in almost none of the OECD countries. Only in some states are the operating and maintenance costs recovered. For instance, even though the water prices for irrigation are relatively high in Greece, the farmers that work in public irrigation units do not even cover their operating and maintenance costs. The Greek authorities believe that irrigation developments contribute to the development of rural areas and thus often grant financial support. Also, the prices charged by the Spanish Hydrographic Confederations for water withdrawal for irrigation purposes are not sufficient for recovering the exploitation costs. In France, the large subsidies being awarded to farmers when they invest money in irrigation equipment render the water price increases useless. Even though the Swedish farmers do not receive any direct subsidies, they can extract groundwaters for free provided that they have a license or authorization to do so (EEA, 1999b)." (Spanish Ministry of the Environment, 2004, page 237).

Let us highlight four elements which, from our perspective, will be at the center of the discussion in the European Union in the coming years: 1) Issues relating to information about the water costs, use rights and consumption levels, particularly in agriculture; 2) the need to raise the average water prices; 3) the criteria to be applied in each geographic region; and 4) the impact of the new CAP.

With regard to the first element, we can say that the available reports of the Spanish Hydrographic Confederations leave much to be desired as regards the necessary background information for the implementation of a heterogeneous price policy (Sevilla, M. 2006). In Spain, the irrigation districts (called "Irrigators' Communities") consume 80% of the water. However, the only available data on their costs and prices come from indirect and incomprehensive surveys.

The EEA considers three types of costs to be taken into account for cost recovery purposes: Operating costs, environmental costs, and resource costs. In practice, only the first type of cost is taken into account, and in a rather partial manner, since there are plenty of subsidies and repayments that are ignored. Therefore, if a real cost-recovery policy is to be implemented, and if we are to face all the existing problems responsibly, increasing water prices will be absolutely necessary.

At present there is great diversity in the water prices charged in different regions of Spain. Even though it is widely admitted that each geographic area should be given different treatment, there is no doubt that moving from abstract theory to the practical and generalized implementation of unequal water prices involves taking considerable political risks, particularly considering that, by virtue of Spain's National Irrigation Plan, there are a number of projects in progress all over Spain to expand and enhance the irrigation infrastructures (Sumpsí Viñas, 1999).

As regards the impact of the new CAP on irrigated agriculture, this issue will depend on a number of factors deriving from the CAP's effects on water prices and the final prices of produce (given the de-coupling of productions and the payments to be awarded to farmers). There is a great variety of crops and the reforms will influence each crop in different ways. According to Atance (García Álvarez-Coque, 2006, page 73):

"Due to the de-coupling principle, now the subsidies are a fixed amount that has no connection with the farmer's activity. Therefore, each farmer or stockbreeder will decide to either continue or to quit their activity on the basis of whether their production-derived income is higher or not than the costs. Undeniably many farmers in the poorer areas will quit. Consequently, implementing the reforms introduced in 2003 in Spain requires managing that quitting risk upfront."

The future impact of all this on irrigated agriculture is still uncertain, because the choice of using or not using irrigation will depend on the market prices of irrigated crops, rather than on the certainty to receive subsidies. Therefore, one can expect a global change in the productions (which is the ultimate goal of said reforms) toward those products that are on high demand.

According to Garrido (Garrido Colmenero, 2003):

"All seems to be gloom and doom in continental irrigated crops. Some question the farmers' capacity to pay for water and many have repeatedly argued that a moderately ambitious pricing policy could eliminate more than a million hectares of irrigated lands (Berbel et al, *1999; Sumpsí* et al., *1998). In this very delicate issue, there are a few unquestionable facts that we should pay attention to. Firstly, farmers of continental crops who use groundwaters for irrigation pay high costs (pumping and equipment in general), as much as 0.15€/m3 (Llamas* et al., *2001; Garrido, 2001). Secondly, the use of groundwaters for irrigation is not decreasing in central Spain in spite of the growing difficulty to operate in many areas: Limited flows (resulting in a less intensive use of land and invested capital) and a certain amount of legal uncertainty until applications are substantiated. This becomes evident in view of the large, ongoing increase in the irrigation of vine and other crops – only in the central Spanish regions of Castilla La Mancha and Castilla León there are 665,000 hectares being irrigated. Third, even though most studies on water demands in the regions of central Spain agree that farmers have a limited capacity to pay for water, the analysis should focus on determining whether irrigated agriculture continues to be more profitable than rainfed agriculture, considering worst-case scenarios in the implementation of the Water Framework Directive, market conditions, and award of subsidies. (page 158)*

In spite of all that, there are vast stretches of new irrigated land that are located in poorer areas which might be abandoned or exploited less intensely, and that would curb the overexploitation of the local aquifers.

5 THE CASE OF IRRIGATED AGRICULTURE IN SPAIN: THE CAP REFORM AND THE POTENTIAL IMPACT OF THE NATIONAL IRRIGATION PLAN

The current irrigated agriculture in Spain was inherited directly from the regenerationist view prevailing in Spain since the early 20th century. Historically, the approach in order to solve the problems of the Spanish agriculture was always based primarily on the transformation of crops (except the ephemeral process of agrarian reform that was carried out between 1931 and 1936), by introducing irrigation in much of the Spanish territory. That led to a spectacular growth of the irrigated area, both from a public and a private point of view.

This was the evolution in the process of public transformation of the Spanish agriculture (from rainfed to irrigated) through the 20th century: 316,000 hectares were transformed under

Table 12. Irrigated areas and how they were introduced.

Origin	Area (thousand Ha)
Historical irrigated areas	1,077
Initiative of the Ministry of Public Works (Act from 1911)	316
IRYDA and INC irrigated areas and coordinated schemes (Ministry of Agriculture, Fisheries and Food and the Ministry of Public Works)	992
Initiative of the regional governments	95
Private initiative	1,280
Total	3,760

Source: Ministry of Agriculture, Fisheries and Food (2000) – National Irrigation Plan

Table 13. Distribution of the Irrigated area in Spain (2002).

Spanish region	Total arable land (1) (Ha)	Total Irrigated Area (Ha)	Irrigated vs. Total farmland (%)
Andalucía	4.854.912	944.435	19,45
Aragón	2.472.792	445.179	18,00
Asturias (Principado de)	339.609	5.773	1,70
Balears (Illes)	180.822	16.642	9,20
Canarias	78.150	23.187	29,67
Cantabria	160.697	1.357	0,84
Castilla – La Mancha	4.722.368	489.808	10,37
Castilla y León	5.178.177	524.199	10,12
Cataluña	1.179.556	271.126	22,99
Comunidad Valenciana	822.894	354.260	43,05
Extremadura	2.224.941	206.973	9,30
Galicia	866.282	82.304	9,50
Madrid (Comunidad de)	347.454	27.702	7,97
Murcia (Región de)	622.725	193.907	31,14
Navarra (C. Foral de)	615.947	87.784	14,25
País Vasco	240.792	10.174	4,23
Rioja (La)	273.973	95.495	34,86
Total Spain	25.182.091	3.780.305	15,01

(1) It covers arable land, fallow, prairies, and pastureland.

the 1911 Act; 992,000 Ha under the 1949 Act (which was later redrafted as the 1973 Agrarian Development and Reform Act); and 95,000 Ha on the initiative of the regional governments.

We should add 695,000 Ha that were turned into irrigated land by private farmers thanks to public subsidies. This represents a high percentage of the total 1,300,000 Ha that were transformed by private entrepreneurs. Similar figures can be found in very few other countries. They reflect the importance that has been attached historically to irrigation in both the Spanish economic policy and the agrarian policy. (Ministry of Agriculture, Fisheries and Food, 2000, the National Irrigation Plan by 2008).

The resulting area is more than three times the historically irrigated areas. This transformation contributed to a constant process of modernization of our agriculture on a national scale and is at the root of the strong growth of the productivity and production of our agriculture (despite the sharp fall in the percentage of the workforce employed by that sector). In this manner, our agricultural sector has greatly contributed to the stability and surplus of our trade balance since Spain joined the European Economic Community in 1986.

Table 14. Areas, yields and production in the Spanish agriculture. Irrigated vs. Rainfed. 1996.

	Area			Yield		
Crops	Rainfed	Irrigated	Total	Rainfed	Irrigated	Production
Cereals						
Barley	3306236	265914	3572150	2887	4337	10697000
Wheat	1806416	206015	2012431	2844	4382	6040454
Maize	75322	364389	439711	3081	9657	3751072
Others	611466	259965	637431			1155726
Total	5799440	862283	6661723			21644252
Oilseeds						
Sunflower	847282	250932	1098214	834	1877	1177757
Soybean	64	5073	5137	917	1987	10141
Rapeseed	16304	81262	97566	1138	1100	107962
Protein						
Pea	53539	28591	82130	791	1460	84110
Bean	7835	2620	10455	779	1949	11206
Lupin	24774	284	25058	689	1266	17435

*Lupin: Rice is not included
Source: Spanish Ministry of Agriculture, Fisheries and Food (1996).

The main benefit of irrigated agriculture is that it generates more direct employment. On average, one irrigated hectare requires 0.141 AWU, whereas one rainfed area just needs 0.037 AWU. In other words, the average workforce potential of irrigated agriculture triples that of rainfed agriculture. These differences become much more dramatic in the agriculture of the South-Eastern coast and the Southern Atlantic coast of Spain, where one irrigated hectare generates up to 50 times more employment than one rainfed hectare, thanks to its higher productivity.

These differences between irrigated and rainfed agriculture are also shown by the different value of the lands in the two systems. The variations depend largely on the crops and their geographic locations, but the land prices are much higher across the board (more than triple on average).

We may observe this behavior in detail in one of the areas with the biggest increase in its irrigated area in Spain: Castilla la Mancha. If one considers the differences in the value of irrigated vs. rainfed lands, it makes sense for land owners to call out for an official transformation of their lands, especially if such transformation is financed by public subsidies. Add to this the chance to increase their productivity by shifting from rainfed to irrigated agriculture, and we find a three-fold interest in this process: Higher land value, higher land productivity, and larger subsidies from the EU (de-coupled production) and from the Spanish public administration that encourage the transformation of rainfed land into irrigated land. The only possible limitations to these trends would be any constraints imposed by higher water prices.

The National Irrigation Plan was launched in the mid 1990s, firstly, in order to regulate and increase efficiency in the management of the water resources included in the National Hydrological Plan (the Popular Party, then in the opposition, did not support the Hydrological Plan in 1993 on the grounds that the Irrigation Plan should be drafted first), and second, in order to bind the agrarian transformation process (expansion of irrigation) to the development and modernization of rural Spain.

It becomes clear, then, that the Spanish governments did not only take into account the historical reasons when preparing and launching the National Irrigation Plan, but also the employment factor and its impact. These are remarkable facts in the new scenario of the Spanish agriculture in the early 21st century.

Table 15. National average prices of Irrigated and Rainfed lands (2001–2002).

	Base weighting 1997 (%)	Prices 01 (Euros/ha)	Prices 02 (Euros/ha)	Price variation		Impact (%)
				(Euros/ha)	(%)	
Rainfed	86.0	5,684	6,014	329	5.8	3.8
Arable land	47.0	4,744	5,146	402	8.5	2.5
Non-citric fruit trees	2.5	5,842	6,051	659	11.3	0.2
Vine	4.2	11,670	12,540	870	7.5	0.5
Olive grove	7.9	15,887	15,950	63	0.4	0.1
Exploitations	24.4	3,135	3,284	149	4.8	0.5
Irrigation	14.0	19,043	20,217	1,174	6.2	2.2
Arable land	9.5	13,875	14,789	914	6.6	1.1
Open-air vegetables	0.5	25,133	30,248	5,114	20.3	0.4
Protected crops	0.2	76,491	78,939	2,448	3.2	0.1
Rice	0.4	24,218	22,920	−1,297	−5.4	−0.1
Strawberry	0.0	35,490	36,962	1,472	4.1	0.0
Citric fruits	1.0	45,644	48,297	2,653	5.8	0.3
Non-citric fruits	0.9	22,813	23,910	1,093	4.8	0.1
Vine	0.3	19,239	21,269	2,030	10.6	0.1
Olive grove	0.8	30,961	31,463	503	1.6	0.1
Exploitations	0.5	9,084	10,133	1,049	11.5	0.1
Total	100.0	7,553	8,001	448	5.9	5.9

Source: Spanish Ministry of Agriculture, Fisheries and Food.

Table 16. Land values in Castilla La Mancha. 2005.

Common prices	Albacete (Euros/ha)	Ciud. Real (Euros/ha)	Cuenca (Euros/ha)	Guadalajara (Euros/ha)	Toledo (Euros/ha)
Labor, uninrrigated	5,703.00	3,264.00	5,608.32	2,725.00	4,244.00
Labor, irrigated	19,410.00	10,573.00	16,411.10	11,872.00	10,504.00
Stone fruit trees, irrigated	27,061.00		29,509.69		
Pip fruit trees, irrigated	24,034.00		24,017.49		
Nut trees	9,359.00		9,673.11		
Rainfed vine	9,967.00	8,528.00	15,916.14		9,850.00
Irrigated vine	17,047.00	15,115.00	24,566.37		
Rainfed pastureland	971.00		3,022.76	703.00	2,490.00
Rainfed olive groves (for oil)	9,134.00	9,916.00	9,782.65	1,721.00	13,790.00
Irrigated olive groves (for oil)	11,584.00	19,518.00	15,359.76		
Natural, rainfed grazing land	1.765,00	1.824,00			3.324,00

Source: Government of the Region of Castilla La Mancha. Annual surveys on land prices. December 2005.

As pointed out in the National Irrigation Plan: "... multifunctional irrigated agriculture, which is characterized by maintaining the local population, regulating the land, and preserving the rural space, constitutes one of the pillars of the new model of European agriculture established in the Agenda 2000. This multifunctional irrigated agriculture is defined in this National Irrigation Plan as "social irrigation". It deserves the support of the authorities because of the valuable services that it renders to society as a whole." (Ministry of Agriculture, Fisheries and Food, 2000, the National Irrigation Plan by 2008, page 5).

The National Irrigation Plan, which should be completed by the year 2008, is an attempt to combine several factors that need to be considered in view of the scenario at the beginning of

Table 17. Distribution of Irrigation systems by Spanish region. 1999.

Spanish region	Gravity (%)	Sprinkling (%)	Localized (%)	Other (%)
Andalucía	29.6	19.5	49.5	1.4
Aragón	70.4	20.7	8.1	0.8
Asturias (Principado de)	58.3	20.5	5.7	15.5
Balears (Illes)	11.9	57.0	29.9	1.2
Canarias	18.3	31.2	47.7	2.8
Cantabria	35.0	54.5	4.5	5.9
Castilla – La Mancha	44.6	52.1	2.0	1.4
Castilla y León	13.0	55.9	28.7	2.4
Cataluña	63.3	12.4	23.6	0.7
Comunidad Valenciana	61.9	1.9	35.7	0.4
Extremadura	57.2	29.1	12.2	1.6
Galicia	79.7	8.0	3.8	8.4
Madrid (Comunidad de)	61.1	28.5	8.6	1.8
Murcia (Región de)	33.2	4.1	62.4	0.3
Navarra (C. Foral de)	79.8	12.2	7.6	0.3
País Vasco	2.6	85.6	4.6	7.2
Rioja (La)	47.5	34.3	17.7	0.4
Ceuta y Melilla	47.5	34.3	26.2	5.8
Total Spain	43.6	27.3	27.6	1.5

Source: National Statistics Institute. Agrarian Census, 1999

the 21st century: The need for an efficient use of water; the role of irrigated agriculture in the development of depressed areas; and the impact on the CAP of new productions introduced by the expanded irrigated agriculture.

At the same time, the National Irrigation Plan has given due consideration to the so-called "new water culture", i.e. new constraints to the use of water. We have moved from an "expansionist" conception of hydraulic infrastructures, whereby resource availability was constrained solely by the lack of new infrastructures, to a new conception that underscores consumption reductions, financial considerations, and water quality. Let us not forget that in Spain nearly 80% of the water is consumed for farming irrigation purposes. Therefore, the bulk of our water savings should come precisely from that sector, as the prevailing irrigation techniques use water in a spendthrift, non-efficient manner.

For all these reasons, more than 70% of the resources allocated to the National Irrigation Plan are assigned to the consolidation and enhancement of the existing infrastructures, many of which are earthen canals and ditches built long ago and in poor condition which are to blame for much of the water losses in Spain.

The so-called "social irrigation" is the second top priority in the Plan. The plan aims to transform vast, disadvantaged areas (in the regions of Aragón, Castilla La Mancha, Castilla León, and others) into irrigated arable land and to exploit their available water resources (even though in some cases these are shared with other regions, e.g. Castilla La Mancha). One of the expected outcomes of this is to promote the economic development of these areas and thus to stop migrations from these regions.

The second goal is linked to the third. The expected increase in agrarian productions (as a result of the transformation) may lead to a clash with the constraints imposed by the CAP. Many of the areas to be transformed will be used to cultivate crops that are limited by the CAP (herbaceous crops, cereals, oilseeds, protein crops, vine, olives, etc.). Therefore, rather than being a blessing, this investment effort might lead into a new problem.

That potential scenario is admitted in the National Irritation Plan (NIP): "Should the entire list of action plans included in the NIP be completed (infrastructure consolidation and enhancement,

Table 18. Actions envisaged under the National Irrigation plan to be completed by 2008. Aggregate investments in million pesetas (1 € = 166 pesetas).

Spanish region	Consolidation and enhancement	Irrigation projects in progress	Social Irrigation	Subsidized Private Irrigation	Other Programs	Total
Andalucía	83,903	41,590	8,580	–	–	134,074
Aragón	55,497	43,030	22,100	–	–	120,626
Asturias	114	–	–	–	–	114
Baleares	3,893	–	5,200	–	–	9,094
Canarias	7,256	–	8,450	–	–	15,706
Cantabria	138	–	2,600	–	–	2,738
Castilla-La Mancha	32,688	15,600	19,600	–	–	67,888
Castilla y León	109,484	57,200	9,100	–	–	175,784
Cataluña	41,901	5,428	8,320	–	–	55,648
Extremadura	21,400	16,510	8,450	–	–	46,360
Galicia	2,790	–	2,600	–	–	5,390
Madrid	3,720	–	–	–	–	3,720
Región de Murcia	43,896	–	–	–	–	43,896
Navarra	19,887	9,750	3,900	–	–	33,538
País Vasco	2,798	–	5,460	–	–	8,258
La Rioja	18,006	–	9,100	–	–	27,106
Comunidad Valenciana	61,191	–	–	–	–	61,190
Sin regionalizar	–	–	–	20,600	4,275	24,875
Total	508,562	189,108	113,460	20,600	4,275	836,005

Source: National Irrigation Plan

Table 19. Irrigated vs. Rainfed area variations (Thousand Ha). Resulting from implementing all of the NIP's programs.

	COP(1)	Maize	Rice	Potato	Beet	Cotton	Forage	FV(2)	Vine	Olive
Irrigation projects in progress	−183.7	112.3	20.4	28.3	0.0	6.8	82.1	85.6	−0.5	0.0
New irrigation (social interest)	−227.1	151.6	3.2	51.3	36.3	2.3	17.0	148.2	−1.7	24.2
New private irrigation	−23.8	3.3	0.4	0.1	0.0	0.2	2.0	14.1	2.0	8.5
Land abandonment	−347.0	0.0	0.0	−3.5	−1.5	0.0	−7.5	−4.0	−40.0	−68.5
Total transformation	−781.6	267.2	24.0	76.2	34.8	9.3	93.6	243.9	−40.2	−35.8
Area variation vs national total (%)	−6.8	55.1	22.5	36.1	21.8	12.0	20.6	24.7	−3.1	−1.6

[1] Cereal/Oilseeds/Protein [2] Fruit/Vegetable
Source: Ministry of Agriculture, Fisheries and Food (2000) – National Irrigation Plan

completion of projects in progress, development of all the arable land where irrigation is possible), by the set deadlines and in compliance with the current EU agrarian policy, it would bring about severe and hardly bearable imbalances in the national agrarian market. For that reason, without detriment to the prior confirmation of hydric availability that is required for each Hydrological Plan, any future transformations of land into irrigated land should be carefully assessed from that critical perspective before they are indeed initiated" (page 33, section 5.11).

Let us consider the estimates on final productions that can be expected as a result of implementing this Plan (Table 19).

If things seemed complicated with the older CAP, the reforms from 2003 may have an even stronger impact on the desired outcome of the NIP investments. The so-called "de-coupling" may lead to the abandonment of those farmlands whose productivity levels are below a certain

Table 20. Irrigation demands (hm3) by 2008.

Spanish region	Current water demand	Increase in demand with new irrigation	Decrease in demand w/enhancement program	Additional resources by 2008	Water demand by 2008
Andalucía	5,025	146	46	444	5,569
Aragón	3,225	278	131	125	3,497
Asturias	25	0	0	1	26
Baleares	136	9	3	7	149
Canarias	210	31	9	30	262
Cantabria	15	0	1	1	15
Castilla – La Mancha	2,267	148	126	133	2,422
Castilla y León	3,352	264	177	141	3,580
Cataluña	2,219	17	96	20	2,160
Extremadura	1,695	139	85	160	1,909
Galicia	619	0	14	3	608
Madrid	268	0	11	10	267
Región de Murcia	1,231	0	49	72	1,254
Navarra	514	43	26	42	573
País Vasco	40	16	0	7	63
La Rioja	342	40	25	11	368
Comunidad Valenciana	2,115	0	139	91	2,067
Sin regionalizar		102			102
Total	23,298	1,233	938	1,298	24,891

Note: The current demand, calculated on the basis of the crops' water demands and the water efficiencies, includes the demands covered currently by the Basin Hydrological Plans and the additional required resources envisaged in the irrigation consolidation and enhancement program.

profitability threshold now that the subsidies are not production-dependent. This process could generate one of two possible trends: 1) a move from rainfed crops to irrigated crops; or 2) a decrease in the farmland area for irrigated crops, if the earned productivity does not compensate for the increased water costs. Maintaining the implicit subsidies through the use of water could be one way out of this, but that would be in conflict with both the Water Framework Directive and the objective of making the production system more efficient.[3]

6 CONCLUSIONS

Agriculture, and particularly irrigated agriculture, despite representing a small percentage of the economic activity of our societies, remains the basis for eradicating hunger and poverty globally. Resources and technologies are available which could be applied successfully in the less developed countries, just like China and India have done, for expanding the areas of irrigated farmland and increasing crop productivity. As a general rule, water is not a limiting factor for such process. However, some regions are suffering from a "hydric stress" which is causing conflicts among the

[3] The changes that are taking place in Spain at the moment as a result of several reform processes (in agriculture, water management, and European subsidies) make it very hard to foresee what transformations will indeed occur in the coming years. In addition, such changes depend largely on political decisions (distribution of subsidies, favoring some areas over other areas, etc.). So, when it comes to glimpsing what the future will bring, all we can say is the authorities will try and avoid dramatic changes, but who knows?

different water users. In those cases, adequate policies are required in order to rationalize water consumption levels.

The liberalization of trade in farm produce and the agrarian reforms in developed countries should contribute to alleviating the situation of the poorer countries through trade and the removal of customs barriers. In the EU, with the CAP reform from 2003 and the gradual implementation of the Water Framework Directive, one can anticipate the adaptation of irrigated farmlands depending on the market profitability of the crops and a decrease in water consumption if the Directive's cost-recovery principle is applied at the same time. These processes will not take place peacefully, especially in southern European countries, e.g. Spain.

In Spain, the above processes have coincided with the launching of the National Irrigation Plan, which is aimed at expanding irrigated areas by following the so-called "social irrigation" approach. Such areas are about the same ones that will be affected by the de-coupling principle of the new CAP.

It is still too soon to see the combined impact of all these processes. However, there is no doubt that there will be a trend toward limiting the growth of irrigated areas and increasing water prices, that is, if the authorities are to promote the efficient use of water and increase the profitability of such uses.

In sum, water and agriculture will remain hot issues in the world in the coming years and will call for reasonable policies in order to meet the new challenges.

REFERENCES

Barbaso, F. (2002) *Perspectiva a largo plazo de la política agrícola de la unión europea.* Director General Adjunto (encargado de las direcciones B, C y D). DG Agricultura, Comisión Europea. Available from: http://www.libroblancoagricultura.com/libroblanco/jtematica/pac/ponencias/f_barbasso/f_barbasso.asp.

CES (2005) *Análisis y perspectivas del sector primario en la Unión Europea.* Madrid, Consejo Económico y Social.

COMMISSION EC (2003) Indicative figures on the distribution of aids, by siza-class of aid, received in the context of directs aids paid to the producers according to Reg. (EC) No. 1259/1999 (Financial Year 2001).

EFE (2005) *Gobierno planea eximir a algunos sectores de sufragar obras agua.* 25-01-2005.

FAO (2002) *World Agriculture: Towards 2010.* Available from: http://www.fao.org/documents/show_cdr.asp?url_file=/docrep/V4200E/V4200E00.htm.

FAO (2002) *Agricultura mundial: hacia los años 2015/2030.* Available from: ftp://ftp.fao.org/docrep/fao/004/y3557s/y3557s04.pdf.

García Álvarez-Coque, J.M. et al. (2006) *La reforma de la Política Agraria Común. Preguntas y respuestas en torno al futuro de la agricultura.* Madrid, MAPA, Eumedia.

García Delgado J.L. & García Grande M.J. (directores) (2005) *Política Agraria Común: balance y perspectivas.* Available from: www.estudios.lacaixa.es.

Garrido Colmenero, A. & Martínez Valderrama, J. (2003) El nuevo marco institucional del agua y la agricultura de regadío. *Papeles de Economía Española*, 96.

Institute for European Environmental Policy and other (2000) A report to the Environment Directorate of the European Commission. The environmental impacts of irrigation in the European Union. London. Available from: http://europa.eu.int/comm/environment/agriculture/pdf/irrigation.pdf.

MAPA (2003) *El Libro Blanco de la Agricultura y el Desarrollo Rural.* Available from: http://www.libroblancoagricultura.com/libroblanco/jtematica/pac/ponencias.aspl.

MMA (2004) *Las aguas continentales en la Unión Europea.* MMA.

MMA (2004) *El agua en Europa: Una evaluación basada en indicadores.* Agencia Europea de Medio Ambiente, MMA.

Massot Martí, A. (2005) *De la crisis de la unión a la crisis de la pac: por un nuevo proyecto para la agricultura europea en un entorno globalizado* (dt). Real Instituto Elcano DT No. 4/2005 Documentos JULIO2005. Available from: http://www.realinstitutoelcano.org/documentos/208.asp.

OCDE–FAO (2005) *Agricultural Outlook*: 2005–2014

ONU (2003) *Population and water.*

Population Reference Bureau (2006) Available from: http://www.prb.org/SpanishTemplate.cfm?Section=Materiales&template=/ContentManagement/ContentDisplay.cfm&ContentID=12597#td.

Sevilla Jiménez, M. & Torregrosa Martí, T. (2006) Las tarifas de utilización del agua como instrumento para la recuperación de costes. *Ponencia. Asepelt. La Laguna 21–24 junio 2006.*

Sumpsi Viñas J.M. et al. (1998) *Economía y Política de Gestión del Agua en la Agricultura.* Prensa, MAPA. Mundi.

UNFPA (2004) Available from: http://www.unfpa.org/swp/2004/espanol/ch4/index.htm and http://www.libroblancoagricultura.com/libroblanco/jtematica/pac/ponencias.asp.

CHAPTER 9

Water and the city in the 21st century. A panoramic vision

Steve Buchberger
University of Cincinnati, USA

Enrique Cabrera
ITA. Universidad Politécnica de Valencia, Spain

ABSTRACT: The demographic explosion of humanity over the last few decades, the unprecedented increase in the standard of living and the marked tendency of the growing population to concentrate in urban centres pose formidable challenges to those in charge of water management in cities. In a completely different framework from the one of only a few decades ago, urban water problems which, until very recently were non-existent, now have to be solved. Solutions which are needed for the future are usually not single ones, require significant investment and, more often than not, because they have to reconcile competing interests, demand a great deal of imagination.

After a brief examination of the relationship between city and water through the ages, a historical journey which shows the speed of change that has occurred, we analyze the major problems, the main actions required to solve them and some of the difficulties that their implantation entails. We shall conclude by outlining some guidelines which, considering the universal and ubiquitous nature of urban water problems, should almost always be implemented.

1 INTRODUCTION

The geographical areas in which water shortages have always been a cause for concern have developed policies based more on the problems of quantity than of quality. Spain, with other Mediterranean countries, is a paradigmatic example of this clear bias, and considering its historical agricultural tradition it could not be otherwise. In these countries irrigated land, although gradually decreasing, today still represents around seventy-five percent of all water consumption. But over the last few decades the situation has been changing very quickly and, consequently, with each passing day the sustainable management of water in cities, where problems of quality and quantity become intertwined, acquires greater relevance. Immediate consequences of this increasing importance are the efforts to reduce agricultural demand which, over the last few years, is receiving much more attention (Cooley et al., 2008). This issue is especially acute where water is scarce, because attempts often are made to meet growing urban demand by releasing water from irrigated land.

Powerful reasons justify these attempts to redistribute traditional water allocations. The main one is priority of use. There is no doubt that today's society, our society of commodities, is incapable of coping with a prolonged failure in urban water supply. But there are other causes, by no means less significant, that also explain it. Among them the increasing population concentration in urban areas, a fact that has given rise to a rapid increase in demand for water in the city at the same time as it puts to the test the huge hydraulic infrastructure (diversion channels, water treatment plants, water flow control tanks, distribution and drainage networks, and, in short, purifying plants) that these systems demand. Historically, the construction and renovation (not so much maintenance) of these water facilities have been very heavily subsidized because water is, let us not forget, a human right. Up to now it was always thought that investments for such a basic necessity as water had to be taken care of by the State at zero direct cost to the citizens. But as subsidies give rise to inefficiency and the need for a rational use of water is increasing, the situation is changing very

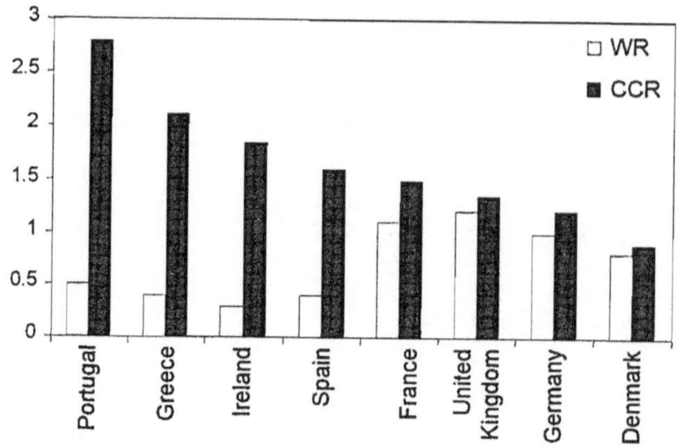

Figure 1. Current Water Tariff and Full Cost Recovery related to per capita income (Merkel, 2003).

quickly. Indeed, in 2010 the principle of cost recovery that the Water Framework Directive, WFD, imposes (EU, 2000) will become obligatory. The WFD will mark a very important turning point for all countries in which water has been subsidized.

This is nothing new for the developed and industrialized countries of northern Europe. Indeed, at the end of the last century, industrialization caused a remarkable deterioration in water quality and it pricked the environmental conscience of a society which, deeply concerned, forced a rapid adjustment of water policy to the new context. In a report of the German Federal Ministry of the Environment (BUNR, 2001) is stated that "During the reconstruction years after World War II, East and West Germany were incapable of integrating the efficient use of water into the expansion of their industrial activities. This fact gave rise at the end of the Sixties and beginning of the Seventies to water contamination reaching levels of social alarm". In order to solve the problem tariff policies were developed and services were regulated. Albeit somewhat later, other developed northern European countries did the same. The WFD will extend this fiscal policy to all EU countries.

Evidence for this is provided in Figure 1 (Merkel, 2003), a snapshot of the state of the issue in some European countries around the year 2000 when the Water Framework Directive came into being. As far as Spain is concerned, little has changed. At that time the Current Water Tariff (shown as CWT in Figure 1) for Spanish users amounted, on average, to 0.4% of their income whereas full cost recovery (FCR in Figure 1) would necessitate increasing the tariffs by a factor of 4 in order to reach 1.6% of the average per capita income. As illustrated in Figure 1, northern European countries at that time were already practically recovering their costs.

The country in Figure 1 closest to full cost recovery is Denmark. This is not surprising when examining, Figure 2 which tracks the dramatic price evolution of drinking water (excluding sanitation) over the last two decades in Copenhagen. In fact, at the end of the Eighties prices began to include full costs and four years later the cost of drinking water had tripled (1 €/m^3 in 1987 compared to 3 €/m^3 in 1991) reaching €4 m^3 in 2002 (Napstjert, 2002).

By contrast, in the last decade the situation in Spain has not changed. In fact, and as a result of the country's spectacular economic growth over the last ten years (although in middle 2008 this cycle has come to an abrupt end) in relative terms one pays less today. Citizens allocate as little as 0.29% of their income (compared to 0.4% at the end of the last century). Indeed, in 2005 average per capita income in Spain (www.ine.es) was 20,838 € whereas, according to that same source, the average water consumption per inhabitant that same year was 166 litres per day, about 60.59 m^3 per year. Considering a total average price of 0.98 €/m^3 (INE, 2005) the total cost per person per year amounted to 59.38 €, less than 0.3% of annual income. This is absolutely logical because in the considered period, the price of water has grown more slowly than average income.

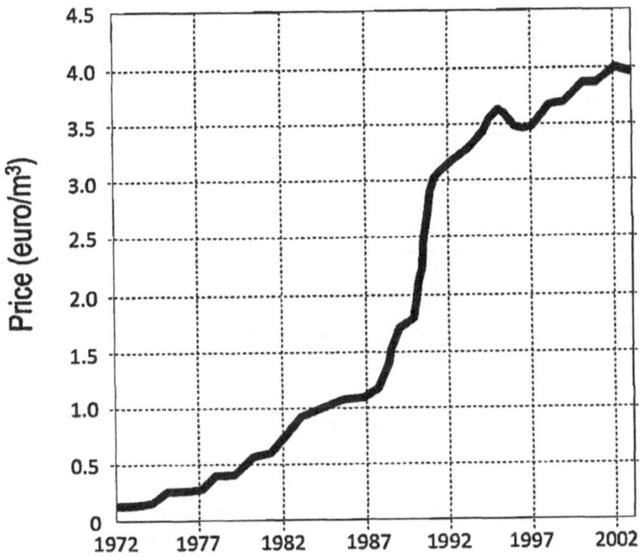

Figure 2. Water price evolution in Copenhagen (Napstjert, 2002).

Table 1. Rates in 1999 for urban drainage in Germany (BUNR, 2001).

	1999 Drainage Rate (divided in two blocks)		Drainage Rate (one block) €/m^3
	Sewage (€/m^3)	Rain water (€/m^2 year)	
Germany	1.79	0.77	2.28
West Germany	1.72	0.78	2.23
East Germany	2.39	0.59	2.54

Of course, without a price policy that reflects full costs and that clearly indicates to the public that sustainable water management in cities has a cost that must be assumed, it is impossible for consumers to change their behaviour to current requirements. Thus, and German urban drainage rates demonstrate this, it must be taken to its ultimate consequences. Indeed, to efficiently approach one of the major problems that the water and city nexus raises nowadays, urban floods, water drainage includes two concepts, a rain term, proportional to the waterproofed surface (it determines the contribution to the total run-off of a building) and the water treatment term (sewage term, linked to the potable water consumed by the customer and recorded by his meter). Table 1 outlines German rates at the end of the last century (BUNR, 2001).

In relation to these we should observe three additional details. In the first place the drainage rates in East German were higher than those in West Germany because East Germany started from lower prices and therefore was further from recovering costs. Secondly separation of rates for sewage and rain water components better reflects the costs that each user generates and helps to explain why they have spread throughout the country and have increased significantly. And so, in 2007 in Berlin these drainage rates stood at 1.637 € per m^2 per year, for the rain water term and 2.487 €/m^3 for the treatment term (http://www.businesslocationcenter.de/de/C/iv/2/se-ite3.jsp). These values are much higher (112% and 39% respectively) than those of the German average for 1999 as outlined in Table 1. The Berlin case is the most significant because before unification the City belonged to East and West. And finally to emphasize the higher pace growth in the rain term, an unequivocal sign of the desire to reflect real costs and, in passing, to pave the way for the construction of domestic tanks to store rainwater. In fact the current drainage rates allow these facilities to be repaid in about ten years (Fewkes, 2006). There is no doubt that operational and environmental reasons advice to make a bet for urban water harvesting.

After highlighting in this introduction cost recovery as a necessary condition for sustainability (although insufficient, because it must be accompanied by professional management), in what follows we briefly review the evolution of the water and city nexus throughout the ages and finally analyze the current complex problem. Even though many symptoms of the urban water problems were already foreseen decades ago, for the most part these problems have continued to grow, because not everything has been done which should have been. This analysis concentrates both on the existing urban water problems and on the activities that need to be undertaken in the future.

2 WATER AND THE CITY THROUGHOUT HISTORY

As water is essential for life, the earliest urban settlements were placed on river banks. To improve and extend access to water man soon showed signs of great creativity which, very probably, reached its zenith during the Roman Empire. Two thousand years ago Rome set up a complex system of aqueducts that linked the more than 100 kilometres between the Apennines and the Eternal City. Roman engineers were able to transport daily 600,000 m^3, a volume which today would be sufficient for a population of three million inhabitants with an allocation of 200 litres per person per day. The Romans had water in abundance (around 500 litres per person per day). The details of these civil engineering works which still surprise us nowadays, are very well documented (Blackman and Hodge, 2004; Bonnin, 1984; Evans, 2000; Viollet, 2000). We also owe to the Romans the term, rivalry, derived from riva (river bank). Ultimately, rivals were those who inhabited opposite sides of the river, because they disputed the same water, a rivalry that today has escalated to neighbouring river basins. Transporting water as far as technology allows has increased the scale, in terms of distance and demographics, of the rivalry.

After the Roman splendour, water distribution in cities, like knowledge itself, went through a dark age. Man requires only some few litres of water per day to satisfy his most vital needs. Paris, one of the booming cities of the Middle Ages, is a good example. In 1553, for a total of 260,000 inhabitants, only 300 m^3 of water were distributed daily. That represents only a litre per person per day (Thirriot, 1987). A century later, around 1669, the 500,000 inhabitants of Paris had 1,800 m^3 of water, about 4 litres per person per day. For many centuries this was the norm. For example in 1740 in Lisbon the 80,000 inhabitants distributed 560 m^3 of water, about 7 litres per person per day (Thirriot, 1987). By the end of the eighteenth century, some decades later, Madrid had no more than 3,600 m^3 of water per day to supply a population of 200.000 habitants (Paz and Paz, 1969). That represents a maximum volume of 18 litres per person per day. All this explains the vigour and the importance in those days of the trade of water carrier immortalized (Fig. 3) by a wonderful painter, Diego Velázquez.

In 1754 the first urban water supply system, as we know it today, was constructed in Bethlehem, Pennsylvania, while still a British colony (Grigg, 1986). As for making water drinkable, London was the pioneering city (Steel, 1972). The first water filter came about in 1829 and in 1908 the first chlorination system. In Spain, the present Madrid city water supply came into being in 1852 and it was Queen Isabel II who contributed most to the financing of the construction of the canal that was to bring water to the city from the nearby mountain range, as the name (Canal de Isabel II) of the company that supplies the whole Region indicates.

Indoor plumbing with running water brought a quantum leap in the quality of life for local citizens and ushered in a new era of unprecedented household convenience with public health benefits. By the beginning of the twentieth century, regardless of the level of development of a country, the value of urban water supply systems was recognized around the world. It was not long before the job of water carrier disappeared. The availability of treated water is now taken for

Figure 3. The water seller of Seville. Diego Velázquez (c.1620). Courtesy of Wellington Collection, London.

granted at urban centres around the globe. This view is so common, that to deny water service, albeit only for a few hours, is considered unacceptable in an urban area.

3 THE WATER AND THE CITY CRISIS IN THE 21ST CENTURY

In Spain the situation in this first decade of the twenty-first century has little in common with the one a hundred years ago when the foundations of the present water policy were laid down. An agricultural country has made way for an industrialized one and one based on services. Although irrigation continues to be the greatest source of water consumption in Spain its contribution to the national economy is in steady decline. Indeed, in 2006 farming's contribution to Gross Domestic Product (GDP) reached a token three percent whereas it employed, only five percent of the population (www.ine.es). And although its environmental and social value continues to be very important (not just for absorbing part of CO_2 that the industrialized world emits but also for maintaining the rural population, conserving landscape and tradition), over the last few decades the migratory flow of residents from the countryside to the city seems unstoppable.

One of the latest reports published by the United Nations on this matter, World Urbanization Prospects. The 2007 Revision (UN, 2008) is compelling evidence of this. Table 2, and Figs. 4 and 5, taken from that report, summarize it perfectly. Table 2 shows the impressive migration of people from the countryside to the city throughout the whole world over the last 60 years. It specifically highlights the case of Latin America and the Caribbean where the percentage of the population living in cities is expected to double from 41.4% in 1950 to 83.5% by 2025. During the tracked period (1950–2050) North America ranks the highest concentration of urban population and is expected to reach up to 92.2%, although in this case the migratory flow as a percentage will be smaller given that it started from a higher threshold (63.9%). All in all, any major area follows the same trend.

Table 2. Evolution by geographical area of the urban population in the last decades (UN, 2008).

	Percentage urban				
Major Area	1950	1975	2007	2025	2050
Africa	14.5	25.7	38.7	47.2	61.8
Asia	16.8	24.0	40.8	51.1	66.2
Europe	51.2	65.7	72.2	76.2	83.8
Latin America and the Caribbean	41.4	61.1	78.3	83.5	88.7
Nothern America	63.9	73.8	81.3	85.7	90.2
Oceania	62.0	71.5	70.5	71.9	76.4

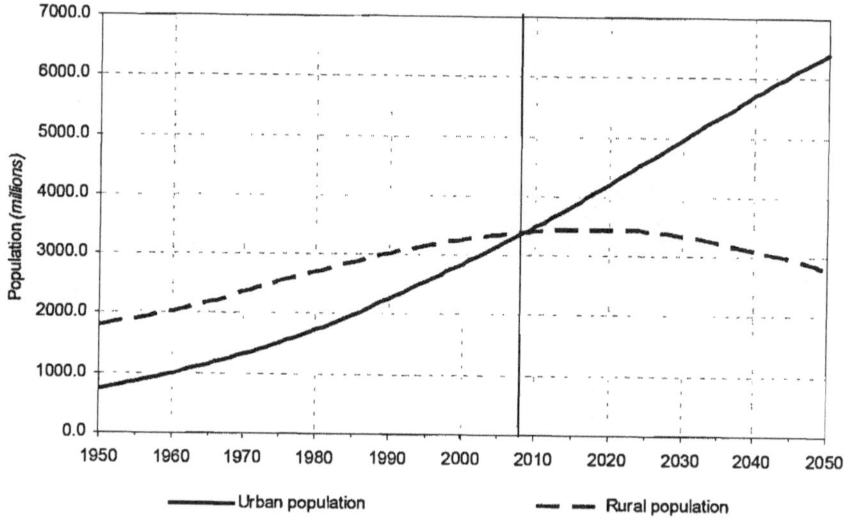

Figure 4. Evolution of the urban and rural populations in the last decades (UN, 2008).

In conjunction with the problem of the overwhelming migratory flow from the countryside to the city it is necessary to superimpose another concurrent demographic issue which, from the point of view of water resources sustainability is even worse. The planet is experiencing an explosive increase in population. In fact, in just one century the world's population will see a spectacular jump from less than three thousand millions at the beginning of the tracked period to the over nine thousand millions projected in 2050. Figure 4, which show concurrent trajectories for global population in rural and urban areas, highlight both facts. Two additional facts are worth to underline. First, the total number of people living in rural areas will peak early in the 21st century and then decline. This trend is consistent with the worldwide migrations displayed in Table 2. Second, the number of people living in urban areas will surpass the number in rural areas for the first time in the history of civilization. This transition from rural to urban majority is happening nowadays.

Figure 5 vividly illustrates that most of the growth in urban population will take place in developing countries rather than in developed ones. Recognizing that sustainable management of urban water in developed countries is already a formidable task, it becomes a more daunting challenge for new cities in developing countries. While basic targets for water and sanitation have been spelled out in the Millennium Development Goals, there is considerable uncertainty about how to achieve

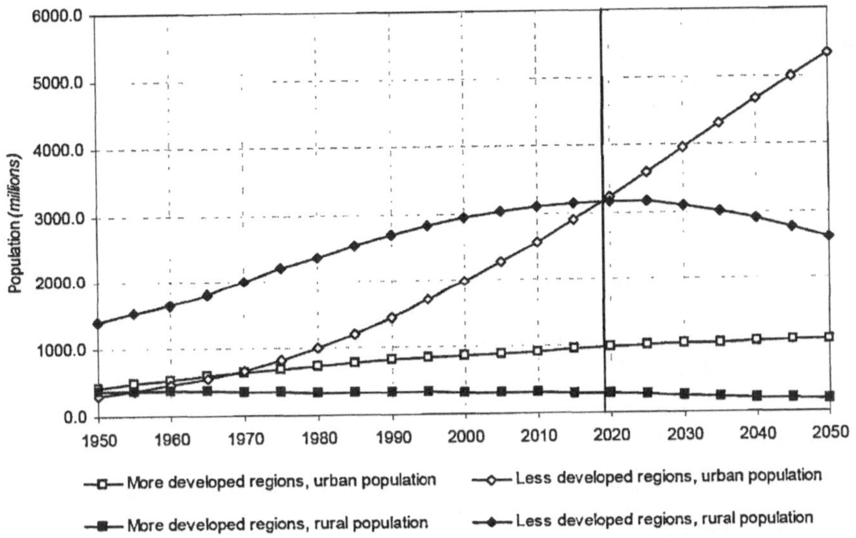

Figure 5. Evolution of the urban and rural populations in developed and undeveloped regions (UN, 2008).

these objectives in the face of dramatic urban growth and concomitant increases in water demand. Additional data on problems specific to the universal right of access to water and sanitation in developing countries are detailed in a recent report of United Nations (UNDP, 2006). However, this is a rather specific issue and, as such, it remains beyond the scope of the present work.

4 THE CHALLENGE OF SUSTAINABILITY

In the preceding sections it has been demonstrated why water policy in the twenty-first century must solve very different problems from those faced in the previous century and earlier times. In the past one hundred years society has witnessed profound demographic and technological changes that now require adaptation of water policies to the new context. However, in countries where the handling of water is intrinsic to the culture of city dwellers (Cabrera, 2008) these adaptations will be complex and difficult to implement. The swift response of northern European countries has already been seen, but on Mediterranean shores history weighs more heavily. Solutions that are suitable for the future are not usually single ones, but instead require significant investment and, more often than not, must reconcile competing interests, and therefore demand a considerable measure of imagination.

As things stand today, it is evident that water is not managed as it should be in any of its uses. And the urban arena, the focus of this work, is no exception. The greatest proof of this is the need to coin new terms such as "sustainability" or "environmental impact". Up to only a few decades ago, neither term was known nor required. They were first mentioned by the Brundtland Commission of the United Nations some two decades ago (Bruntland, 1987). Since then there has been great awareness of the existence of the unsustainability problem. However, the solutions that have been adopted are not sufficient because the proposed remedies do not properly account for the high growth rate of the problems.

As a consequence, a true sustainability policy remains elusive. In order to put one into practice we need to find the complex breakeven point in the three-dimensional space defined by the social, economic and environmental axes. This point is difficult to identify because these competing interests are almost always conflicting, thus it is not easy to find "win-win" type solutions.

It is, nonetheless, instructive to identify the main difficulties that exist on each of the axes in order to balance the water-city equation correctly in the three-dimensional space in which sustainability is situated. And irrespective of whether the axes can be ranked by importance (because they are all equally weighted) they allow us to structure the major problems that actually appear with a certain degree of logic. However, we should emphasize that all the difficulties to be overcome that are outlined in general are found in all the axes, although it is always possible to assign a specific problem to one of them.

And after that first classification, a second ranking can be established within each axis. We referred to the more or less universal nature of the problem. The clearest example is the millennium goals which, very likely, are the greatest problems which, within the water and the city framework, society faces at present; fortunately they only affect the developing countries.

Therefore, given all the reservations already set out, the list of problems can be ranked around the three axes that follow:

4.1 *Economic axis*

From a strictly economic standpoint, the main problems to be solved in this axe are:

- Infrastructure renovation.
- The eternal dilemma: public management versus private management.
- Achievement of the full cost recovery principle
- The millennium challenges

Each of these are commented on below, albeit succinctly.

4.1.1 *Infrastructure renovation*

Regardless of the extent of the infrastructures that growing cities may require, renovation of existing facilities poses a complex argument. Justification of new works to serve future needs is always much easier than convincing users that the infrastructure has a limited life and that, once this period comes to an end, it must be renewed. Further complicating the picture is the fact that much of the urban water infrastructure is buried underground, where it is out of sight and, consequently, out of mind of most consumers. The situation is so complex, or at least it was seen as so distant, that it is only in the most developed countries that the issue of infrastructure rehabilitation has started to be of concern in the last few years. Let us not forget, and this may be gleaned from section 2, that in modern urban water distribution and drainage systems many miles of mains are more than 100 years old.

Nevertheless all too often water bills do not include a charge for the inevitable progressive replacement of all these installations. Indeed the explanation for the huge differences in the unitary price of a cubic metre of water between countries with similar standards of living may be sought, for the most part, in the inclusion of renewal costs, although they are also due to the different quality provision standards and to the possible inclusion of environmental costs. In any case it does not seem reasonable that citizens of Berlin should pay 2.487 € per m^3 to purify their water (an amount to which must be added, see Table 1, the cost of rainwater drainage) while in Spain the total amount for sanitation (drainage and purification), according to the latest published data of the National Statistics Institute (INE, 2005), is up to eight times less (0.3 € per m^3).

This is a general problem which has been growing over the years. Current generations are using and exhausting an infrastructure built and bequeathed by previous generations. If provisions are not made soon to replace and restore the aging water infrastructure future generations will inherit a heavy mortgage. And we absolutely must insist on this point. To get an idea of the enormous magnitude of this looming problem note that in the United States the EPA has estimated that the investment required to renew the water supply (drainage and separate purification not included) over the next twenty years will be 276,800 million dollars. Two thirds of this total would be set aside for replacing mains (EPA, 2005). Further evidence of the need to renew these aging urban systems is reflected in the growing costs being paid to evaluate the current state and remaining life of the existing water infrastructure. This reconnaissance is a critical first step. Resources being limited, we need to know what to renew first, which requires assessing the state these infrastructures are in, the most valuable assets of these systems (Cabrera and Pardo, 2008).

4.1.2 *The eternal dilemma: public management versus private management*

The debate over the advantages and disadvantages of public or private management is intrinsic to urban supply and, doubtless, it is a dilemma that will never end. In fact, this duality arose with development of the modern water supply system. Indeed already in 1875 an impassioned debate between those in favour of both systems took place in the city of Birmingham, England. At that time the argument (still effective today) was made: "The quantity and quality of water to be supplied to the public, are matters of greater importance than mere profit and should be controlled and managed by representatives of the people and not by private speculators" (Thackray, 1990). Those who defended public management won the argument. Things have changed very much since then. Nowadays in England and Wales public water management, simply, does not exist.

Today, as then, the debate is still current. Therefore the attention that the subject continues to deserve proves the point (Hukka and Katko, 2003; WSTB, 2002; Wolf and Hallstein, 2005; Hall, 2008) and recently (June 2008) the web page remunicipalization, with ample information on the matter, (www.remunicipalisation.org) has just appeared. And so while time passes, the issues remain the same as highlighted with the following argument: Water privatization has spread rapidly throughout the world over the last decade, particularly in the South. But the tide now seems to be turning. Increased tariffs and a failure to deliver promised improvements, have left water multinationals facing increasing opposition. In any case such a lively debate is an unequivocal sign of the complexity of the subject. The page analyzes many highly topical cases, even one of most unexpected, that of the city of Paris. We are speaking about the capital of the country where the private urban water management borne. Its City Council has recently decided not to renew contracts with the private sector.

4.1.3 *Achievement of the full cost recovery principle*

The importance of the principle of cost recovery has been emphasized in the introduction to this article. Consequently it is not necessary to return to this question except to remind the reader that cost recovery tends to promote economic efficiency while achieving high standards of quality Nonetheless, we must simply highlight one curiosity. In Europe, and contrary to what one might initially think, those who best recover costs are the northern countries' public companies (remember the case of Copenhagen). Amongst other reasons, a likely explanation is that increases in tariffs arouse fewer suspicions among the general public.

4.1.4 *The millennium challenges*

Much ink has been spilt, and quite naturally so, on emphasizing the basic right of all humans to have access to water and sanitation, a right included in the Millennium Development Goals approved in September 2000 by the plenary session of the United Nations. The ambitious Millennium Goals are to be applauded for recognizing the basic human need for access to water and sanitation and for setting a strict timetable of 2015. However, worldwide monitoring by international organizations (WHO, 2005) and participating governments (GG, 2006) show variable progress toward achieving the Millennium Goals. It seems clear that in many countries these goals are not going to be achieved, particularly in some Ibero-American ones (RAC, 2007). Even now when we are over half way into the period between the launching of the Millennium Goals and the established horizon, it seems reasonable to ask if, in spite of the efforts that have been made, is the distance to be travelled getting shorter, staying the same, or growing longer? The answer is required because within our context of water and the city the reliable provision of safe water is, beyond a shadow of a doubt, the most important of all the problems confronting humanity in the twenty-first century.

4.2 *Social axis*

From an essentially social standpoint, the main problems facing water and the city depend on geographical area. The influence of geography is so profound and complex, it is nearly impossible to attribute a universal character to any of the seven problems listed below. They are:

- The weight of history
- The need to adapt management and control structures. Service regulation.
- The subsidy culture.
- Income inequality
- Deficient rating
- Conflicts of interest
- Qualification improvement.

These problems affect each country in a very different way as can be seen from the brief description for each that follows.

4.2.1 *The weight of history*

The influence of past precedent on current thinking cannot be underestimated. This problem has already been analyzed in the second point of the present article where it was shown that the weight of history can be a significant burden, especially in countries where the water problem is seen more from the point of view of quantity than quality. These countries tend to be plagued with problematic management structures poorly adapted to handle contemporary difficulties. In addition, their water management institutions are generally fragmented. In these conditions integrated water management is very complex if not futile. Historical private ownership water rights also weigh very negatively and, in short, many laws were drafted in a very different context from the present one. All these are important obstacles when adapting water policy at the present moment. Countries with a shorter historical background in water matters have hardly any inertia which may allow them to introduce the necessary policy changes without too many difficulties.

4.2.2 *The need to adapt management and control structures. Service regulation*

Among the series of problems included under the preceding general heading, the weight of history, probably the most important of them is the unsuitability of management and institutions for solving today's urban water challenges. Considering the degree of coordination required among many diverse constituents and competing interests, it seems that a single centralized administration for urban water management offers the most efficient and effective approach to implement and enforce the necessary water policy measures. For this reason the need for an overarching institutional control is specifically highlighted. At present it is common to find countries with fragmented competences, and that is the worst possible scenario for some of these administrations to assume responsibility for change which is always complex.

Fortunately, some countries are reacting with the creation of regulatory agencies for water that bring together the hitherto dispersed competencies. Britain did so two decades ago and more recently other countries, such as Australia or Portugal, have followed suit. In others in which the water companies themselves are already efficient and are coordinated amongst themselves (as in the cases of Holland or Germany) this action does not seem necessary. However, regardless of how it is actually structured, there is no doubt that a proper ordering of the sector is required.

4.2.3 *The subsidy culture*

As water is a vital resource, many states took advantage of the spectacular advances in civil engineering at the beginning of the twentieth century to develop large-scale hydraulic works and financed them with public money. In this way the user by paying a symbolic price would have water. However, when the service costs are not fully repaid, one loses sight of its value which, inevitably leads to irresponsible use. The WFD (EU, 2000), conscious of the fact that the subsidy culture does not promote efficiency, has included cost recovery among its main directives (See its article 9). It will be binding as of 2010.

Subsidies for water are inherent in countries with an agricultural tradition and for that reason their politicians are reticent to include all costs in the price of water. But 2010, just around the corner, signals the end of the social compensation funds which flowed to the majority of southern European countries, since they had the lowest incomes in the Union. Mediterranean countries used these funds to finance a good part of their urban water infrastructures. The loss of this subsidy will translate into a substantial increase in the price of water and, although unpopular, this economic correction will promote sustainable water management in the Mediterranean countries.

A last reflection in this respect is required. The proper handling of water has a cost which, if society takes on pollution, is not in question. Therefore the object of the debate is how to distribute the costs. If these costs continue to be partially subsidized (and, for example, the State pays them indirectly through general taxation) those who use water rationally are partly financing the costs of those who squander it. This is wholly unacceptable. Perhaps only irrigation water, for reasons which are beyond this discussion, should be subsidized, but in a way that encourages saving. At the moment, traditional water rights of the irrigation bloc are related to surface area cultivated and this provides no incentive to conserve. In any case this is a question that is beyond the scope of this work.

4.2.4 *Income inequality*

In the developing countries, unequal distribution of wealth can adversely impact water policy. For example, citizens on very low incomes must hand over a disproportionately high percentage of their income (say over 5%) to pay that part of the water costs which corresponds to them. Such an exorbitant fee could upset social stability. This fact, which is beyond the realm of water policy, is one of the most complex factors to overcome. Only with a very carefully graduated social rating, and which we shall deal with below, can this problem be alleviated.

4.2.5 *Deficient rating*

The principle of cost recovery is simple: the total expenditure that the management system entails, must balance with the payments made by users. But it does not specify how the costs should be distributed. And so, in the same way as a country balances expenditure and income while the exchequer distributes the charges by taking into account citizens' income, a scheme must be devised to allocate the total costs of the system fairly among all users and such that it favours saving. The solution that has been adopted is a pricing structure using progressive blocks of consumption so that as unitary demands increase, the price per cubic metre rises significantly. Under such a plan the first block of consumption (let us say 8 m^3per month for a family of four) can have a very low cost, but above a certain level of consumption the price to the user, regardless of income, of each cubic metre must rise significantly.

4.2.6 *Conflicts of interest*

With urban growth comes increasing demand for water on the part of cities. Ideally, this additional demand could be met by developing new water supplies. However, when it is not possible to develop new water supplies to meet growing municipal demand, a common practice is to divert water from one use to another. Here, for instance, it is not unusual to transfer water traditionally used in agricultural production to urban applications. Switching water to uses with considerably higher economic capacity generates very unfair competition. The case of Mexico City is well known. Its growing water demand has had to be satisfied with contributions from resources which are very distant from the city, to the detriment of those who have always used it. These situations generate social problems whose solution is very complex but which, at least, forces us morally to use water in the most rational way possible so that, to a certain degree, we compensate those who have had to relinquish the use of this resource. The importance of this subject explains the attention several authors have dedicated to it (Molle and Berkoff, 2006).

However, the conflicts of interests appear in frameworks that go beyond strictly urban boundaries. For example in the USA the water disputes between the native Indians and the new users are well known (Galoway, 2005). Sometimes the problems that arise are not so much of quantity but of quality. And thus, for example, the pollution of water tables generates externalities (Rogers et al., 1998) which companies supplying drinking water must assume. It is not surprising, then, that initiatives (Heinz, 2004) are contemplated that establish compensations between cities and farmers. For example, some farmers commit to not using polluting fertilizers for which the municipality offers them an economic compensation for their loss of agricultural production. It is, therefore, a win-win type of agreement, a clear example of the advantage of implementing global approaches.

4.2.7 *Qualification improvement*

In contradistinction to the twentieth century where water policy essentially co-evolved with the technical capabilities of civil engineering, water policy of the twenty-first century will require skills from a more interdisciplinary standpoint, particularly from the fields of management and sustainability. It also demands learning from experience, especially past mistakes (Frederiksen, 2007). It seems, then, necessary to adapt university curricula to the new knowledge needs, a matter pending in countries which, like Spain, have seen a remarkable boom in great hydraulic works and, consequently, education related to water have been orientated essentially towards civil engineering. As a result of this orientation, very necessary at the time, present programmes of study do not usually include content, such as water economy, that today is necessary for its sustainable handling.

4.3 *Environmental axis*

Within the three-dimensional space in which sustainable water management must find its breakeven point, the environmental axis is the most recent. Indeed it has come *hand in hand* with progress which at the outset it was unaware of, it did not know the majority of its secondary effects. For this reason, and opposite to what happened with the previous social axis, almost all the problems that from this standpoint are visible have a marked universal character, albeit their gravity is, depending on place, quite variable.

The set of the main environmental problems are:

- Water pollution
- Urban floods
- Water–land nexus
- Water, energy and climate trilogy in an urban framework
- The exhaustion of resources (seen, essentially, from the standpoint of quantity)
- Local problems

In what follows we shall discuss, albeit briefly, each of them in turn.

4.3.1 *Water pollution*

Any use of water entails, to a greater or lesser extent, a degradation of its quality. It so happens, however, that the capacity of self-purification by Nature was generally sufficient until the arrival in the twentieth century of large-scale industrial and economic development. Man's intensive use of water degraded it beyond its capacity to self-regenerate. Irrigated land demonstrates this, the oldest use after that by humans. When, in order to increase productivity, in the earlier second half of the last century farmers resorted to fertilizers and chemical pesticides, underground water began to become polluted. In some cases, the unintentional pollution to the water table (due to excessive levels of nitrates) is so severe that today many aquifers are no longer able to provide urban areas with the drinking water they had always supplied.

But water pollution is not just the consequence of the use of water. Rapid urbanization in detriment of former rural lands presents serious water pollution effects as well. Consider urban stormwater runoff. During a rain event, stormwater flows from paved streets and cascades through engineering drainage systems directly to receiving surface waters. The first rainwater that arrives at the culverts drags with it a good fraction of the pollutants existing on the land surface, including fuels that urban vehicles use and fertilizers and pesticides applied to urban landscapes. It is now well recognized that stormwater runoff contains a very high polluting load, with BOD, COD and TSS concentrations comparable to domestic sewage. This is a major problem to which most modern cities devote considerable attention (Schreder and Pawlowsky-Reusing, 2004); the importance of the subject justifies it. This new contamination is in addition to the already existing one, that of urban effluents.

Water pollution arouses deep social alarm and, consequently, is a key element of the engine that drives society to advance along the complex road that leads to sustainability. On the other hand, the cost of solving water pollution problems acts as a kind of economic brake. So along the uncharted path to sustainability, there is a delicate balance between desire to be responsible and the willingness to pay. For this reason the environmental axis also underlines the importance of the full cost recovery principle. This trial of strength between environmental and economic considerations demonstrates the need to look for a break-even point. Of course, pollution is a problem that does not admit half measures. In order to prevent pollution, the necessary economic resources need to be put in place. Otherwise the natural environment will be exhausted or ruined thus jeopardizing the availability of essential resources for future generations.

4.3.2 *Urban floods*

To some extent all building developments involve waterproofing the land and/or infringing on floodplains. This, from the standpoint of the hydrological cycle, has two negative effects. In the first place, impervious buildings reduce infiltration and, with it, the replenishment of water tables. Non-infiltrated water runs along paved streets until it finds its natural or artificial course. Secondly the proliferation of paved areas leads to a reduction in the time it takes a drop of water to enter the sewer (calculated from the moment it hits the ground until it enters the artificial drainage system), a parameter known as time of concentration. Both effects are common in urban areas and when they overlap, often produce dramatic increases in the runoff response (peak flows and volumes) for an urban catchment compared to the virgin pre-development conditions.

Waterproofing of land, which sometimes reaches areas next to the natural courses of rivers prone to periodic flooding, is the main reason for ever more frequent urban flooding. This has become a widespread problem and one whose solution is very complex especially when it has not been anticipated. The construction of large rainwater retention tanks and ponds, an extended version of individual domestic cisterns to which we have already referred, as well as the use of various compensatory techniques (such as porous materials for pavements or decentralized rain gardens) are examples of best management practices which appear promising as effective mitigation measures for urban flooding.

The fact that urban flooding is becoming ever more frequent and the economic (and sometimes human) damage that it causes is enormous, has attracted the attention of municipal authorities (NYS, 2003), researchers (Loucks et al., 2006) and academia (Gómez, 2006) who are developing strategies to better manage stormwater and, hence, reduce economic damages in urban settings.

4.3.3 *Water–land nexus*

Land management directly influences water policy and vice versa (Cabezas et al., 2008) because a land-use decision is also a water decision (Falkenmark and Rockström, 2006). Indeed, the use assigned to land directly determines its water needs and for that reason land planning must be integrated into water policy. This is an essential issue which affects the water and city nexus very directly and which, because of the numerous conflicting interests that exist in this respect, hinders many of the activities of the decision-makers. And of course, if the problem is not tackled, over the years its complexity grows.

The interaction between water and land must be considered both in terms of quantity and quality. Recent works (Goonetilleke A. et al., 2005) have already proposed guidelines for quantifying the water–land relationship. And one last question which against this background must not be overlooked is jurisdiction which, when water use is changed, usually arises. Urban growth usually comes at the expense of rural areas. Cities are gaining ground on farmland and this it is an issue that must be assessed not only in the short term as usually happens. The problem must be approached from the widest possible context (Molle and Berkoff, 2006).

Without a shadow of a doubt, due to its enormous complexity and to the various sensitivities that it arouses, the land-water nexus is one of the greatest challenges that Man must resolve as soon as possible because time, as has already been mentioned, is against it.

4.3.4 *The water, energy and climate change triangle*

The current global energy crisis is another chapter in the continuing saga of an insatiable demand for oil combined with limited refinery production and growing awareness of finite petroleum reserves. While high energy costs pose an obvious economic hardship, they also encourage creative efforts to improve operational efficiencies and conserve natural resources. This is very important because the cost of energy associated with sustainable water use is significant. Indeed, each stage of the urban water cycle (from its catchment to its final discharge, including the purification required to restore its quality to what it was when caught) involves a high energy cost. Sometimes this charge is overshadowed by both the huge investment that hydraulic works require and because the energy cost is diluted among a very large number of domestic and industrial users. Of course, the sum of many small terms can add up to an impressive total. A recent study in the State of California (CEC, 2005) estimates that the energy cost relating to water is 19 percent of all electricity consumption and 32 percent of all gas use. In Spain, with no official data on the matter, an initial estimate indicates (Cabrera et al., 2008) that the water industry may consume about 10 percent of the nation's energy.

The solution to the problem lies in managing water in the most efficient way possible and promoting conservation. Because when natural environment withdrawals decrease, in addition to favouring biodiversity (there is more water in the natural environment) energy consumption is significantly reduced. In fact it is very simple to quantify, in energy terms, the benefit of a more efficient use of water. On the one hand the oversizing of installations (mains, water treatment plants, etc.) is avoided and with it the associated energy cost linked to all major civil engineering works. On the other hand, and much more importantly, energy is saved on a daily usage basis. In principle, this energy savings can be quantified across the urban water cycle. The amount of energy saved will depend on the point in the overall cycle in which water usage is optimized. In fact, in energy terms, repairing a leaking water main does not have the same repercussion as installing a low capacity toilet cistern; nor is saving water from a spring the same as saving it from a desalination plant.

The water-energy nexus as set out here is a relatively new approach. Traditionally, water managers have faced energy savings from the perspective of process improvement. For example, process and energy breakthroughs in the last few years in desalination have been spectacular. However, energy savings obtained from a more rational use of water have hitherto been ignored. This promising new avenue has tremendous potential as can be seen from a recent study (McMahon et al., 2006) which demonstrates that the best cost-benefit ratio (in Kw-hr saved) that can be obtained in a domestic energy saving programme is obtained by using water-saving devices.

Finally, the water-energy nexus has a direct connection to climate change. It is not difficult to estimate reductions in greenhouse gasses expected from various water conservation practices. Reducing greenhouse gas emissions, of course, is a primary strategy in the fight against global warming. It is only necessary to know the energy cycle that the saved water goes through and the origin of the energy used in handling it. Recent works by the NRDC and the PI (NRDC and PI, 2004) have designed spreadsheets (PI, 2004) to facilitate these calculations for both urban and agricultural applications.

4.3.5 *The exhaustion of resources (seen, essentially, from the standpoint of quantity)*

The growth of many cities in regions with a traditional shortage of water resources forces engineers and planners to resort to alternative water supplies, either as transfers from other river basins, or if urban sprawl is coastal, as desalination of seawater. The final decision on which is the solution that best solves the problem is, once again, the one that diminishes the set of environmental, social and economic costs. While such an analysis is never simple, the inherent complexity increases because political interests often remove it from the rationality that the case requires. We have a good example in Spain. In this first decade of the twenty-first century, discussion has concentrated, with no intermediate positions, on deciding which new resource, (transfer or desalination) is most suitable. This is an absurd debate that only makes the search for the best solution in each case more difficult (Cabezas et al., 2008).

5 THE WAY FORWARD

All over the world the challenges with respect to water policy that society must confront in the twenty-first century are enormous. A recent study carried out in the USA demonstrates this (NAS, 2004). Before highlighting some guidelines for sustainable water management, two key considerations should be noted. Firstly, there is no magic remedy to solve all the problems. Secondly, the solutions to be adopted must be appropriate for both the framework in which they are found and for the initial circumstances. With these two guiding premises here are some recommendations to nudge progress along the road to more sustainable urban water management:

- *Remove the world of water from the political arena: water is a question of State.* Water is a vital resource and, besides, it is essential for any life form on the planet. And, moreover, future generations need it. For these reasons water cannot be, as all too frequently happens in countries whose atavistic character is all too present, at the centre of political debate. Water is simply a question of State.
- *Adapt the decision-making structures to current problems.* When managing water efficiently and sustainably, one of the major difficulties to overcome is the frequent fragmentation of competencies. The ultimate responsibility for supplying water to the public is usually that of municipalities, whereas management of this resource normally depends on a higher authority. As integrated water management becomes more necessary, the limitations and disadvantages of fragmented management structures are more apparent. The advisability of integrating water and land management policies only increases the evidence of the need to centralize both responsibilities in spite of the evident difficulties that its implementation entails. But without a doubt this is a capital issue when pressing ahead steadily along the road of sustainability.
- *Seek agreed solutions, the only valid ones in the long term.* Public participation and, in short, governance, are subjects which are talked about a lot nowadays. It is another way of transferring and, because of its novelty, is in the throes of the processes of running-in and of learning. The European Union project, *Harmonicop* under the motto *"Learning together to manage together – Improving participation in water management"* has been one on the first attempts to implement this novel and advisable way of going forward (www.harmonicop.info).
- *Educate and make the public environmentally aware.* Both water pollution and the deterioration of the natural environment at large scale are relatively recent, just some decades, problems. A large section of the public are not yet aware of it, or perhaps they have not understood that to preserve the environment has an economic cost which we all must take on. For this reason it is necessary to explain clearly what is at stake, thus counteracting the enormous weight of history. In this item we are not speaking about old drinking water quality problems in cities that generate epidemics, as the cholera in London in 1854, and that modern water supply systems have definitively solved.
- *Implement the cost recovery principle, by establishing a suitable rating system* fostering efficient use of water, as has been demonstrated throughout this work, is an essential means of action.
- *Foster transparent management.* The public is very sensitive to the price it must pay for its water service. Rates cannot be increased without a clear justification of the reasons for these rises. Consequently, transparent management is an essential issue.
- *Setting out clear rules of the game, especially as far as control of private operators are concerned.* The way to resolve this issue is by creating regulatory agencies to which mention has already been made.

- *Exchange experiences*. The majority of current problems are very complex because the solutions do not satisfy all the parties involved to the same extent. And as human behaviour is extremely unforeseeable and the most complex to model, difficulties when finding a satisfactory solution are manifold. For this reason, recognizing that the circumstances of each case can be very diverse, learning by analyzing the experiences and decisions that those who have previously faced similar problems and those who best manage water, will always be a wise strategy.
- *Boost economies of scale*, wherever possible. Because there are many small and medium-sized municipalities in which, from a strictly economic point of view, it is much more complex to manage water efficiently. For this reason to set up associations is, from the economic point of view, a sensible and laudable decision. In any event this is not an easy task in many cases. The Italian experience bears witness to this. Because the Galli Act that Italy promulgated in 1994 to this regard, to that end has produced mixed results (Citroni et al., 2007). In that sense, before any initiative is adopted in that direction, we need to strengthen, by learning from what has happened and bearing in mind the particular circumstances in each case, what in Italy has been successful and, at the same time, try to diminish the impact of the difficulties encountered there.
- *Develop and apply new technologies*, to optimize investment and to apply the most robust solutions. To do this it is necessary to undertake a cost benefit analysis of the various alternatives, as well as to be aware of the different options available for solving specific problems. From this objective comparison a final solution should emerge.

6 CONCLUSIONS

The passage of time demonstrates the growing importance of managing this most precious natural resource, water, in a sustainable way. The challenge, as much due to the demographic explosion over the last few decades as to the dramatic migration of the rural population to urban centres, acquires a special dimension with a solemn urgency within the framework of water and the city. This is especially pronounced in the large cities of developing countries. The greatest challenge is the need to introduce significant changes into many of the facets of water management, a sphere with a huge inertia and for this reason very resistant to change. Consequently the tendency to a laisser faire attitude only aggravates the problem, given that one enters a vicious circle from which is very difficult to break free. Only when the majority of society is totally aware of the need to act without delay, will politicians begin to introduce the necessary changes to promote adaptive urban water management. And, in order to do so, an educated public which backs them resolutely is essential. That is the great challenge confronting society today if it wants to show the slightest degree of solidarity with future generations.

REFERENCES

Blackman, D.R. & Hodge, A.T. (2004) *Frontinus' Legacy. Essays on Frontinus' de aquis urbis Romae*. Ann Arbor, Michigan.

Bonnnin, J. (1984) *L'eau dans l'antiqueté. L'hydraulique avant notre ère.* Collection de la Direction des Etudes et Recherches d'Electricite de France. Paris, Eyrolles.

Bruntland, G. (ed.) (1987) *Our Common Future: The World Commission on Environment and Development.* Oxford, Oxford University Press.

BUNR (Bundesminiterium für Umwelt, Naturschutz und Reaktorsicherheit) (2001) *Water Resources Management in Germany.* Bonn, Germany, Federal Ministry for the Environment.

Cabezas, F., Cabrera, E. & Morell, I. (2008) *El agua, cuestión de Estado. Perspectiva desde la Comunidad Valenciana.* AVE (Asociación Valenciana de Empresarios). Valencia junio de 2008.

Cabrera, E. (2008) *El agua en España.* Madrid, Ministerio de Medio Ambiente.

Cabrera, E., Cobacho, R., Espert, V. & García-Serra, J. (2008) Agua y Energía: Una relación que conviene comprender. *Congreso Latinoamericano de la IAHR. Cartagena de Indias.* Septiembre 2008.

Cabrera, E. Jr. & Pardo, M.A. (2008) *Performance Assessment of Urban Infraestructure Services.* London, IWA Publishing.

CEC (California Energy Comisión) (2005) *California's Water-Energy Relationship*. Final staff report. CEC 700 – 2005 – 011 SF. California Energy Comisión. State of California. November 2005.
Citroni, G., Gianelli, N., Lippi, A. & Profeti, S. (2007) Chi governa l'aqua?: Regolazione, potere locale e arene della rappresentanza nella governance del servizio hídrico integrato. *Convegno SISP 2007*, Catania, Sicily, 20–22 settembre.
Cooley, H., Christian-Smith, J. & Gleick, P. (2008) *More with less: Agricultural water conservation and efficiency in California*. Oakland, California, Pacific Institute. September 2008.
Evans, H.B. (2000) *Water Distribution in Ancient Rome: The Evidence of Frontinus*. Michigan, The University of Michigan Press.
EPA (Environmental Protection Agency) (2005) Drinking Water Infrastructure Needs Survey and Assessment. Third Report to Congress. US Environmental Protection Agency, Office of Water, Washington, DC 20460, June 2005.
EU (European Union) (2000) Directiva 2000/60/CE del Parlamento Europeo y del Consejo de 23 de Octubre de 2000. Diario Oficial de las Comunidades Europeas, de 22.12.2000.
Falkenmark, M. & Rockström, J. (2006) *The New Blue and Green Water Paradigm: Breaking new Ground for Water Resources Planning and Management. Journal of Water Resources Planning and Management*, May–June, 2006.
Fewkes, A. (2006) The technology, design and utility of rainwater catchment systems. In: Butler, D. & Fayaz, F.A. (eds) *Water Demand Management*. IWA Publishing.
Frederiksen, H.D. (2007) *Water resources management: Stewardship and services. Journal of Water Resources Planning and Management*, January–February, 2007.
Galoway, G. (2005) 2004 Julian Hinds Water Resources Development Award Lecture. *Journal of Water Resources Planning and Management*, July–August, 2005.
GG (Gobierno de Guatemala) (2006) *Hacia el cumplimiento de los Objetivos de Desarrollo del Milenio en Guatemala*. Gobierno de Guatemala. Secretaría de Planificación y Programación de la Presidencia. Ciudad de Guatemala. Julio 2006.
Gómez, M. (2006) *Curso de Hidrología Urbana, 6a edición*.Grupo Flumen. Universidad Politécnica de Cataluña. Barcelona 2006.
Goonetilleke, A., Thomas, E., Ginn, E. & Gilber, D., 2005. Understanding the role of land use in urban stormwater quality management. *Journal of Environmental Management* 74 (1), 31–42.
Grigg, N.S. (1986) *Urban Water Infraestructure. Planning, Management and Operations*. New York, John Wiley & Sons.
Hall, D. (2008) Water privatisation. European Union. Europe's water. Hotel Metropole, Brussels. February 2008.
Heinz, I. (2004) Sustainable Farming as a result of negotiations: An analysis at European level. *2004 GIGR International Conference*. Beijing, October 2004.
Hukka, J.J. & Katko, T.S. (2003) *Water Privatisation Revisited. Panacea or Pankake?*. Delft, The Netherlands, IRC International Water and Sanitation Centre.
INE (Instituto Nacional de Estadística) (2005) *Encuestas del agua*. Madrid, Available from: www.ine.es.
Loucks, D.P., Stedinger, J.R. & Stakhiv, E.Z. (2006) Indiviudal and Societal Responses to Natural Hazards. *Journal of Water Resources Planning and Management*, September–October, 2006.
McMahon, J.E., Whitehead, C.D. & Biermayer, P. (2006) *Saving Water Saves Energy*. Berkeley, California, Lawrence Berkeley National Laboratory.
Merkel, W. (2003) El Futuro de la Industria de Agua en el mundo. *Ingeniería del Agua*, 10(3), 337–353.
Molle, F. & Berkoff, J. (2006) *Cities versus Agriculture: Revisiting Intersectoral Water Transfers, Potential Gains and Conflicts*. International Water Management Institute, Report 10. Colombo, Sri Lanka.
Napstjert, L. (2002) *Water savings in Copenhagen*. Watersave Network Fourth Meeting, Loughborough University, UK.
NAS (National Academy of Sciences) (2004) Confronting the Nation's Water Problems. Washington, The National Academies Press.
NRDC and PI (National Resources Defense Council and Pacific Institute) (2004) *Energy Down the Drain. The Hidden Costs of California's Water Supply*. Agosto, National Resources Defense Council.
NYS (New York State) (2003) *Stormwater Management Design Manual*. Department of Environmental Conservation, New York State.
Paz, J. & Paz, J.M. (1969)*Abastecimiento y Depuración de Agua Potable*. Madrid, Publicaciones de la ETSICCP.
PI (Pacific Institute) (2004) *User Manual for the Pacific Institute Water to Air Models*. Oakland, California, Pacific Institute.

RAC (Red Agua y Ciudad) (2007) *Agua y ciudad en Iberoamérica. Una valoración.* Valencia, España, Red CYTED Agua y Ciudad.

Rogers, P., Bathia, R. & Huber, A. (1998) *Water as a Social and Economic Good. How to Put the Principle into Practice.* Stockholm, Global Water Partnership.

Schreder K. & Pawlowsky-Reusing, E. (2004) Integrated sewage management to reduce pollution load in Berlin. *W&W International Journal*, October 2004.

Steel, E.W. (1972) *Abastecimiento y Saneamiento urbano.* Barcelona, Gustavo Gili.

Thackray, J.E. (1990) *Privatization of Water Services in the United Kingdom.* Urban Water Infrastructure, K.E. Schilling & E. Porter. Kluwer Academic Publishers, pp. 33–42.

Thirriot, C. (1987) *Pouvoir politique et recherche hydraulique en France aux XVII et XVIII siècles. Hydraulics and Hydraulic Research. An Historical Review.* Holanda, Balkema.

UN (United Nations) (2008) *World Urbanization Prospects. The 2007 Revision.* New York, United Nations.

UNDP (United Nations Development Programme) (2006) *Human Development Report. Beyond Scarcity: Power, Poverty and the Global Water Crisis.* New York, United Nations.

WHO (World Health Organization and Unicef) (2005) *Water for Life. Making it Happen.* Geneva, Switzerland.

Wolf, G. & Hallstein, E. (2005) *Beyond Privatization. Restructure Water Systems to Improve Performance.* Oakland, California, Pacific Institute.

WSTB (Water Science and Technology Board of the National Academy of Sciences) (2002) *Privatization of Water Services in the United States: An Assessment of Issues and Experience.* Washington, DC, National Academy Press.

Viollet, P.L. (2000) *Hydraulique dans les Civilisations Anciennes.* Paris, Presses de l'Ecole National des Ponts et Chaussés.

CHAPTER 10

European water research: From past to future trends

Avelino González
European Commission, Research Directorate General. Brussels, Belgium

ABSTRACT: Water research is promoted and financially supported by the European Communities since the beginning of the successive Research and Technological Development (RTD) Framework Programmes (FPs). Traditionally water is one of the core aspects addressed by the Environmental research programme, but not exclusively. Other EU research programmes such as Industrial Technologies, International cooperation or specific SMEs related activities are also addressing complementary aspects of water research. The description of the main subjects addressed is complemented with the identification of major drivers justifying a coordinated action at European level including the appropriate allocation of resources. Moreover an overview of the related EU legislation and initiatives of interest is presented. The historical continuity of European environmental research through the different FPs is completed with a description of the on going Seventh Framework Programme (FP7 2006–2013) including the structure, budget and components.

1 INTRODUCTION TO EU WATER RESEARCH

The water sector is almost unique in the sense that it combines aspects related to environment, economy, industry, welfare and health. Sustainable management of fresh water RESOURCES is a keystone priority that embraces preservation, protection and sustainable use. Water is certainly more than a product, good or service, discussions about the right for water culminated at the last World Water Forum in Mexico with a EU statement considering that "water is a primary human need and that water supply and sanitation are basic social services".

The sector is confronted to diverse needs and pressures derived from their use in industry, food production, energy, transport, and human consumption or more simply, but not negligible, in the environment. The intrinsic transboundary character and the evidence of the global impact on the water cycle is now well recognised, justifying coordinated activities at all levels including European Union level.

It is estimated that the European Commission's budget for supporting research represents the equivalent of about 6% of the overall research investment in Europe at national, public or private, level. However the influential impact of the EU research policy, in terms of both priorities and funding mechanisms, on national programmes is notably relevant. This leading role is a major consideration when designing research policies and programmes by the European Commission.

2 MAJOR SUBJECTS FOR WATER RESEARCH

Water research is a relatively large area that includes a variety of aspects, actors and potential beneficiaries. Research and technological development in this area necessarily includes approaches aiming to increase basic knowledge and/or contributing to the development and application of innovative tools, methodologies and technologies.

The detailed structure of the FP that reflects possible approaches is depending on the chosen strategy. However, and independently of the mentioned structure, the relevant subjects for research can be grouped in few categories that are usually present in a way or in another, in water research projects:

- Water quality/quantity (resources, ecosystems, impacts of climate change . . .)
- Water management (IWRM, modelling, DSS, scenarios development . . .)
- Water technologies (treatment, distribution/collection, pollution prevention, monitoring and control . . .)
- Water policies and socio-economic aspects (needs/impacts of regulation, health . . .)
- Water-related extreme events (floods, droughts . . .)

Soil, or sediments, and water relationship as well as functioning and interacting mechanisms are also usually considered in the subjects for water research.

Sector's based approaches are present in relevant thematic research programmes such as food production (water and agriculture), industrial technologies (water and industry) and energy (water and energy production).

European research is frequently about facilitating and consolidating cooperation among multi-disciplinary teams. Therefore approaches responding to specific requirements, including users or consumers' needs, can nowadays integrate some of the mentioned subjects within the same project.

3 DRIVERS

When addressing issues linked to research policy it is well recognised that setting priorities is a real challenge that shall consider what are the major drivers for research, including its impacts and added value.

Legislation, and particularly environmental regulation, is frequently mentioned as a major driver for research. This may be discussed but a clear distinction should be made between environmental research policy and environmental policy. Notably in terms of time frame. Ideally research results may lead to the development of new or amended pieces of legislation, and consequently, when a policy is drafted, the required tools and methodologies should be available for its implementation. Policy implementation can be considered as a driver but more for the application of research results rather than for the development of a research policy. Moreover environmental policies are designed in view to preserve and protect our environment and to promote a sustainable development, in other words legislation is made for the society. Therefore the major driver for research is the civil society.

To analyse rapidly society wishes some figures are needed. A large survey (Special Eurobarometer 224, 2005) has been conducted to assess European citizens' perception, information and attitude towards scientific and technological issues (Fig. 1). European society has demonstrated highest concern on environmental issues, the results of the mentioned survey show that citizens are most interested in new themes on environmental pollution (38%). Surprisingly this subject is ranked higher than 'new medical discoveries' or 'new inventions and technologies'. About 57% of Europeans perceived science and technology as responsible for most of the environmental problems the world is facing today, but 50% agreed that S&T can play a positive role in improving

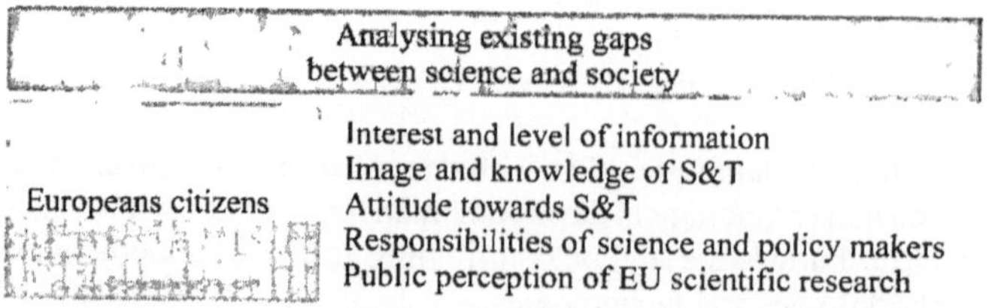

Figure 1. Survey conducted to assess European citizens' perception, information and attitude towards scientific and technological issues.

Key figures
20% of EU surface water threatened with pollution
65% of EU drinking water is supplied by groundwater
60% of EU cities overexploit their groundwater resources
50% of wetlands are endangered by overexploitation
20% increase of irrigated land since 1985
18% of EU population is suffering water stress

Figure 2. Some key figures on EU waters.

the environment. Close to 50% of respondents confirmed their higher interest in environmental science and technology development.

Another European survey was conducted recently to assess the attitudes of citizens towards environment (Special Eurobarometer 217, 2005). Among main environmental issues, including climate change, man made disasters, air pollution, impact of chemicals on health, natural hazards, waste..., 47% of Europeans are most worried about water pollution (seas, rivers, lakes, groundwater, etc).

Both surveys evidenced that our society is extremely concerned by environmental aspects and particularly by water pollution.

Since the signature of the Amsterdam Treaty the European Research policy and programmes' implementation is a legal obligation. Research is to date very high in the political debate and recognised essential in contributing to economic growth and employment. The Lisbon Strategy, which is aiming to make the EU "the most competitive and dynamic knowledge-based economy in the world, capable of sustainable economic growth with more and better jobs and greater social cohesion", is strongly supporting environmental RTD actions.

As part of its mission European Research must also support other Union's policies such as environment, energy, health, agriculture, regional development etc. Water policy is certainly one of the largest pieces of EU environmental legislation. Many factors are contributing to this situation and particularly:

- international character
- chemical and physical properties
- human health
- key factor for development
- economic sector
- natural resource

Certain aspects related to the overall water sector are also ruled by EU legislation regulating the internal market and competition or addressing health and consumers protection. Other linked issues are related to policies covering areas such agriculture, fishery, energy, industry, ...

3.1 *Brief outline of EU water legislation*

The intention is to present briefly the current situation in terms of EU water policy. More detailed information about the legislation or on-going related activities is available from the web[1].

The *EU Water Framework Directive* (WFD)-(2000/60/EC) is aiming to protect all waters–rivers, lakes, coastal and groundwater, focussing in integrated river basin management and based on ecological status of water bodies. The main implementation coming deadlines are the following:

- 2006 Establishment of monitoring networks. Start public consultation (Art. 8 & 14)
- 2008 Present draft river basin management plan (Art. 13)

[1] Information about environmental EU policies is available at http://ec.europa.eu/environment/index_en.htm

- 2009 Finalise river basin management plan including programme of measures (Art. 13 & 11)
- 2010 Introduce pricing policies (Art. 9).
- 2012 Make operational programmes of measures (Art. 11).
- 2015 Meet environmental objectives (Art. 4).
- 2021 First management cycle ends (Art. 4& 13).
- 2027 Second management cycle ends, final deadline for meeting objectives (Art. 4 & 13).

EU WFD – Daughter Directives:

3.1.1 *List of Priority Hazardous Substances (COM(2001) 17 final)*

Addresses the hazardous substances that are of particular concern for the freshwater, coastal and marine environment. These substances will be subject to cessation or phasing out of discharges, emissions and losses within an appropriate timetable that shall not exceed 20 years.

3.1.2 *The new Groundwater Directive (2006/118/EC)*

Establishes a set of water quality standards as well as measures for preventing and limiting groundwater pollution that considers local situations. It complements the WFD in relation to the requirements for chemical status assessment of groundwater as well as provides means for identification and reversal of current trends in pollutants concentrations.

The first Directive's deadline requires establishing groundwater quality parameters by the end of 2008 beside carrying out studies on pollution trends.

3.1.3 *Urban waste water treatment directive (91/271/EEC)*

The objective of the Directive is to protect the environment from the adverse effects of discharges of urban waste water and of waste water from industrial sectors of agro-food industry.

Main deadlines for implementation (EU15) are presented in the Table 1.

3.1.4 *Drinking water Directive (98/83/EC)*

The Directive addresses the quality of water intended for human consumption. The last version contains a review of parametric values, and where necessary strengthen them in accordance with the latest available scientific knowledge. The point of use is the point of compliance with the quality standards (water quality at the tap) and it makes reference to ISO/CEN standards, including the obligation to report on quality and the obligation to inform the consumer on drinking water quality and new measures to be taken. The revision of the Directive was conducted in view to

Table 1. Urban waste water directive implementation deadlines.

(p.e.)	0–2000	2000–10.000	10.000–15.000	15.000–150.000	+150.000
Sensitive areas	if collection 31/12/2005 appropriate treatment	collection 31/12/2005 secondary* treatment	colletion 31/12/1998 more advanced treatment	colletion 31/12/1998 more advanced treatment	colletion 31/12/1998 more advanced treatment
Normal areas	if collection 31/12/2005 appropriate treatment	collection 31/12/2005 secondary* treatment	collection 31/12/2000 secondary treatment	collection 31/12/2000 secondary treatment	collection 31/12/2000 secondary treatment
Less sensitive areas (coastal waters)	if collection 31/12/2005 appropriate treatment	collection 31/12/2005 appropriate treatment	collection 31/12/2000 primary or secondary treatment	collection 31/12/2000 primary or secondary treatment	collection 31/12/2000 primary (exceptional) or secondary treatment

*appropriate treatment if discharge to coastal waters

streamlining the legislation to parameters essential for health and environment (48 parameters). The main changes in parametric values from 1980–1998 are:

- Lead: reduced from 50 μg/l to 10 μg/l.
- Copper: value reduced from 3 to 2 mg/l
- Values for individual and for total pesticides retained (0.1 μg/l/0.5 μg/l)
- More stringent ones introduced for certain pesticides (0.03 μg/l)
- Standards introduced for new parameters like trihalomethanes, trichloroethene and tetra-choloroethene, bromate, acrylamide etc.

Implementation deadlines: The Member States have until 25/12/2000 to transpose the Directive into national legislation and additionally 5 years, i.e. until 25/12/2003, to ensure that the drinking water complies with the standards set, except for Bromate (10 years), Lead (15 years) and Trihalomethanes (10 years).

3.1.5 *Bathing water quality (76/160/EEC)*

The bathing water quality should be monitored and tested in order to protect bathers from health risks and to preserve the environment from pollution. The 1976 Bathing Water Directive has set binding standards for bathing waters throughout the European Union. The annual Bathing Water Report and Tourist Atlas show substantial progress in the quality of bathing waters and large public awareness. On 24 October 2002, the Commission has adopted the proposal for a revised Directive (COM(2002)581) that is based on the outcome of recent epidemiological research and managerial experiences with the previous Directive. The revised proposal provides long-term quality assessment and management methods in order to reduce both monitoring frequency and monitoring costs.

3.1.6 *Sewage Sludge Directive (86/278/EEC)*

This Directive is aiming to encourage the use of sewage sludge in agriculture and to regulate its use in such a way as to prevent harmful effects on soil, vegetation, animals and man. To this end it prohibits the use of untreated sludge on agricultural land unless it is injected or incorporated into the soil. Specifies rules for the sampling and analysis of sludge and soils have been adopted for its implementation. In addition it sets out requirements for the keeping of detailed records of:

- the quantities of sewage sludge produced
- the quantities used in agriculture
- the composition and properties
- the type of treatment
- the sites where the sludge is used

A proposal for the revision of the Sewage Sludge Directive, linked to the development and implementation of the Thematic Soil Strategy, is in preparation. It is addressing in particular:

- an harmonised and better definition of sewage sludge
- the concept of advanced and conventional treatment
- the introduction of two land type uses (agricultural and non-agricultural)
- the interception of pollutants at source
- the reduction of the allowable heavy metals concentration
- the definition of limit values for persistent, toxic and bioaccumulative organic compounds and certain surfactants
- the description of sampling and analytical standard methods

3.1.7 *A new EU Floods Directive (2007/60/EC)*

The Directive requires the assessment of flood risks in all water courses and coastal lines in the European Union with the aim to manage and reduce the risks that flooding poses to human health, the environment, cultural heritage and economic activities.

A first assessment shall allow identifying by 2011 the river basins and associated coastal areas at risk. Flood risk maps will be established for those zones by 2013 with the objective of setting flood management plans by 2015 addressing approaches for risk prevention, protection and preparedness.

The implementation shall be coordinated among Members States within the context of the river basins management plans promoted by th WFD and shall include public participatory approaches and information.

3.1.8 *Other pieces of water-related EU legislation*

The Nitrates Directive (91/676/EEC) is aiming to protect waters against pollution caused by nitrates from agricultural sources. At least 30–40% of rivers and lakes show eutrophication symptoms or bring high nitrogen fluxes to coastal waters and seas. The agricultural origin of these N fluxes accounts for 50 to 80% of total N inputs to EU waters.

The Integrated Pollution Prevention and Control Directive (96/91/EC) address the mimization of point sources pollution from major polluting industrial installations. A system based on permits must take into account the whole environmental performance of the plant, i.e. emissions to air, water and land, generation of waste, use of raw materials, energy efficiency, noise, prevention of accidents, risk management, etc. The permits must be based on Best Available Technologies (BATs) that have been defined in collaboration with the major polluting industrial sectors. Industrial activities have been divided into some 30 sectors following the Annex I of the Directive. For each sector it takes around two years to complete the work and to produce a so-called BREF (BAT reference document). A revision of the Directive is in preparation, the BREFs documents will also follow an updating and revision process. Available BREFs can be downloaded from the European Integrated Pollution Prevention and Control Bureau website[2].

The deadlines for IPPC Directive implementation at new industrial installations was October 1999, for existing installations the final deadline is October 2007.

Water-related research links to existing EU legislation is well recognised and particularly proved resulting from its contribution to undergoing revisions of Directives. Some examples of RTD projects financially supported by the EU Sixth Framework Programme are given in the table 2.

3.2 *Other initiatives of interest*

Water issues are globally calling for many actions contributing to foster sustainable development strategies. The European Union has launched different initiatives that can beneficially build on research activities and results such as:

- The European Water Initiative (EUWI)[3] that is aiming to contribute to the achievement of the Millennium Development Goals , notably by promoting the access to safe water and basic sanitation. The European Water Initiative includes a component addressing RTD issues and scientific cooperation.
- The EU Environmental Technologies Action Plan (ETAP)[4] aims to foster the development and adoption of environmental technologies. It addresses different priority action grouping RTD activities, financial incentives and international cooperation measures.
- The EC Horizon 2020 initiative[5] is focussing on water, energy and waste with the aim of de-polluting the Mediterranean region by 2020 in collaboration with partners' countries. RTD activities contributing to pollution prevention, sustainable management of resources and environmental technologies development in major polluting areas (industrial emissions, urban wastewater and municipal waste) are of beneficial interest.

---

[2] http://eippcb.jrc.es/pages/FActivities.htm

[3] Detailed information is available at http://ec.europa.eu/research/water-initiative/index_en.html

[4] Detailed information is available at http://ec.europa.eu/environment/etap/

[5] More information is available at http://ec.europa.eu/environment/enlarg/med/horizon_2020_en.htm

Table 2. Example of RTD projects supporting directly EU environmental policies.

Project Acronym	Project Title	EU linked Directive
REBECCA	Relationships between ecological and chemical status of surface waters	WFD
SWIFT-WFD	Screening method for Water data Information in support of the implementation of the Water Framework Directive	WFD
AQUAMONEY	Development and Testing of Practical Guidelines for the Assessment of Environmental and Resource Costs and Benefits in the WFD	WFD
EAQC-WISE	European Analytical Quality Control in support of the Water Framework Directive via the Water Information System for Europe	WFD
BRIDGE	Background cRiteria for the IDentification of Groundwater thrEsholds	Groundwater
VIROBATHE	Methods for the concentration and detection of adenoviruses and noroviruses in European bathing waters with reference to the revision of the Bathing Water Directive 76/160/EEC	Bathing Water
HORIZONTAL-ORG	Horizontal Standards on Organic Micropollutants for Implementation of EU Directives on Sludge, Soil and Treated Bio-waste	Sewage Sludge
HORIZONTAL-HYG	Horizontal Standards on Hygienic parameters for Implementation of EU Directives on Sludge, Soil and Treated Bio-waste	Sewage Sludge

– The EC initiative on water scarcity and droughts (COM(2007) 414 final)[6] addresses the related challenges by a number of measures and actions to be undertaken at regional, national or EU level. The proposed options are favouring measures for water efficiency and water savings and promote means for improving drought risk management plans. Enhancing research activities in the area of water scarcity and droughts, and namely in the FP7 context, is part of the identified policy options.

4 EU FUNDED WATER RESEARCH SO FAR

The Framework Programme is to date the major instrument dedicated to develop and support the European Union's policy for research and technological development and aiming to strength the competitiveness of the European industry. In pursuing this objective the European research shall be coordinated with research activities implemented at Member States' level in view to support all European Union's policies.

The research activities supported at Community level started in 1957 with the signature of the EURATOM treaty and the establishment of the Joint Research Centre. The adoption of the ESPRIT programme (R&D in the field of information technologies) in 1984 was a preliminary step for the First Framework Programme. A revision of the Treaty of Rome and aiming to re-launch European integration, the Single European Act was signed in 1986 and came into force in 1987, the starting year of the Second Framework Programme. With this new treaty science and technological development became a Community responsibility.

The Second Framework Programme was followed by 4 consecutive FPs as detailed in the Table 3 that indicates the period for implementation and the overall and specific budget for environmental research. The FP6 was completed by 2006 in terms of programme implementation and projects' selection, some of them lasting till 2010. The continuation and reinforcement of Community RTD activities is secured for the period 2006–2013 with the FP7.

[6] Detailed information is available at http://ec.europa.eu/environment/water/quantity/scarcity_en.htm

Table 3. Overview of FPs' budget and the Environmental research share.

	FP 1	FP 2	FP 3	FP 4	FP5	FP 6
Duration	1984–1987	1987–1991	1990–1994	1994–1998	1998–2002	2002–2006
Overall budget	3.750 M ECU	5.396 M ECU	6.600 M ECU	11.879 M ECU	13.700 M EURO	17.883 M EURO
Specific theme	Protecting the environment	Environment	Environment	Environment[(1)]	Environment and Sustainable	Global Change and Ecosystems
Specific budget	195 M ECU	261 M ECU	587 M ECU	809,5 M ECU	1.083 M EURO	769 M EURO

[(1)] Includes Environment and Climate and Marine Sciences and Technology

Table 4. Structure of Environmental research within the FP4.

FP4 Environment and Climate

A. Research into the natural environment, environmental quality and global change
   1. Climate change and impact on natural resources
   2. Atmospheric phsysics and chemistry, interaction with the biosphere and mechanisms of the impact of environmental change

B. Environmental technologies
   1. Instruments, techniques and methods for monitoring the environment
   2. Technologies and methods for assessing environmental risk and for protecting and rehabilitating the environment
   3. Technologies to forecast, prevent and reduce natural risks

C. Space techniques applied to environmental monitoring and research
   1. Methodological research and pilot projects
   2. Research and development work for potential future operational activities
   3. Centre for Earth Observation (CEO)

D. Human dimensions of environmental change
   1. Socio-economic causes and effects of environmental change
   2. Economic and social responses to environmental problems
   3. Integration of scientific knowledge and economic and social considerations into the framing of environmental policies
   4. Sustainable development and technological change

FP4 – Marine Sciences and Technology

I. Marine science
   1. Marine systems research
   2. Extreme marine environments
   3. Regional seas research

II. Strategic marine research
   1. Coastal and shelf sea research
   2. Coastal engineering

III. Marine technology
   1. Generic technologies
   2. Advanced systems

4.1 *Participating in EU funded research*

The FP7 rules for participation[7], and additionally the related specific programme and/or work programmes, provided the minimum conditions that shall be fulfilled.

[7] The document is available at http://eur-ex.europa.eu/JOHtml.do?uri=OJ:L:2006:391:SOM:EN:HTML

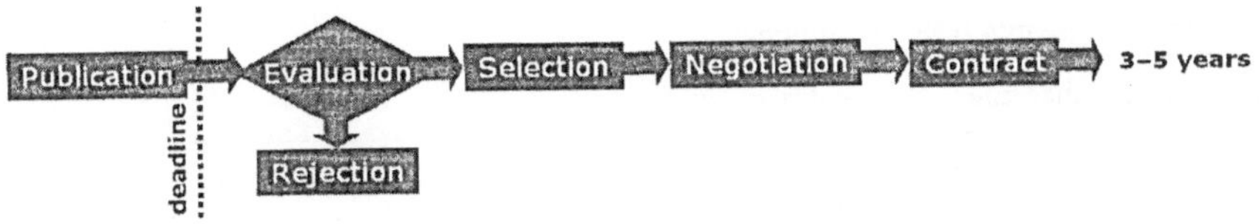

Figure 3. Main steps between call for proposal publication and contract execution.

Any legal entity (company, university, research centre, organisation or individual) established in different EU Member States (Austria, Belgium, Bulgaria, Cyprus, Czech Republic, Denmark, Estonia, Finland, France, Germany, Greece, Hungary, Ireland, Italy, Latvia, Lithuania, Luxembourg, Malta, Netherlands, Poland, Portugal, Romania, Slovakia, Slovenia, Spain, Sweden and United Kingdom) or Associated States (Switzerland, Israel, Norway, Iceland, Liechtenstein, Turkey, Croatia, the Former Yugoslav Republic of Macedonia, Serbia, Albania and Montenegro) can participate in a consortium submitting a proposal.

Partners from third countries[8] can also participate on a similar basis and receive financial support[9] provided the enhanced contribution is justified.

4.2 *FP implementation*

The Framework Programme is implemented by the European Commission on the basis of call for proposals addressing different topics for research derived from the Specific Programmes adopted by the European Union. The thematic calls for proposals are published within the Official Journal of the European Union and are publicly available from a dedicated server on the internet (http://www.cordis.lu) together with the Work Programmes describing the research subjects and other guidance documents. After the deadline for submission the eligible proposals are evaluated by the Commission services with the assistance of panels of independent experts. The top ranked proposals will be proposed for selection and negotiation according to the available budget for funding.

After a successful negotiation the consortium composed by different partners and the European Commission will enter into a contract defining the entities, responsibilities and description of work as well as the financial envelope and project duration. The EU contribution is usually rated as 50% of the overall estimated cost of the project that is taking the form, in a majority of the cases and types of contracts, of a reimbursement of the actual eligible costs exposed by the partners to carry out the project. The contractual modalities for participation, dissemination of results, financial Community contribution and other rights and obligations are derived from an EU decision linked to the Framework Programme, the so-called "Rules for participation".

4.3 *FP preparation and orientation*

Orientations and focus of the successive FPs were and are reflecting the strategies endorsed by the European Union. As examples the European Union Strategy for Sustainable Development was clearly influencing the FP5 scope and design. The Lisbon Strategy objective of making the European Union "*the most competitive and dynamic knowledge-based economy in the world*" is integrated within the FP6 context and the current FP7. Initiatives such as the "Environmental Technologies Action Plan (ETAP)" and the "Environmental and Health Action Plan" have been considered when preparing calls for proposals during FP6 and FP7. Climate change concerns and impacts as well as

[8] The list of target countries, the so-called International Co-operation Partners Countries (ICPC) is available at ftp://ftp.cordis.europa.eu/pub/fp7/docs/icpc-list.pdf.

[9] The situation can evolve in the future, updated rules and conditions are publicly available at the CORDIS server (http://cordis.europa.eu/)

Table 5. EU water research structure in FP5.

FP5 – Sustainable Management and Quality of Water
1.1 Integrated management and sustainable use of water resources at catchment scale
1.1.1 Strategic planning and integrated management methodologies and tools at river basin scale
1.1.2 Socio-economic aspects of sustainable use of water
1.1.3 Operational management schemes and decision support systems
1.2 Ecological quality of freshwater ecosystems and wetlands
1.2.1 Ecosystem functioning
1.2.2 Ecological quality targets
1.3 Treatment and purification technologies
1.3.1 Management of water in the city
1.3.2 Wastewater treatment and re-use
1.4 Pollution prevention
1.4.1 Abatement of water pollution from contaminated land, landfills and sediments
1.4.2 Combating diffuse pollution
1.5 Surveillance, early waming and communication systems
1.5.1 Pollution surveillance and control
1.5.2 Improved flood forecasting
1.6 Regulation of stocks and technologies for arid and semi-arid regions and generally deficient regions
1.6.1 Improving know ledge on water resources use and management
1.6.2 Prevention and mitigation of saline water intrusion
1.7 Pre-normative, co-normative research and standardisation

the EU energy policies are as well affecting the FP7 implementation. International collaboration with third countries is a research component strengthened in the current FP.

The mentioned ETAP set up measures facilitating the take up of appropriated environmental technologies while contributing to competitiveness and growth. One of the priority actions dealing with research has identified water as a predominant sector whether a technology platform would favour the definition and adoption of a set of strategic research targets. A technology platform, composed by a variety of stakeholders sharing a long term vision on a particular issue, is a mechanism aiming to identify common research objectives in view to develop and promote a specific technology. The Water Supply and Sanitation Technology Platform (WSSTP)[10] delivered its first vision document, strategic research agenda and deployment plan in 2006.

When developing a research policy two complementary aspects must be considered. From one side we have the thematic approach including the subdivision of the programmes according to the areas for research and technological development as derived from the corresponding Specific Programmes. The second aspect is related to strategic objectives and implementation means including contractual rules and financial provisions.

The presentation of the thematic subdivision of the environmental research programme may differ notably when comparing the Specific Programmes structure. For example water was clearly identified as a heading for research within the FP5 (see Table 5) and FP6 (see Table 6). However water research activities were not considered as a separate area within FP4 (see Table 4) and FP7 (see Table 9) but more integrated on climate change impacts, management of natural resources or environmental technologies. Comparing the areas for environmental research as defined for implementation in FP4 and FP7, similarities in terms of subdivision can be found.

The second aspect is of major relevance as it defines the means to achieve the strategic objectives. Most of the rules for participation and implementation are translated into a variety of funding mechanisms, reimbursement rates, intellectual property rights, contractual obligations, rules for dissemination and exploitation of results, etc. In a first instance all those issues are certainly less

[10] The WSSTP Vision Document and the Strategic Research Agenda are available at http://www.wsstp.eu/

Table 6. Water research structure in FP6.

FP6 – Water cycle and soil-related aspects

1 Hydrology and Climate processes
  1.1 Climate modelling at catchment-regional scale
  1.2 Climate variability, floods and droughts
2 Ecological impact of global change, soil functioning and water quality
  2.1 Ecological impact of global change on surface water bodies, ecosystem, health indicators and remediation strategies
  2.2 Water – soil systems functioning and management
3 Integrated management strategies and mitigation technologies
  3.1 Integrated water management at catchment scale
  3.2 Integrated urban water management and mitigation technologies
  3.3 Management of water under scarcity and mitigation technologies
4 Scenarios of water demand and availability
  4.1 Scenarios of water demand and availability at 25–50 y perspective

Table 7. Number of FP6 related projects and EC grant distributed per area of research.

FP6 – Water cycle and Soil related aspects

Area	# of RTD projects	EC grant (€)
Hydrology and climate processes	3	13.538.000
Ecological impact of global change, soil functioning and water quality	8	33.912.000
Integrated management strategies and mitigation technologies	31	111.042.000
Scenarios of water demand and availability	1	6.993.000
Specific Support Activities	5	2.508.000
Scientific Support to Policies	10	20.186.000
TOTAL	58	188.179.000

attractive for the scientific community. However the mentioned rules are essentially governing the modalities for FP7 implementation. In this area major changes have been introduced along FPs' definition and implementation. And possibly this adaptation is the suitable approach to achieve major objectives contributing to the European Research Area (ERA)[11]

The results of the FP6 area "Water cycle and Soil related aspects" in terms of contracts and funding are summarised in the Table 7. Visibly the numbers are less substantial compared to the overall figures of FP5 (more than 180 contracts and about 250 M€) and it could be noted that some activities are partially or not really covered (monitoring, sensor development, industrial water treatment . . .). A number of FP6 projects will be completed by 2009 and 2010 and therefore it is difficult estimating the overall outcome and impact. It is however expected a valuable contribution to the water sector and a relevant structuring impact in the water research policies in the European Union.

5 CURRENT AND FUTURE TRENDS FOR EU WATER RESEARCH

The Seventh Framework Programme for RTD is aligned to the EU Financial Perspectives for the period 2007–2013. The 7 years duration, exceeding the usual 4 years, is introducing a new variable

[11] ERA is a structure aiming to support better coordination and convergence of research and innovation policies and strategies (http://ec.europa.eu/research/era/index_en.html)

Table 8. FP7 Cooperation themes.

	Themes		Budget
COOPERATION	1. Health		6.100
	2. Food, Agriculture and Fisheries, and Biotechnology		1.935
	3. Information and Communication Technologies		9.050
	4. Nanosciences, Nanotechnologies, Materials and new Production Technologies		3.475
	5. Energy		2.350
	6. Environment (including Climate Change)		1.890
	7. Transport (including Aeronautics)		4.160
	8. Socio-economic Sciences and the Humanities		623
	9. Security and Space	Space	1.430
		Security	1.400
	Total (€ million)		32.413

demanding first a visionary concept along with a flexible, dynamic and reacting programme. With an overall budget exceeding 52€ billion the FP7[12] is being implemented by the European Commission. Aiming to strengthen links with education and innovation programmes, the so-called "Triangle of Knowledge", the FP7 is drafted around 4 main Specific Programmes for its implementation, in addition to the Joint Research Centre and Euratom activities:

- Cooperation – Collaborative research
- Ideas – Frontier research – European Research Council
- People – Human potential, Marie Curie fellowships
- Capacities – Research capacities (infrastructure, SMEs, regions, internat. Cooperation)

The specific programme "Cooperation", the more conventional compared to former FPs, is subdivided into themes as described in the Table 8.

The theme "Environment (including Climate Change)" is subdivided in activities and sub-activities. The environmental research FP7 structure is shown in the Table 9. Water related research is embedded in the presented activities, such as natural resources management (ecosystems, IWRM, modelling, DSS ...), natural hazards (floods, droughts ...), environmental technologies (water treatment technologies, sensors and monitoring ...).

The overall budget for the theme "Environment, including Climate Change" is about 1.890€ million for the 7 years duration of the FP7. As detailed in the Figure 4 the available annual budget is gradually increasing starting in 2007 with an annual budget slightly higher than the last year of FP6 situated around 224€ million. A sharp increase of the annual budget is planned for 2011 that will culminate in 2013 with a budget 60% higher that the available in 2007. It should be noted that there is no prescribed budget per activity, which brings flexibility and possibilities for adaptation.

So far two main calls have been implemented in 2007 and 2008 with a number of selected projects. The catalogue of water related projects selected so far is available from the Circa server[13], totalling an overall EC funding of about 63€ million for 19 RTD projects. In addition a joint call with the theme 4 Nanosciences, Nanotechnologies, Materials and new Production Technologies allowed in 2008 the selection of 5 projects addressing 'Nanotechnologies for water treatment' counting with about 9.5€ million EC grant.

[12] All documents related to FP7 legal basis are available at http://cordis.europa.eu/fp7/find-doc_en.html#legal-basis

[13] FP7 2007–2008 Water research http://circa.europa.eu/Public/irc/rtd/eesdwatkeact/library?l=/projects_information/catalogue_2008pdf/_EN_1.0_&a=d

Table 9. Activities and sub-activities within the FP7 Environment theme.

FP7 Environment (including Climate Change)
Climate change, pollution and risks
Pressures on environment and climate
Environment and health
Natural hazards
Sustainable Management of Resources
Conservation and sustainable management of natural and man-made resources and biodiversity
Management of marine environments
Environmental Technologies
Environmental technologies for observation, simulation, prevention, mitigation, adaptation, remediation and restoration of the natural and man-made environment
Protection, conservation and enhancement of cultural heritage including human habitat
Technology assessment, verification and testing
Earth observation and assessment tools for sustainable development
Earth and ocean observation systems and monitoring methods for the environment and sustainable development
Forecasting methods and assessment tools for sustainable development, taking into account differing scales of observation

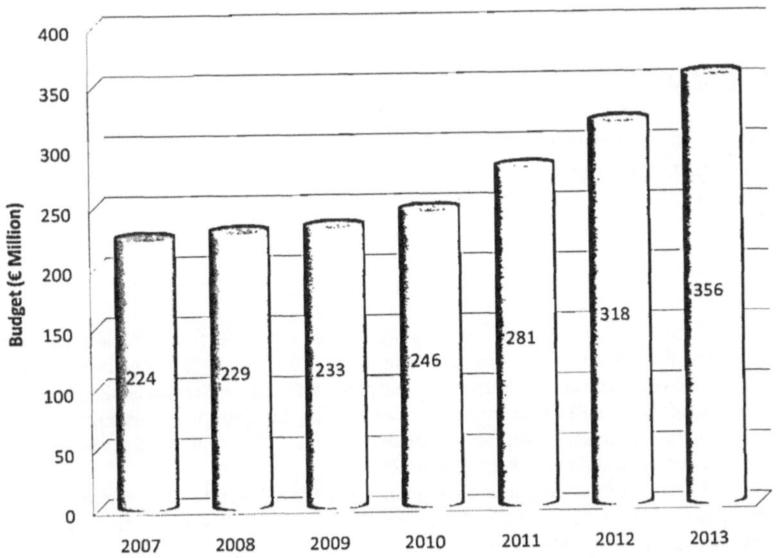

Figure 4. Planned evolution of the annual budget (€ million) for the FP7 theme Environment (including Climate Change).

The FP7 implementation is going on, the call for proposals in 2009 includes also topics dealing with water research and it is expected that a number of projects will be selected. The future calls of the theme "Environment (including Climate Change)" will continue addressing the area of water research. Sector-based and cross-thematic approaches tackling water related issues will expand notably in other themes of the FP7 Cooperation Specific Programme. The aspects identified in the Strategic Research Agenda of the Water Supply and Sanitation Technology Platform will be considered besides attracting increased industrial and SMEs participation.

Policy and strategic drivers will affect the definition of the future research activities, including international cooperation. Main issues that can be anticipated at this stage are related to expected increasing competition for environmental services and depleting natural resources that might arise due to societal and environmental changes.

Some specific aspects expected to be addressed in a close future could be:

- water scarcity, including conservation, use optimisation, land use impacts and urbanisation in coastal areas;
- safe and sustainable water reuse, recycling and recovery of valuable products including energy;
- water for food and productivity, including scenarios for resource distribution
- integrated water resources management in changing environments

The list of issues at stake might be rather long and would require prioritisation in consultation with EU Member States. However the expected limiting factor will be the potential budget available for supporting water research.

6 DISCLAIMER

The information and views set out in this paper are those of the author(s) and do not reflect necessarily the official opinion of the European Communities. Neither the European Communities institutions and bodies nor any person acting on their behalf may be held responsible for the use which may be made on the information contained therein.

ACKNOWLEDGEMENTS

The views expressed are the result of a combination of publicly available information combined with a valuable exchange of experiences with colleagues of the European Commission and RTD project participants.

REFERENCES

Catalogues of EC funded water research projects within the environmental research programmes (FP4, FP5, FP6 and the FP7 projects selected after the calls 2007 and 2008). Available from: http://circa.europa.eu/Public/irc/rtd/eesdwatkeact/library?l=/projects_information&vm=detailed&sb=Title.

Council Decision 2006/971/EC of 19 December 2006 concerning the Specific Programme Cooperation implementing the Seventh Framework Programme of the European Community for research, technological development and demonstration activities (2007–2013).

Decision No. 1982/2006/EC of the European Parliament and of the Council of 18 December 2006 concerning the Seventh Framework Programme of the European Community for research, technological development and demonstration activities (2007–2013).

European Commission (2003) Environmental Technologies Action Plan – Discussion paper – "A report from the Water Issue Group".

European Commission (2005) Special Eurobarometer 217/Wave 62.1 "The attitudes of European citizens towards environment".

European Commission (2005) Special Eurobarometer 224/Wave 63.1 "Europeans, Science and Technology".

European Commission, COM(2004) 38 final, "Stimulating Technologies for Sustainable Development: An Environmental Technologies Action Plan for the European Union".

CHAPTER 11

The interdisciplinary challenge in water policy: The case of "water governance"[1]

J.E. Castro
School of Geography, Politics and Sociology, Newcastle University, UK

ABSTRACT: It is widely ackowledged that the world water crisis is mainly a crisis of governance. However, there is no shared understanding of what "governance" means, how it works, who are its actors. The prevailing conceptions of governance in mainstream water policy documents tend to be instrumental and idealistic. Perhaps the most important consequence of instrumental and idealistic understandings of governance is the rhetorical depoliticization of what is, paradoxically, a political process. The main mechanism of this "depoliticization" of governance is the exclusion of the ends and values informing water policy from the debate. Instrumental and idealistic understandings of governance constitute a major obstacle for the scientific understanding of the process and for achieving success in policy interventions directed at tackling the water crisis. The paper argues for the development of a balance between the techno-scientific, socio-economic, political, and cultural aspects of water management activities, which may help in superseding the artificial separation of water research and practice in disciplinary and corporatist feuds.

"The water crisis is largely a crisis of governance"
UNESCO, 2006: 1

1 INTRODUCTION

The challenges facing water management have become increasingly global in scope since the 1970s. This reflects the rising awareness about the uncertainties posed by the worsening situation of the hydrosphere, and particularly freshwater, and the unsustainability of water management practices in many areas. It is also a reflection of the conflicts flaring up from the protracted social inequalities affecting the access to water for essential human uses and from the inefficiency, ineffectiveness, and inefficacy characterizing water management in many regions, not just in the poorer countries. In this regard, since the 1970s the international community has launched significant and far-reaching policy initiatives in response to the challenges. These include tackling desertification, controlling water pollution, developing conflict prevention measures in the light of ongoing and potential water conflicts, monitoring and preventing water-related threats and hazards (ranging from the impact of floods and other disastrous climatic events to the persistence, revival and emergence of water-related diseases), to overcoming the deficiencies and inequalities in the allocation and distribution of water for essential human use in developing countries (for a synthesis of the main international initiatives since the 1970s, see "Milestones 1972–2003: from Stockholm to Kyoto" at UNESCO's Water Portal, http://www.unesco.org/water/wwap/milestones/index.shtml).

However, despite the important efforts made in recent decades, there is a growing awareness that the struggle for reducing ecological unsustainability and limiting the negative impact of water-related hazards and defficiencies in water management is being lost in many countries. As an example, let us consider the goal of guaranteeing universal access to essential water and sanitation

[1] An earlier version of this text was published in Brazil as "Water governance in the twentieth-first century", in Ambiente e Sociedade, Vol. 10, no 2, pp. 97–118, University of Campinas, Brazil.

services, which continues to be a main target of the international community. The goal of universalizing these services was restated in the late 1970s, when the aspiration to provide essential volumes of safe water to every human being on earth by 1990 was endorsed by the United Nations. The 1977 UN Water Conference in Mar del Plata, Argentina, which led to the International Drinking Water Supply and Sanitation Decade (1980–1990), declared that everyone has "the right to have access to drinking water in quantities and of a quality equal to their basic needs". The Decade was officially closed by the Global Consultation held in New Delhi in 1990, which produced the New Delhi Statement calling for "some [water] for all rather than more for some" (UN, 1980; 1990). Unfortunately, and although significant progress has been made in some areas, that goal was not achieved. As a matter of fact, current estimates show that at the beginning of the twentieth first century 1.1 billion people, around 17 per cent of the world population, still lacks access to safe water while around 2.4 billion, or 40 per cent, has no access to adequate sanitation (EC, 2002a,b). Moreover, while the objectives for 1990 had been to guarantee universal access to essential volumes of water, the current targets as expressed in the UN Millennium Development Goals (MDGs) adopted in 2000–2002 are limited to halving the proportion of the world population without access to these services by 2015 (UN, 2000, 2002). Although from a certain perspective the new goals may be more "realistic", in practice this means that the international community is prepared to accept that a large proportion of human beings will continue to suffer disease and death owing to the lack of essential water services perhaps for decades to come. In this connection, a recent evaluation of the progress made in relation to the MDGs shows that even these limited objectives will not be achieved in many of the poorest countries, which are characterized by "fragile states [...] with weak governance and institutions" (WHO, 2005: 27, 71).

There is increasing recognition that, to a large extent, the main causes for this unacceptable state of affairs are neither technical nor "natural" but rather are, broadly speaking, of a social and political nature. The water crisis, we are told, is mainly "a crisis of governance" (UNESCO, 2006: 1). But, what does "governance" mean in this context? Although the prevailing uses of this concept in the literature dedicated to water seem to suggest a shared understanding of the meaning of governance, in fact the answer to this question is not straightforward. For some, governance is an instrument, a means to achieve certain ends, an administrative and technical toolkit that can be used in different contexts to reach a given objective, such as enforcing a particular water policy. For others, governance is a process involving not the instrumentalization of decisions taken by experts and powerholders, but rather the debate of alternative, often rival projects of societal development, and the definition of the ends and means that must be pursued by society, through a process of substantive democratic participation. In addition to the contrasting conceptions of governance discussed here, there are also different intellectual and political traditions, some of them defending irreconcilable positions, which inform dissimilar understandings and practices of governance. Thus, for instance, while certain traditions understand that water governance must be structured around the principles that water is a common good and that essential water services are a public good that cannot be governed through the market, other traditions defend the entirely opposed view that water must be considered as an economic resource, essential water services as a private good, and that in consequence the governance of water and water services must be centred on market principles. These are just a few examples to demonstrate that the question about what exactly "governance" means requires careful consideration. We come back to this later.

The need to achieve a shared understanding of the "water crisis" has also important implications for water-related academic and techno-scientific endeavors, emphasizing the call for meaningful, not just rhetorical, interdisciplinarity in water research. In this regard, although a high degree of sophistication has been reached in the techno-scientific fields related to water, such as hydrogeology, hydraulic engineering, or biotechnology applied to water management, we are still very far from plainly understanding the historical, socio-economic, cultural and political processes underpinning the "water crisis". This gap between the techno-scientific and socio-political fields of knowledge, we claim, may contribute to explain why the enormous technological progress made in relation to water in recent decades has not been reflected in more sustainable, efficient, effective and efficacious practices of water management. Therefore, there is a need for establishing a balance between the techno-scientific, socio-economic, political, and cultural aspects of water management activities, and superseding the artificial separation of water research and practice in disciplinary and corporatist feuds. Correspondingly, the development of genuinely interdisciplinary approaches that contribute towards developing water governance and management practices grounded on the principles of sustainability and social justice is one of the most urgent challenges facing water governance in the twentieth-first century.

2 THE "WATER CRISIS"

Writing "water crisis" with inverted commas denotes that the very notion that there is a water crisis is a contested matter. The intensity of the debate, and its propensity to become marred in circular arguments, are well reflected in the following statement from the First UN World Water Report: "the water crisis that exists is set to worsen despite continuing debate over the very existence of such a crisis" (UNESCO, 2003: 5). To make things worse, the camp of those who accept the existence of a global water crisis is divided, often irreconcilably, when it comes to defining the dimensions, meanings, and extent of such crisis or, more importantly, to proposing the actions that are needed for overcoming the crisis or at least for mitigating its negative impacts. For instance, let us briefly explore the most recent edition of UNESCO's World Water Report, which follows on the steps of the 2003 report in defending the argument that a global water crisis exists (UNESCO, 2006).

The first thing that must be said is that the report presents overwhelming evidence of the existence of a global water crisis, and it is an excellent effort to reflect the multidimensional character of such crisis. It reminds us that from the total water volume on earth only 2.5 per cent is freshwater, and that only a fraction of this freshwater "in storage" is usable for human consumption. This freshwater is unevenly distributed in geographical terms, and is subject to severe and adverse pressures from naturally occurring and human-driven processes. The report also identifies the main human drivers of these impacts: "population growth, particularly in water-short regions, major demographic changes as people move from rural to urban environments, higher demands for food security and socio-economic well-being, in-creased competition between users and usages, pollution from industrial, municipal and agricultural sources" (Id. pp. 121–136). It also engages with arguably all the major themes characterizing the water crisis, including the problem of essential water and sanitation services, the water-related risks and threats to human health, the links between water management and poverty, water for industry, agriculture and energy, water for environmental sustainability, and the growing number of environmental refugees displaced by climatic and human-driven processes (Id., 9, 316). Moreover, and of greater relevance for this article, "governance is an overarching theme" of the report and it certainly provides powerful insights into some of the crucial challenges affecting water governance worldwide (Id., p. 45). However, the report is also an excellent example of the protracted difficulties facing water experts, specialists, and practitioners to overcome such obstacles to scientific knowledge as artificial disciplinary boundaries, and continued lack of conceptual frameworks to develop truly interdisciplinary coordinations, especially between the techno-sciences and the social sciences. Let us consider some examples.

Firstly, in relation to the permanence of artificial boundaries, the continued use of concepts such as "water sector" or "water resources" throughout the report, suggests that the dimensions and concepts of traditional disciplines have disproportionate prevalence over other approaches. A similar report where, for instance, ecological economists or political ecologists play a more central role would certainly frame the analysis with a different conceptual apparatus that incorporates the interconnectedness that exists between water management and other human endeavors, which is lost in the traditional treatment of activities as "sectors".

Secondly, the report tends to define water almost invariably as a "resource", including a chapter on "The state of the resource" (Id., p. 119). The document also pays attention to the ecosystemic character of water issues, but the prevalence in the report of a language that reduces water to one of its many dimensions, that of being a resource for humans, illustrates the persistence of disciplinary enclosures preventing cross-fertilization in the production of scientific knowledge about water. The repeated conceptualization of water as a resource, used more than 1400 times in the document, would be strongly criticized by ecologists and ecological economists, among others, as being tributary to a resource-oriented model of water management that is actually responsible to a large extent for the current "water crisis". There is a growing body of literature dealing with these problems, including a number of studies focusing on "water security" that highlight the implications and contradictions inherent in treating water as a "natural resource", as a "commodity", as an "entitlement", and so on (Webb and Iskandarani, 1998; see also EUWATER, 2005).

Thirdly, the treatment of water values, to which the report dedicates a whole chapter, reflects the existing contradictions and confrontations between irreconcilable positions on this subject. It also adopts one of the main competing arguments without paying sufficient attention to alternative positions in the highly contested debate about valuation. Let us examine first the contradictions. The report states that

"As a physical, emotional and cultural life-giving element, water must be considered as more than just an economic resource. Sharing water is an ethical imperative as well as an expression of human identity and solidarity [...]. Valuing water, including sustaining and fostering water-related cultural diversity, heritage and knowledge, is critical to enhancing our ability to adapt in a changing world. Economic valuation of water resources must be recognized as existing within this larger and more complex context of valuing water" (Id., p. 403, 405).

This is a well-thought statement which raises the reader's expectations about the propositions that the report may have to offer in terms of developing systems for capturing this multidimensional and complex universe of water values. However, what comes next is a conventional lesson on economic valuation of water resources and services that fails to live up to the rhetorical recognition that economic valuation is just one among other dimensions of the problem. Moreover, the approach to economic valuation that is given central stage in the document is just one among a number of different rival positions competing in the field, but this is not adequately explained. For instance, the report classifies "residential water supply" and "residential sanitation" under "Consumer Goods" within the category "Commodity (or Private) Goods" and not under "Public Goods", a category reserved in this document for the protection of the "aquatic environment", "wild lands", and "biodiversity and endangered species" (Id. p. 409). Thus, an ongoing debate taking place globally about the need to consider essential water services such as water and sanitation as public goods, a social right, and a universal human right, and not a private good or commodity is entirely neglected (see, among others, Ward, 1997; Petrella, 2001; Strang, 2004; EUWATER, 2005). Intentionally or not, the report has abandoned here the scientific approach to support one of the rival positions in the debate, without adequate justification. Once this positioning of the authors has been identified, other apparent contradictions in this crucial section of the document become more intelligible. For instance, it states next that "Governance strategies should be selected to optimize the achievement of societal goals. In this context, valuation can be viewed as a fairly neutral and objective process by which social goals and trade-offs can be identified and debated and the optimal governance strategies chosen" (Id. p. 410).

Although in some passages of the document there is a clear recognition that governance cannot be reduced to a policy instrument (Id., pp. 46–49), the key section of the report "Responding to the challenge of valuing water" is grounded on this instrumental understanding of governance as a strategy to achieve certain goals. A number of questions arise from this statement. How are these "societal goals" defined? Who defines these goals? Why a particular language of valuation, economic valuation, has been preferred over others? Who has the power to decide that this is the relevant language of valuation for water management issues (on value diversity and languages of valuation, see Martínez Alier, 2002). What principles inform this "governance strategy" based on economic valuation? The instrumental understanding of governance adopted in this crucial section of the document is, unfortunately, prevalent in the specialized water literature, which has tended to depoliticise water management processes by treating them as mainly (or even merely) "technical", "objective and neutral" (we come back to this later).

We have taken advantage of some gaps and internal contradictions in what is otherwise a state of the art review of the situation affecting the hydrosphere. Our main reason for discussing the above examples is to cast light on some of the crucial challenges affecting the governance of water in the twentieth-first century. We believe that the increasing rhetorical recognition of the need for a more complex analysis of the water crisis, as exemplified by the 2006 UNESCO World Water Report commented above, can stimulate genuine attempts to develop a more comprehensive, interdisciplinary understanding of water governance.

In this regard, one of the common themes that can be identified in the diverse international initiatives directed at tackling the water crisis is the widespread recognition of the centrality of "good", "effective" or "sound" governance (i.e. ADB, 1995; EC, 2000, 2002b; GWP, 2003; Camdessus, 2003; Cosgrove, 2003; UNDP, 2004; UNESCO, 2006). However, as already mentioned, despite the apparent agreement on the crucial importance of "governance", the debate is marred by conceptual ambiguity and is subject to the tensions inherent in the very nature of the process of democratic governance. Let us briefly review some aspects of this debate relevant to our discussion.

3 GOVERNANCE

The debate on governance is subject to underlying confrontations between rival and at times even incompatible intellectual and political traditions, which defend often irreconcilable opposing principles and values. Although this is often blurred by the assertive use of the concept in mainstream public policy documents, the fact is that different actors have diverse, often contradictory, understandings of governance. This, consequently, informs very different, frequently incompatible, policy strategies and decisions, given that governance or, to be more precise, democratic governance is a political process characterized by the confrontation of rival political projects grounded on different values and principles. The case of water governance lends itself as an excellent ground to illustrate these nuances. Rather than being just a matter of pure academic disquisition, the contradictions between competing intellectual and political frameworks underscore much of the institutional and political transformations undergone in the field of water policy and management.

In this connection, from a general perspective, the concept of governance aims at conceptualizing evolving forms of government and regulation that trascend those based on traditional state hierarchies and market systems (Hirst, 1994; Held, 1995; Amin, 1997). In the field of development policy, for instance, the concept of governance has become central to the argument that the traditional forms of management based on "state monopoly" over decisions and institutional arrangements are been replaced by new forms characterized for "pragmatic pluralism" (Esman, 1991; see also UNESCO, 2006: 48). Thus, "governance" would be a process resulting from the articulation of the classic forms of authority embodied in the state (hierarchical organization) with those characteristic of the private sector (driven by market competition) and the voluntary sector or "civil society" (characterized by citizens' voluntary action, reciprocity, and solidarity) (e.g. UNDP, 1997, 1998; Picciotto, 1997; see also Streeck and Schmitter, 1985). For instance, in reference to the situation in the European Union, governance has been described as a multi-layered, multi-scale, and multi-sector ensemble characterised by a combination of hierarchical structures, participatory dynamics, associative action, and market mechanisms, and would be based mainly on a culture of dialogue, negotiation, active citizenship, subsidiarity, and institutional strengthening (Heinault et. al., 2002).

Far from being an abstract academic discussion, this debate has far-reaching consequences for public policy in general, including water policy. As already mentioned, despite rhetorical recognition to the contrary, in the water policy literature governance is often understood instrumentally, as a mean to achieve certain objectives, as a policy strategy, rather than as a complex process of democratic dialogue, negotiation, and citizen participation that includes the discussion about what objectives must be pursued by society. Also, and closely related to the previous point, the conceptualization of governance that tends to prevail in this literature often presents an idealized vision of the interrelations between the main spheres involved: the state, the market, and "civil society". This idealized version of governance presents the state, the market and "civil society" as partners participating in symmetric, triangular interaction, as in the notions of "public-private partnership" and "tri-partite partnership", which have become central in mainstream public policy (e.g. Picciotto, 1997; UNDP, 2006; World Bank, 2006). We argue that there is a need to critically examine these instrumental and idealized understandings of governance that can be identified in the policy literature.

For instance, key concepts comprised in the notion of governance, such as "civil society", have different, even opposing, meanings for different intellectual and political traditions (see, for instance, Cohen and Arato, 1994; Kaviraj and Khilnani, 2001). Thus, for the free-market liberal tradition "civil society" is coterminous with the market: a sphere of action characterized by the free concurrence of self-interested, egoistic individuals pursuing their own ends. For free-market liberalism, a tradition that has arguably exercised a major influence in global public policy, and certainly in water policy, since the 1980s, there is no triangular interaction because there are in fact only two partners in the picture: the state and the market. Moreover, for this intellectual tradition the only role of the state should be to guarantee the free operation of market forces, minimizing or, preferably, cancelling state control and regulation over private actors (e.g. Brooke Cowen and Cowen, 1998; Newbery, 1999). Let us emphasise here that although this minimalist understanding of governance in the free-market liberal tradition is not widely shared in the water-policy community, it has nevertheless exercised significant influence in shaping public policy, including water policy, worlwide since the 1980s. As stated by Joseph Stiglitz, former Chief Economist at the World Bank and 2001 Economics Nobel Prize, in his evaluation of the influence of free-market liberalism in global public policy:

In setting the rules of the game, commercial and financial interests and mind-sets have seemingly prevailed within the international economic institutions. A particular view of the role of government and markets has come to prevail – a view which is not universally accepted within the developed countries, but which is being forced upon the developing countries and the economies in transition (Stiglitz, 2002: 224–5; see also Leys, 2001).

As Stiglitz's statement suggests, the free-market notion of governance, that is, "the particular view of the role of governments and markets" held by this tradition, is not widely accepted. It certainly differs in substantial ways with the understanding of governance held by rival intellectual and political traditions. For instance, contrary to the identification of "civil society" with the market held by free-market liberals, the pluralist and communitarian traditions tend to understand "civil society" as the realm of voluntary action, reciprocity, and solidarity, a buffer space between the market and the state. This understanding of civil society as a separate sphere of action vis a vis the state and the market has played a crucial role in the worldwide social and political struggles against dictatorships and authoritarian regimes since the 1960s, and gained momentum since the 1980s with the fall of the Berlin Wall and the collapse of military dictatorships in Latin America and elsewhere. From another angle, this notion of civil society reflects the expanding role of Non Governmental Organizations (NGOs), social movements, and other actors that have become increasingly influential in public policy, and certainly in water policy. On the one hand, this understanding of civil society contributes to a more complex concept of governance that captures the multi-actor, multi-dimensional, multi-sector character of public policy decisions and actions. On the other hand, however, as already discussed much of the water policy literature tends to adopt an idealized notion of civil society as the realm of reciprocity, voluntary action and solidarity, and this notion informs and idealized understanding of governance as a balanced partnership between the state, the market and "civil society". This idealized notion, in turn, provides the rhetorical framework for the adoption of an instrumental understanding of governance, as a neutral and objective tool or strategy for policy implementation, which is devoid of any political content. Thus, in an apparent paradox, governance, which is esentially a political process, becomes depoliticised in the water policy literature. We come back to this in a moment, but let us briefly discuss first another aspect of the complex nuances characterizing the understanding of governance: the diverse notions and practices of governance in different political cultures.

The diversity in the understandings of governance across different political cultures can be illustrated, for the sake of brevity, by reference to the rival notions and practices characterizing the notion of "citizenship" (see, for instance, Delanty, 2000; van Steenbergen, 1994). "Active citizenship" is one of the main drivers of action within "civil society" according to the understanding of governance prevailing in the water policy literature. However, what are the notions and practices of "citizenship" and "citizen participation" underlying these discourses? Again, we are confronted with rival, even irreconcilable notions of citizenship, although this fact is obscured in the policy literature which tends to assume a shared understanding of this concept. For instance, free-market liberalism has a particular understanding of citizenship that is limited to the realm of civil and political rights. In a nutshell, the free-market notion of citizenship is centred on the protection of individual rights, particularly the right to own property, to formal judicial procedures, and to exercise the political right of electing or being elected for government. The essence of this tradition is the protection of individual freedom against state intrusion, which includes freedom from state controls and excessive regulation in the pursuit of market interests. Contrastingly, to give another example relevant for water policy, for the social-democratic tradition, in its different national varieties, the individual rights of citizenship are complemented by "social rights", such as the right to have universal access to essential public services like as education and public health, which includes the access to affordable and safe water and sanitation services. Social rights of citizenship in this tradition are deemed to ensure the abatement of market-based social inequalities to provide all citizens with a status that is independent of their market position and thus enabling the less favoured members of society to excersise their citizenship rights more fully. This notion of social rights is rejected in the free-market liberal tradition, which considers social rights as an obstacle and not as vehicle for individual freedom and citizenship. Moreover, these tensions at the heart of one of the most cherished notions in modern western political theory, citizenship, adopt a diversity of configurations in the different countries and political cultures of the western hemisphere. As before, this is not merely an academic disquisition that lacks relevance for the earthly concerns of those involved in practical policy and managament activities. The influence of the rival positionings about citizenship informing different political cultures can be clearly identified in the current water policy documents, debates, and practices.

These considerations are even more relevant when we address the situation of non-western and, particularly, developing countries, given that notions such as "governance", "civil society" or "citizenship" emerged from the specific historical experience of Western Europe and the US and their empirical reference may be completely absent in other societies. For instance, let us focus for a moment on the notion of governance as a "partnership", which as discussed earlier pressupposes a balanced, symmetrical association between "the state", the "market", and "civil society". In practice, this notion has no empirical correlate in many countries, which are characterized by a frail public sector with low or null capacity for regulation and law enforcement, and where "civil society" is often limited to a small local elite, given that the bulk of society cannot afford to participate meaningfully in the social and political life or take part in the decision-making process. Unfortunately, this is the situation in a large number of countries that are among the worst affected by the "water crisis" and where the need for "good water governance" is consequently more urgent. A recent report forecasts that many of these countries will not be able to achieve the MDGs precisely because of the fragility of the public sector and the resulting poor "governance" (WHO, 2005: 27, 71). Thus, in many developing countries the notion of governance as a "partnership" is meaningless, as citizens have no capacity to exercise democratic control over public or private actors in charge of water management and is often defenceless in the face of water-related risks and hazards. However, this situation is by no means limited to developing countries, given that citizen participation in the process of environmental governance tends to be very limited in developed countries too (Dryzek, 1997; see also Beck, 1992, 1998).

Although many of these caveats about the meaning of "governance" are well-known and form part of the wide-ranging debates taking place around the world on this subject (e.g. GWP, 2003), in practice the prevailing understanding of governance as an instrument or as an idealized system of shared responsibility continues to permeate public policy decisions and practices, including those involving water management. In our perspective, one of the most crucial problems is that the mainstream water policy literature tends to present a depoliticized understanding of governance, although it is essentially a political process. The main mechanism of this depoliticisation of "governance" is the exclusion of ends and values from the debate, thus reducing it to a merely instrumental, technical, supposedly neutral management process or policy strategy. For instance, let us consider the suggestion for water reform offered in a recent study commissioned by the World Bank. The authors argued that:

The major thrust of institutional reforms within the water sector is to enhance the functional capabilities, operational strength, and institutional readiness to handle water challenges both at present and in the future. Given this thrust, the main objectives of institutional initiatives are rather transparent. These objectives are to: make water as an economic good, strengthen allocation capabilities, increase the reliance on market forces, revive the payment culture, ensure financial self-sufficiency, promote decentralized decision structure, and encourage the adoption of modem technology and information inputs (Saleth and Dinar, 1999: 36).

In this statement we are presented with a number of objectives for institutional reform. Leaving aside the discussion about the suitability of these objectives, the main questions in relation to water governance would be: who are the actors that decide that these are the main objectives for reforming water institutions? What is the process through which this decision is taken? What is the role of the citizens in this process? Are they consulted? What mechanisms are available for them to participate in this process? Moreover, what are the ultimate ends and values informing the adoption of such objectives? And what understanding of water governance underlies the study's approach to the reform of water institutions? The reference to this study is just an example of the contradictions inherent in the prevailing technocratic approaches to water management. In this case, a highly political process such as that required for reforming water institutions tends to be depoliticized in the analysis and presented as a neutral, "transparent", policy instrument.

However, there exist alternative understandings of governance that provide elements for thinking beyond instrumental action, as the following example illustrate:

The core of governance has to do with determining what ends and values should be chosen and the means by which those ends and values should be pursued, i.e. the direction of the social unit, e.g. society, community or organization. Governance includes activities such as efforts to influence the social construction of shared beliefs about reality; the creation of identities and institutions; the allocation and regulation of rights and obligations among interested parties; and the distribution of economic means and welfare services. Governance, in other words, is the shaping and sustaining of the arrangements of authority and power within which actors make decisions and frame policies that are binding on individual and collective actors within different territorial bounds (Hanf and Jansen, 1998: 3).

In this perspective, governance cannot be reduced to an instrument for the implementation of policy decisions taken, presumably, by experts in the relevant fields (see, for instance, Dryzek, 1997). Governance is not a strategy, and is not an idealized scheme of interaction between also idealized actors. Governance, always in this perspective, is a political process involving the exercise of political power by political actors who seek to define the ends and values that must inform social development. It also comprises the identification of means to pursue those ends and values, and the adoption of suitable arrangements for the exercise of authority and power in the process. This understanding of governance inmediately elicits a number of questions, in the light of the previous discussion. What are the ends and values that inform water policy and management? Who participates in the determination of these ends and values? Who determines the means by which those ends and values should be pursued? How are these decisions taken? How do (do they) common citizens participate in the determination of those ends and values, and in the identification of the means for pursuing them?

In this connection, the determination of the ends and values in relation to water management, and the selection of the means to pursue those ends and values, does not happen in a social vacuum. Rather than being the result of a balanced partnership, the process of water governance resembles a highly asymmetric and evolving structure where the actors tend to have dissimilar proportions of political power and knowledge. In practice, water policies that have often a significant political content are designed and implemented with disregard for the values, opinions, and preferences of the citizens and in the absence of democratic governance arrangements. In practice, water governance consists in the interaction between governments, large businesses, political parties, civil and other organizations representing sectoral interests (e.g. workers' unions, religious organizations, peasant movements, etc.), international agencies (e.g. international financial institutions and other agents of the process of "global governance"), NGOs, and other relevant powerholders. These actors are involved in continuing debates and in social and political confrontations around how water and essential water services should be governed, by whom, and for whom. These confrontations are at the heart of the process of democratic water governance, which is characterized not only by dialogue and negotiation but also, unfortunately, by growing uncertainty and protracted social and political conflicts. To this we turn next.

3.1 *Water uncertainty and conflict*

One particular area that requires urgent efforts towards enhancing inter-disciplinary coordination between the techno- and the social sciences concerns the study of the uncertainties and conflicts emerging around the management of water and water services. Regarding water uncertainty, debates on risk and "manufactured uncertainty" have emphasised environmental threats and hazards among which water-related extreme events and human defficiencies in the management of water have a central place (e.g., Beck, 1992; McGranahan et. al., 2001). International concern on these issues has led to a wide variety of efforts aimed at assessing the dimension and scale of these risks in the search for adequate approaches to limit their negative impacts (Kasperson et. al, 1995; Kasperson and Kasperson, 2001; UNEP-UNICEF-WHO, 2002; WHO, 2003a,b; WHO-Europe, 2003; UN-Habitat, 2003; UNESCO, 2003, 2006; UNICEF, 2005). Similarly, existing and potential conflicts over water at the international level have elicited an ongoing academic and political debate and a number of important initiatives oriented at preventing conflict and promoting water sharing and cooperation (e.g. Cosgrove, 2003). We will come back to water conflicts but let us first consider briefly the notion of water uncertainty and risk.

Arguably, the ultimate water uncertainty concerns the very survival of the hydrosphere, and particularly its freshwater component. Pressures on available freshwater are driven by contradictory forces such as the rising water volumes extracted for human uses and the need to slow down and reduce water abstractions to restore and protect the fragile equilibrium of ecosystems and water bodies. In particular, water needed for agriculture, which currently accounts for about 70 per cent of the world's freshwater consumption (estimates indicate that in some developing countries, but also in certain developed countries, irrigation uses up to 85 per cent of freshwater abstracted), poses a crucial challenge (Bruinsma, 2003: 138; World Bank, 2004: 5, 14). For instance, the UN Food and Agriculture Organization forecasts that developing countries will need an average increase of 14 percent in irrigation water withdrawals until the year 2030, which according to FAO will not have a significant impact on the agreggate available freshwater (Bruinsma, 2003: 140–142; the document admits that individual countries are already in a critical situation). However, environmentalists claim that to stop the generalized overpumping of aquifers, falling water tables, and rapid deterioration of aquatic ecosystems water abstractions should be significantly reduced to restore sustainable water levels (Brown, 2005: Chapter 6). The critics point at dramatic examples such as the Dead Sea (Friends of the Earth, 2006) and the Aral Sea in Central Asia (Altyev, 2006), which have shrunk to a fraction of their original sizes as a result of extensive irrigation and water-consuming industrial activities, and these are just two examples in a long list of dying rivers, lakes, aquifers, wetlands and water bodies (Brown, op. cit.). In this context, it is difficult to foresee how we could possibly achieve simultaneously food security and sustainable water management. Similar dilemmas are faced in other areas of water management owing to competing demands on freshwater sources coming from rising living standards in urban areas of developing countries and from the expansion of cash crops and tourism in water-scarce regions, or from the worldwide destruction of mangroves through the expansion of shrimp farming, to mention just a few areas of concern. Other authors have also examined how social cleavages grounded on poverty, gender, and ethnicity, among other factors, impinge on the water insecurity affecting large sectors of the world's population (Webb and Iskandarani, 1998). These and other water uncertainties, in turn, are intimately related to existing or potential conflicts over water, which we examine next.

3.1.1 *Water conflicts*

The prospect that social and political conflicts over the distribution and allocation of water will increasingly "become a key part of the 21st-century landscape" is regularly restated by international leaders (e.g. van Ginkel, 2001). For instance, in February 2006 the British government issued a dramatic warning about the increased likelihood of "wars over water" and announced that its military forces must be prepared to intervene in "humanitarian disaster relief, peacekeeping and warfare" related to dwindling natural resources, particularly water (The Independent, 2006). This is not entirely surprising given that over the last few decades international security experts have warned that water was becoming more important than oil as a potential source of conflicts around the world (Gleick, 1993, 2000). Some authors have pointed out that the fact that global freshwater sources are unevenly and irregularly distributed, that some regions of the world are extremely water-short, and that water bodies are often shared by two or more countries is a looming source of conflicts, and the situation would be set to worsen as we progress into the twentieth-first century. These warnings seem to have good ground when we consider that 263 river basins, where about half of the world population is located, are shared by two or more countries (Cosgrove, 2003: 1). It is also estimated that fewer than 10 countries control about 60 percent of the world's freshwater sources, and a large number of groundwater aquifers are shared by two or more countries (Ohlsson, 1992; Samson and Charrier, 1997). Nevertheless, this notion that international water wars are inminent is fiercely contested by authors who argue that there is scarce historical evidence in favour of the hypothesis that transboundary waters tend to be the cause of war between countries and that rather peaceful cooperation in water sharing would have been the main international pattern for millennia (Allan, 2001; Cosgrove, 2003: 10–11; Yoffe et. al., 2004).

This highly relevant debate on the potential for international water conflict and cooperation is far from being settled. However, there is a second dimension of water conflicts that continues to receive relatively less attention in the mainstream water policy literature: intra-national water conflicts. This characterization may be misleading, as in fact in many cases water conflicts have both an inter- and an intra-national dimension. Nevertheless, the focus here is particularly on social struggles over water that range from confrontations over the control of water bodies and water infrastructure to urban conflicts over the inequalities and inefficiencies in the access to essential water services. On this subject, there is solid historical evidence showing that the control of water and water systems has played a significant role in the emergence of social and political conflicts, and continue to do so. Thus, water control has been a major factor in the establishment and consolidation of asymmetrical power relations often leading to structural conditions of inequality and injustice in the access to water, not just in the classical "hydraulic civilizations" studied by Karl Witffogel (Wittfogel, 1956, 1959) but also in recent centuries and to the present time. Among other cases it can be mentioned Bolivia (Crespo Flores et. al., 2003), India (Shiva, 1992), Italy (Santino, 1994, 2003), Mexico (Musset, 1991; Bennett, 1995; Perló Cohen and González Reynoso, 2005; Castro, 2006), Spain (Arrojo Agudo and Martínez Gil, 1999; BCFS, 2004), and the United States (Meyer, 1984; Worster, 1985; Hundley, 1992; Berry, 1998), just to mention some examples.

In more recent years, the record of intra-national water conflicts include from peaceful demands to the authorities, judicial litigation, demonstrations, mass parades, and other forms of civic protest including civil disobedience such as non payment of taxes or water bills, to direct confrontations involving in the extreme the destruction of property (e.g. destruction of water infrastructure) and often the loss of human lives. Although these forms of water conflict have become widespread around the world (see, for instance, Shiva, 2002; Bouguerra, 2003; Barraqué and Vlachos, 2006), they tend to receive less attention in the mainstream water policy literature. However, this is arguably one of the most difficult challenges facing water governance in the twentieth-first century: while it may be possible that the predictions about future international water wars are exaggerated, the occurrence of intra-national social struggles fuelled by water inequality and injustice is unlikely to diminish in the foreseeable future.

3.2 *Water conflict as an object of knowledge*

As suggested in the previous discussion, water conflicts are part and parcel of wider social and political confrontations between alternative, often antagonistic societal projects, confrontations that are at the heart of the process of governance. However, the aim of this chapter is not to explore the confrontations themselves but to contribute towards the development of interdisciplinary coordinations in the production of scientific knowledge about water conflicts, which requires the exploration of how physical-natural and social processes interweave. In this regard, the evidence shows that the emergence of water conflicts is seldom the sole result of "natural" causes such as freshwater scarcity in arid and semi-arid regions. Cooperation, solidarity and successful bottom-up "water governance" arrangements have been developed in very adverse conditions of natural water scarcity, as in the classical example of medieval Valencia in Spain (Glick, 1970), but also in places as pre-colonial Bali in Indonesia (Geertz, 1980), Ceylon (Leach, 1959), or the Philippines (Ostrom, 1990) to mention a few typical cases. Conversely, there are obvious examples of protracted social conflicts over water in the context of very favorable hydrological conditions such as for instance in Guayaquil, Ecuador (Swyngedouw, 2004) or in the state of Chiapas in Mexico (Castro, 1992).

Unfortunately, on the one hand, the production of scientific knowledge about water conflicts, and in general about water, is characterized by high fragmentation along the lines of entrenched epistemic cultures that continue to develop largely unconnected from each other. On the other hand, however, the existing fragmentation in the knowledge about water conflicts offers an excellent opportunity to develop genuine interdisciplinary approaches that bring together the expertise developed in the techno- and the social sciences, and other epistemic fields. In this regard, relevant suggestions for the study of water conflicts can be found in the interdisciplinary field of political ecology, which is concerned with the study of "ecological distribution conflicts" (Guha and Martínez Alier, 1997: 31). Political ecological perspectives have inspired an expanding body of water research (Swyngedouw et. al., 2002) on a number of problems ranging from the links between conflicts over the provision of urban water services and the process of global capital accumulation (Swyngedouw, 1999, 2004), the multidimensional character of water struggles arising from neoliberal water reform policies (Laurie et. al., 2002; Laurie 2007), to the interrelations between intra-national water conflicts and the long-term development of citizenship (Castro, 2006), just to give a few examples.

However, the development of interdisciplinary strategies for the production of knowledge across the techno- and the social sciences continues to be difficult and progress is slow. Among other aspects that require further consideration is the fact that knowledge about water is produced from a number of distinctive, often unconnected epistemic perspectives, and the resulting fragmentation of knowledge tends to become structural owing to entrenched disciplinary and institutional power configurations, a problem which is not limited to the field of water research (e.g., Knorr Cetina, 1999). For instance, in our studies on contemporary social conflicts over water in Mexico we identified a number of distinct epistemic subjects involved in water management activities who understand and explain water conflicts from very different, often unconnected perspectives (Castro, 1995; 2006). For the sake of the analysis we derived from the empirical research the existence of three epistemic subjects: the water expert, mainly water engineers and others directly involved in the techno-scientific aspects of water management, the water functionary, who are members of the bureaucratic and policy-institutional apparatuses in charge of water management activities, and the critical social scientist, referring broadly to the work of social scientists producing knowledge about water from a critical perspective such as contemporary political ecology. The evidence suggests that these different subjects construct their knowledge about water conflicts on the basis of different rationalities and epistemic structures, which underpin the identification of very different observables for the identification and explanation of "water conflicts" (on the concept of observable

Table 1. Water conflict and epistemic subjects.

Epistemic subject	Rationality	Observables	
Water expert (Geo-hydrologists; hydraulic engineers, etc.)	Techno-scientific	Quantitative indicators Physical-natural and technical conditions and drivers Water resources	
Administrative-financial experts	Market	Quantitative indicators Economic efficiency Market criteria	
Water functionary	Policy-administrative	Bureaucratic norms Electoral and party-political considerations	"Water conflict"
Ecologist	Ecological	Indicators of sustainability-insustainabiity Ecosystems	
Critical social scientist	Socio-political	Power configurations Structural inequalities Social identities Languages of valuation	

see Piaget, 1978: 43–6; 1977: 342–6.). Table 1, where we have added additional examples of epistemic subjects involved in water research, illustrates schematically the diverse approaches of these subjects to "water conflicts".

For instance, in the early 1980s Mexican water experts elaborated a map of "conflicts over water in the main Mexican cities" to predict the ocurrence of such events between 1980 and the year 2000 (SARH, 1981: 50). A close examination showed that they grounded their analysis on quantitative observables, such as the interactions between water availability, demand, supply, consumption, cost and population, urban and industrial growth over the period under analysis. They conceptualized urban water conflicts from a techno-scientific perspective, and therefore conflict in their analysis would be the result of the lack of expected correspondence between quantitative variables, such as a geometrical increase of water demand in the arid areas of the country where water availability was already compromised in 1980. In contrast, for the "water functionary" the notion of water conflicts places the emphasis on a different array of observables, which can also be illustrated from our research on Mexico. Besides the techno-scientific rationality (after all many water functionaries are techno-scientists by training) they are subject to policy-bureaucratic, and often also party-political, interests such as concerns about the impact of water conflicts on electoral prospects. Therefore, their observables are for instance the recurrent events of urban social protest over the poor quality of the water services or the civil disobedience of water users who have decided not to pay their bills in protest for a recent hike in the tariff. In general, the water functionary must deal with processes that fall outside the technical domain of the expert, such as "popular discontent", "the social and economic characteristics of the population" that create conditions for water troubles, or the inherent contradictions between "the economic, social, psychological and environmental values of water" (SARH, 1981: 14). In turn, the critical social scientist is concerned with the task of making observable the intertwining between the social regularities and physical-natural processes that are at the heart of water conflicts. For instance, and remaining with the Mexican example, the socio-political rationality of this subject provides a framework for inquiring into the socio-economic and political mechanisms that underpin the exclusion of a large fraction of the population from access to safe and affordable water services, a major cause of water conflict in Mexico.

A similar scheme of analysis could be applied for the treatment of water conflicts in other areas of activity, such as the widespread struggles against "water privatization" or the opposition to large scale hydraulic works like dams and inter-basin water transfers. However, the scheme in Table 1 is only a simplification to cast light on the distinctive rationalities operating in water research and policy, which may help to better understand some of the key obstacles for interdisciplinary coordination in the study of water conflicts. It is important to clarify that the epistemic subjects represent bodies of knowledge and traditions of thought, not individuals or collective actors, who in practice may embody one or more epistemic cultures. We believe that the identification of the conceptual frameworks, rationalities, and observables operating in the field of water research, as we have attempted to sketch here, is an esencial exercise to strengthen the foundations of meaningful interdisciplinary in this field.

4 CONCLUSIONS

There is increasing recognition that the "water crisis" is mainly a crisis of governance. Unfortunately, although the use of the concept of "governance" often assumes a shared understanding, in fact there exist underlying confrontations between rival theoretical bodies of knowledge and political and cultural traditions for which governance has entirely different meanings. Moreover, much of the mainstream debate on the topic has been aimed at depoliticising the processes under discussion and presenting them as mainly (or even merely) "technical" in nature, probably in the belief that depoliticising water management activities would provide opportunities for abating or at least controlling water uncertainty and conflict. An important aspect of this debate concerns the question of social participation in relation to problems of water uncertainty and risk, which is a central component of the process of democratic governance. How are the risks associated with water management communicated to the wider public? How do citizens participate in the process? What mechanisms are available for them to participate? How are the societal goals informing water policy identified? What ends and values are prioritized in these goals? What means are chosen to pursue those ends and values? What languages of valuation are chosen in the process? Who takes these decisions? Who are the actors that these decisions intend to benefit? What mechanisms of democratic control exist to monitor decision takers and implementors of water policy? These and other similar questions are at the heart of the process of democratic governance, and we know that this process is undergoing a severe crisis worldwide. Unsurprisingly, this crisis of water governance is being increasingly expressed in the form of inter-, and particularly intra-national social and political conflicts over water, which present one of the most formidables challenges for the scientific community involved in water research and practice.

Our conclusion draws on the perspective of one of the epistemic subjects sketched above, the critical social scientist, which stems from a long-standing tradition in the social sciences concerned with developing the appropriate cognitive structures for making observable such structural regularities as cyclical social conflicts —whether in relation to water or not. However, the task of elaborating adequate explanations of the causes and consequences of water uncertainty and inequality requires the development of further interdisciplinary coordination between the intellectual domains of, for instance, water engineers, hydrologists, and social scientists, which to date has been a slow and relatively fruitless endeavour. The existing gap between the intellectual domains developed by techno-scientists and critical social scientists concerned with social inequality and struggle remains a major obstacle to achieve this goal. The persistence of this obstacle continues to hamper our full understanding of "water conflicts", and consequently diminishes the chances we may have to avoid their negative consequences, which almost systematically affect the most vulnerable sectors of the population.

In this connection, there is a need for adopting a critical perspective of the understanding of water governance as an instrument, a supposedly neutral policy tool, which aims at depoliticising what is essentially a political process. The idealized and instrumental approaches to water governance tend to neglect in their analysis, despite rhetorical recognition to the contrary, the existence of fundamental social divisions underpinning water insecurity, injustice, and inequality, which are major drivers of water conflict. Thus, a truly inter-disciplinary approach to the problem must strive to make observable those processes that create and reproduce the structural socio-economic and political inequalities that continue to preclude a large sector of the world's population not only from participating in the governance of water, but even from accessing essential volumes of safe water for daily survival. This kind of approach requires addressing "water conflicts" as an object of knowledge on its own right, which constitutes a crucial step towards transforming the unacceptable conditions characterizing the "water crisis". Our work seeks to make a contribution towards this daunting venture by calling for efforts to develop higher levels of coordination between the different cognitive structures and epistemic cultures involved in the production of knowledge about water.

5 ACRONYMS

ADB Asian Development Bank
CBI Council for Biotech Information
DNA Deoxyribonucleic acid
EC European Commission
EUWATER European Network for a New Water Culture
FA Food and Agriculture Organization
GWP Global Water Partnership
MCMA Mexico City Metropolitan Area
MDGs Millennium Development Goals
MSSRF M.S. Swaminathan Research Foundation
OECD-WPB Organization for Economic Co-operation and Development – Working Party on Biotechnology
SARH Secretariat of Agriculture and Hydraulic Resources (Mexico)
SEMARNAT Secretariat of Environment and Natural Resources (Mexico)
UNCED United Nations Conference on Environment and Development (The Earth Summit 1992)
UNDP United Nations Development Programme
UNEP United Nations Environment Programme
UNESCO United Nations Educational, Scientific, and Cultural Organization
UNICEF United Nations Children's Fund
USAID United States Agency for International Development
WCW World Commission on Water for the 21st Century
WHO World Health Organization
WWF World Water Forum

REFERENCES

Allan, T. (2001) *The Middle East Water Question. Hydropolitics and the Global Economy.* London, I. B. Tauris & Co.

Altyev, T. (2006) *Address of the Chairman of the Executive Committee of the International Fund on the Aral Sea.* Available from: http://enrin.grida.no/aral/aralsea/english/obr/obr.htm.

Amin, A. (1997) *Beyond Market and Hierarchy: Interactive Governance and Social Complexity.* Cheltenham, Elgar.

Annan, K. (2002) *UN Secretary General's Message for the World Water Day.* Available from: http://www.unesco.org/water/water_celebrations/water_day_2002.shtml.

Arrojo Agudo, P. & Martínez Gil, F.J. (Coords.) (1999) *El Agua a Debate desde la Universidad. Hacia una Nueva Cultura del Agua.* Zaragoza: Institución Fernando el Católico.

Asian Development Bank (ADB) (1995) *Governance: Sound Development Management.* Manila, ADB.

Barraqué, B. & Vlachos, E. (eds) (2006) *Urban Water Conflicts: An Analysis on the Origins and Nature of Water-related Unrest and Conflicts in the Urban Setting.* Paris, UNESCO Working series SC-2006/WS/19.

BCFS – Boletín Ciudades para Un Futuro Más Sostenible (2004) *El Plan Hidrológico español: un estudio de caso*. Available from: http://habitat.aq.upm.es/boletin/n27/lista.html.

Beck, U. (1992) *Risk Society. Towards a New Modernity*. London, Sage.

Beck, U. (1992, 1998) Politics of risk society. In: Franklin, J. (ed.) *The Politics of Risk Society*. Cambridge, Polity Press, pp. 9–22.

Bennett, V. (1995) *The Politics of Water: Urban Protest, Gender, and Power in Monterrey, Mexico*. Pittsburgh, University of Pittsburgh Press.

Berry, K. (1998) Race for water? Native Americans, Eurocentrism, and Western water policy. In: Camacho, D.E. (ed.) *Environmental Injustices, Political Struggles*. Durham, Duke University Press.

Bouguerra, M.L. (2003) *Les Batailles de l'Eau. Pour un Bien Commun de l'Humanité*. Paris, Les Editions de l'Atelier.

Brook Cowen, P.J. & Cowen, T. (1998) Deregulated private water supply: a policy option for developing countries. *Cato Journal*, 18(1), 21–41.

Brown, L.R. (2005) *Outgrowing the Earth: The Food Security Challenge in an Age of Falling Water Tables and Rising Temperatures*. New York, W. W. Norton & Co.

Bruinsma, J. (2003) *World Agriculture: Towards 2015/2030. An FAO Perspective*. London, Earthscan.

Camdessus, M. (Chair) (2003) *Financing Water for All. Report of the World Panel on Financing Water Infrastructure*. World Water Council, Global Water Partnership. Available from: http://www.worldwatercouncil.org/download/CamdessusReport.pdf.

Castro, J.E. (1992) *El Conflicto por el Agua en México. Los Casos de Tuxtla Gutiérrez, Chiapas y Ciudad Juárez, Chihuahua, 1986–1991* (unpublished thesis, Master in Social Sciences). Mexico City, Latin American Faculty of Social Sciences (FLACSO).

Castro, J.E. (1995) Decentralization and modernization in Mexico: the management of water services. *Natural Resources Journal*, 35(3), 1–27.

Castro, J.E. (2006) *Water, Power, and Citizenship. Social Struggle in the Basin of Mexico*. Houndmills, Basingstoke, Palgrave-Macmillan.

Cohen, J.L. & Arato, A. (1994) *Civil Society and Political Theory*. Cambridge, Mass. and London, The Massachusetts Institute of Technology (MIT) Press.

Cosgrove, W.J. (Comp.) (2003) Water Security and Peace. A Synthesis of Studies Prepared under the PCCP–Water for Peace process (An UNESCO–Green Cross International Initiative). Technical Documents in Hydrology, PC-CP Series, # 29. Paris, UNESCO. Available from: http://unesdoc.unesco.org/images/0013/001333/133318e.pdf.

Crespo Flores, C., Laurie, N. & Ledo, C (2003) *Cochabamba case study*. Research Project Report. In: Castro, J.E. (Coord.) PRINWASS Project. Oxford, University of Oxford. Available from: http://users.ox.ac.uk/~prinwass/.

Delanty, G. (2000) *Citizenship in a Global Age. Society, Culture, Politics*. Buckingham, Open University Press.

Dryzek, J. (1997) *The Politics of the Earth. Environmental Discourses*. Oxford, Oxford University Press.

European Commission (EC) (2000) *The European Community's Development Policy. Communication from the Commission to the Council and the European Parliament*. Available from: http://europa.eu.int/eur-lex/en/com/cnc/2000/com2000_0212en01.pdf.

European Commission (EC) (2002a) *EU Water Initiative: Water for Life. Health, Livelihoods, Economic Development, Peace, and Security*. Brussels. Available from: http://europa.eu.int/comm/research/water-initiative/index_en.html.

European Commission (EC) (2002b) *Water Management in Developing Countries: Policy and Priorities for EU Development Cooperation*. Communication from the Commission to the Council and the European Parliament. Available from: http://europa.eu.int/eur-lex/en/com/cnc/2002/com2002_0132en01.pdf.

EUWATER (2005) *European Declaration for a New Water Culture*. Saragossa, New Water Culture Foundation – EUWATER Network. Available from: http://www.unizar.es/fnca/euwater.

Esman, M.J. (1991) Management Dimensions of Development: Perspectives and Strategies. West Hartford, Kumarian Press.

Food and Agriculture Organization (FAO) (2003) *Review of World Water Resources by Country* (aquastat). Rome, FAO. Available from: http://www.fao.org/DOCREP/005/Y4473E/Y4473E00.HTM.

Friends of the Earth (2006) *Projects – the Dead Sea*. Available from: http://www.foeme.org/projects.php?ind=21.

Geertz, C. (1980) *Negara. The Theatre State in Nineteenth-Century Bali*. Princeton, NJ, Princeton University Press.

Gleick, P.H. (1993) Water and conflict: Fresh water resources and international security. *International Security*, 18(1), 79–112.

Gleick, P.H. (2000) *Water Conflict Chronology*. Oakland, Pacific Institute for Studies in Development, Environment and Security.

Glick, T.F. (1970) *Irrigation and Society in Medieval Valencia*. Cambridge, Mass., Harvard University Press.

Global Water Partnership (GWP) (2003) *Effective Water Governance. Learning from the Dialogues*. Available from: http://www.gwpforum.org/gwp/library/Effective%20Water%20Governance.pdf, GWP.

Guha, R. & Martínez-Allier, J. (1997) *Varieties of Environmentalism. Essays North and South*. London, Earthscan.

Hanf, K. & Jansen, A.-I. (eds) (1998) *Governance and Environmental Quality: Environmental Politics, Policy and Administration in Western Europe*. Harlow, Addison Wesley Longman.

Heinalt, H., Getimis, P., Kafkalis, G., Smith, R. & Swyngedouw, E. (eds) 2002. *Participatory Governance in Multi-level Context*. Opladen, Leske & Budrich. pp. 107–131.

Held, D. (1995) *Democracy and the Global Order: From the Modern State to Cosmopolitan Governance*. Cambridge, Polity Press.

Hirst, P. (1994) *Associative Democracy: New Forms of Economic and Social Governance*. Cambridge, Polity Press.

Hundley, N. Jr. (1992) *The Great Thirst. Californians and Water, 1770s–1990s*. Berkeley, University of California Press.

Kasperson, J.X. & Kasperson. R.E. (2001) *Global Environmental Risk*. Tokyo, United Nations University Press and Earthscan.

Kasperson, J.X., Kasperson, R.E. & Turner, B.L. II (eds) (1995) *Regions at Risk. Comparisons of Threatened Environments*. Tokyo, United Nations University Press.

Kaviraj, S. & Khilnani, S. (eds) (2001) *Civil Society: History and Possibilities*. Cambridge, Polity.

Knorr Cetina, K. (1999) *Epistemic Cultures. How the Sciences Make Knowledge*. Cambridge, Mass., Harvard University Press.

Laurie, N. (ed.) (2007) *Special Issue on "'Pro-poor' water: Past present and Future Scenarios"*. Geoforum, 38(5).

Laurie, N., Radcliffe, S. & Andolina, R. (2002) The new excluded 'indigenous'?: The implications of multi-ethnic policies for water reform in Bolivia. In: Seider, R. (ed.) *Multiculturalsim in Latin America. Indigenous Rights, Diversity and Democracy*. Houndmills, Basingstoke, Palgrave-Macmillan. pp. 252–276.

Leach, E.R. (1959) Hydraulic society in Ceylon. In: *Past and Present*, pp. 2–26.

Leys, C. (2001) *Market-Driven Politics. Neoliberal Democracy and the Public Interest*. London, Verso.

Martínez-Alier, J. (2002) *The Environmentalism of the Poor. A Study of Ecological Conflicts and Valuation*. Cheltenham y Northampton, Edward Elgar.

McGranahan, G., Jacobi, P., Songsore, J., Surjadi, Ch. & Kjellén, M. (2001) *The Citizens at Risk. From Urban Sanitation to Sustainable Cities*. London, Earthscan.

Meyer, M.C. (1984) *Water in the Hispanic Southwest. A Social and Legal History, 1550–1850*. Tucson, Arizona, The University of Arizona Press.

Musset, A. (1991) *De l'Eau Vive à l'Eau Morte. Enjeux Techniques et Culturels dans la Vallée de Mexico (XVIe-XIXe Siècles)*. Paris, Éditions Recherche sur les Civilisations.

Newbery, D.M. (1999) *Privatization, Restructuring, and Regulation of Network Utilities. The Walras-Pareto Lectures*. Cambridge, Mass., The MIT Press.

Ohlsson, L. (ed.) (1992) *Regional Case Studies of Water Conflicts*. Göteborg, Peace and Development Research Institute (PADRIGU), University of Göteborg.

Ostrom, E. (1990) *Governing the Commons: The Evolution of Institutions for Collective Action*. Cambridge, Cambridge University Press.

Perló Cohen, M. & González Reynoso, A.E. (2005) *¿Guerra por el Agua en el Valle de México? Estudio sobre las Relaciones Hidráulicas entre el Distrito Federal y el Estado de México*. Mexico City, National Autonomous University of Mexico, Autonomous Metropolitan University, and Friedrich Ebert Foundation.

Petrella, R. (2001) *The Water Manifesto. Arguments for a World Water Contract*. London, Zed Books.

Piaget, J. (1977) *The Grasp of Conciousness*. London, Routledge and Kegan Paul.

Piaget, J. (1978) *The Development of Thought. Equilibration of Cognitive Structures*. Oxford, Basil Blackwell.

Picciotto, R. (1997) Putting institutional economics to work: From participation to governance. In: Clague, Ch. K. (ed.) *Institutions and Economic Development: Growth and Governance in Less-developed and Post-socialist Countries*. Baltimore, John Hopkins University Press.

Saleth, R.M. & Dinar, A. (1999) Evaluating water institutions and water sector performance. World Bank Technical Paper 447, Washington DC, World Bank.

Samson, P. & Charrier, B. (1997) *International Freshwater Conflict. Issues and Prevention Strategies*. Geneva, Green Cross International.

Santino, U. (1994) *Il ruolo della mafia nel saccheggio del territorio*. Palermo, Casa Europa, Centro Siciliano di Documentazione "Giuseppe Impastato".

Santino, U. (2003) *L'acqua rubata. Dalla mafia alle multinazionali*. Palermo, Centro Siciliano di Documentazione "Giuseppe Impastato". Available from: http://www.centroimpastato.it/publ/online/Acqua.php3.

Secretaría de Agricultura y Recursos Hidráulicos (SARH), Comisión del Plan Nacional Hidráulico (1981) *Plan Nacional Hidráulico*. Mexico City, SARH.

Shiva, V. (1992) *The Violence of the Green Revolution. Third World Agriculture, Ecology and Politics*. Mapusa, Goa, India, The Other India Press.

Shiva, V. (2002) *Water Wars. Privatization, Pollution, and Profit*. Cambridge, MA, South End Press.

Stiglitz, J.E. (2002) *Globalization and its Discontents*. London, Penguin.

Strang, V. (2004) *The Meaning of Water*. Oxford, Berg Publishers.

Streeck, W. & Schmitter, Ph. (1985) *Private Interest Government*. London, Sage.

Swyngedouw, E.A. (1999) *Flows of Power: Nature, Society and the City*. Oxford, Oxford University Press.

Swyngedouw, E. (2004) *Social Power and the Urbanization of Water. Flows of Power*. Oxford, Oxford University Press.

Swyngedouw, E.A., Maria Kaïka & Castro, J.E. (2002) Urban water: A political-ecology perspective. *Built Environment*, 28(2), 124–137.

The Independent (2006) *Armed forces are put on standby to tackle threat of wars over water*. London, 28 February 2006.

United Nations (UN) (1980) *International Drinking Water Supply and Sanitation Decade: Report of the Secretary General*. New York, UN.

United Nations (UN) (1990) *The New Delhi Statement: Some for all rather that more for some*. Global Consultation on Safe Water and Sanitation for the 1990s. New Delhi (10–14 September 1990), UNDP and the Government of India.

United Nations (UN) (2000) *Millennium Declaration*. New York, UN. Available from: http://www.un.org/millennium/summit.htm.

United Nations (UN) (2002) *Key Commitments, Targets and Timetables from the Johannesburg Plan of Implementation*. World Summit on Sustainable Development, Johannesburg. Available from: http://www.johannesburgsummit.org/html/documents/summit_docs/2009_keyoutcomes_commitments.doc.

United Nations Development Programme (UNDP) (1997) *Governance for Sustainable Human Development*. New York, UNDP. Available from: http://magnet.undp.org/policy/default.htm.

United Nations Development Programme (UNDP) (1998) *Reconceptualizing Governance*. New York, UNDP. Available from: http://magnet.undp.org/Docs/!UN98-21.PDF/Recon.htm.

United Nations Development Programme (UNDP) (2004) *Water Governance for Poverty Reduction. Key Issues and the UNDP Response to Millennium Development Goals*. New York, UNDP.

United Nations Development Programme (UNDP) (2006) *Public-Private Partnerships for the Urban Environment (PPPUE)*. New York, UNDP. Available from: http://pppue.undp.org.

United Nations Environment Programme (UNEP), United Nations Children's Fund (UNICEF), and World Health Organization (WHO) (2002) *Children in the New Millennium: Environmental Impact on Health*. Available from: http://www.who.int/water_sanitation_health/hygiene/settings/millennium/en/index.html.

United Nations Educational, Scientific and Cultural Organization – UNESCO, World Water Assessment Programme (2003) *Water for People, Water for Life*. The United Nations World Water Report. Paris, UNESCO and Berghahn Books.

United Nations Educational, Scientific and Cultural Organization – UNESCO, World Water Assessment Programme (2006) *Water, a Shared Responsibility*. The United Nations World Water Report 2. Paris, UNESCO and Berghahn Books.

United Nations – Habitat (2003) *Water and Sanitation in the World's Cities: Local Action for Global Goals*. London, Earthscan Publications Ltd.

United Nations Children's Fund (UNICEF) (2005) *Water, Environment, and Sanitation*. Available from: http://www.unicef.org/wes/.

van Ginkel, H. (2001) *Discourse of the United Nations Under-Secretary General at the Stockholm Water Symposium, 13 August 2001*. Quoted in the Financial Times, 14 August 2001, p. 6.

van Steenbergen, B. (1994) *The Condition of Citizenship*. London, Sage.

Ward, C. (1997) *Reflected in Water, A Crisis of Social Responsibility*. London, Casell.

Webb, P. & Iskandarani, M. (1998) *Water Insecurity and the Poor. Discussion Papers on Development Policy*. Bonn, Center for Development Research.

Wittfogel, K.A. (1956) The hydraulic civilizations. In: Thomas, W.L. (ed.) *Man's Role in Changing the Face of the Earth.* Chicago, The University of Chicago Press, pp. 152–164.

Wittfogel, K.A. (1959) *Oriental Despotism. A Comparative Study of Total Power.* Third reprinting. New Haven, Yale University Press.

World Bank (2004) *Water Resources Sector Strategy. Strategic Decisions for World Bank Engagement. Washington, The World Bank.* Available from: http://www-wds.worldbank.org/external/default/WDSContentServer/IW3P/IB/2004/06/01/000090341_20040601150257/Rendered/PDF/28114.pdf.

World Health Organization (WHO) (2003a) *The Right to Water.* Geneva, WHO. Available from: http://www.who.int/water_sanitation_health/en/righttowater.pdf.

World Health Organization (WHO) (2003b) *Water Supply, Sanitation and Hygiene Development.* Geneva. WHO. Available from: http://www.who.int/water_sanitation_health/hygiene/en/.

World Health Organization (WHO) (2005) *Health and the Millennium Development Goals.* Geneva. OMS. Available from: http://www.who.int/mdg/publications/MDG_Report_08_2005.pdf.

World Health Organization – Regional Office for Europe (WHO-Europe) (2003) Dealing with uncertainty: how can the precautionary approach help protect the future of our children? In: *Fourth Ministerial Conference on Environment and Health. Stockholm, 26–27 June 2003.* Available from: http://www.who.dk/document/eehc/2ndigoprepmeet_11.pdf.

World Bank (2006) *Public-Private Partnership in Infrastructure.* Washington: The World Bank. Available from: http://web.worldbank.org/WBSITE/EXTERNAL/WBI/WBIPROGRAMS/PPPILP/0,,menuPK:461139~pagePK:64156143~piPK:64154155~theSitePK:461102,00.html.

World Water Council (2004) *Analysis of the 3rd World Water Forum*, Marseille, WWF.

Worster, D. (1985) *Rivers of Empire. Water, Aridity, and the Growth of the American West.* New York-Oxford, Oxford University Press.

Yoffe, S., Fiske, G., Giordano, M., Giordano, M., Larson, K., Stahl, K. & Wolf, A.T. (2004) Geography of international water conflict and cooperation: Data sets and applications. *Water Resources Research*, 40(5), 1–12.

CHAPTER 12

The future of water management: The case for long-range hydraulic interconnections

M. Fanelli
Consultant, Former Director of the Hydraulic & Structural Research Centre of ENEL – Milan (Italy)

ABSTRACT: The increasing anthropisation of the environment poses an urgent problem of improvement in the management of water resources as well as in the control of hydro-geological hazards. Imbalances in spatial and temporal distribution of water availability (as well as of climatic excesses) need be redressed. Consideration of these requirements at a merely local scale can be insufficient, uneconomical or ineffective. The case can be made for the analysis of the feasibility of long-range hydraulic interconnection systems, aiming at making it possible that the widely unbalanced availabilities of distant regions with complementary needs could be shared to mutual advantage. Such systems have been seriously studied, and their construction has indeed begun, in some important extra-European countries (the paper briefly reviews the examples of Lybia, China and Brazil); the huge investments and the long time spans needed for their realisation highlight the necessity of a careful appraisal of the cost/benefit balance as well as of a sound financial planning. The subject of the social acceptance of such projects is also mentioned, insofar as the likely occurrence of public opinion opposition (on grounds of perceived needs of ecological protection and of perceived priority of competitive infrastructures) is not to be underestimated.

In the last part of the paper concerning the Italian situation a recent preliminary feasibility study for the definition of such an hydraulic interconnection system laid across the dorsal mountain range of the Apennines is presented. More in-depth studies will however be necessary before final conclusions about the advisability of advancing a firm proposal for such a system can be made.

1 THE GLOBAL BACKGROUND

The whole world over, water is becoming a scarce resource, so much so that its price to the public is rapidly growing and some political analysts even do not hesitate to raise the dark prospect of future 'wars for water'. The issue of water resources management is therefore taking on a critical role not only in the general framework of the agenda of sustainable growth, as a key factor in the fight for survival of endangered populations or in the aspirations of less favoured nations to a more acceptable life quality, but even in developed countries. Indeed, the media news reflect ever more frequently the fact that the water-related problems are gaining centrality and priority not only in the underprivileged countries, where water shortages hinder progress, but more and more even in the industrialised ones, where rising consumptions and climate excesses put the existing infrastructures under heavy stress and foreshadow a possible impending impairment of living standards. Therefore both rising demands and the perceived evidence of climatic changes for the worse, causing every year extensive damage, lost production and loss of lives, cause the pressure to mount – at the public opinion as well as at the decisional levels- for better hydrological hazards management.

In this context it is interesting to note that in the last decades the central planning Authorities of some important countries (Lybia, China and Brazil being at the moment the most conspicuous examples, see § 3) have actively promoted the design, and even initiated the building, of complex, costly systems of long-range hydraulic networks, their extended lines of conveyance conduits being punctuated by reservoirs and pumping stations where dictated by topographic circumstances. These systems of interconnecting tunnels, reservoirs and canals fulfil the function of making possible the sharing of the abundant natural hydraulic inflows of regions experiencing above average precipitations (or from zones endowed with other relevant water sources, as e. g. fossil underground reserves, see § 3.1) with distant arid or semi-arid zones. More specifically, the overall capacity of the system reservoirs, to be created through the erection of dams, fulfil the function of transferring the surplus inflows of rainy seasons to the times of drought (the stored volumes avoiding at the same time most of the more damaging floods, and the distributed capacity of not-too-distant reservoirs improving the dynamic response of the system, which otherwise would suffer from excessively large time and space changes in hydraulic head). Side benefits provided by such systems would include the possibility of increasing the production of hydroelectric power, provided the interconnected system be situated at a suitable elevation.

Interconnected systems of this kind can be many hundreds of kilometres long (even more than 1,000 km) and their cost may range in the tens of billions of € or of $. Their technical and economic feasibility has just now begun to appear as conceivable, as a consequence of the growing value attributed to the resource 'water' [and of the growing import of economic losses due to hydrological events] on one hand, and of the advancing technology of large infrastructure building on the other hand[1].

Besides the recent extra-European initiatives alluded to in the preceding lines (and illustrated in § 3), also in Europe, in the Author's opinion, some countries could reasonably consider the cost/benefit balance of such hydraulic long-distance interconnections. Italy is beyond doubt one of these countries (see § 4.1), and for this reason the Italian Company CESI SpA has recently undertaken the initial steps in order to carry out a pre-feasibility research study about the technical and economic possibility of creating a system of pressure tunnels interconnecting a sizable number of reservoirs distributed all along the Apennines mountain range that runs near the NW-SE axis of the Italian peninsula. Such a system would be more than 1,000 km long and comprise about 60 small or medium-sized reservoirs for a total capacity of nearly 2,000 million cubic metres (see § 4.2); it would be able to store most of the peak flood discharges of the Northern rivers, which cause annually very damaging floods and landslides, and to transfer (with the necessary time delay) significant volumes of water to the Southern regions, which suffer on the average from severe summer droughts. The storage capacity of the reservoirs distributed along the system would also help to alleviate the hydro-geologic disasters recurring during autumn or winter in the geologically unstable Southern territories as well as in some of the Northern regions.

The mentioned CESI research exercise will be succinctly described in the following § 4.2; the initial conception of this system assumes that both reservoirs and interconnecting tunnels will be situated at a uniform elevation (tentatively put at around 500 m above sea level, see § 4.2), thus creating a network of freely communicating vessels and hence dispensing with the necessity either of pumping stations or of regulating valves, in favour of the maximum simplicity of functioning and management of the system. More in detail, the crest of each spillway will be put at the same elevation and the pressure tunnels will be excavated at a uniform depth.

It goes without saying that in every study of this kind a thorough, exhaustive analysis of the cost and benefit balance of the enterprise must be carried out hand in hand with the optimisation of layout as well as with the appraisal of technical feasibility and of the options for financial planning; a reliable estimate of the costs and benefits being, admittedly, a difficult enough exercise in itself. In any case, in the Author's opinion the growing scarcity and value of the water resources, together with the increasingly destructive extremes of the climate, will probably force the planning Authorities of many a country to consider more and more seriously, in the near future, the prospect (if not indeed the convenience) of creating such large infrastructures, not unlike many other equally extended networks already accepted – and routinely taken for granted- into the mainstream fabric of modern society (long-range electric power transportation networks, oil and methane regional and trans-national pipelines, global communication networks...). The very large capital outlays

[1] History shows, anyway, that the construction of long-range hydraulic connections has occurred, albeit as exceptional, once-in-a-while achievements, even in ancient times; suffice it to mention, just to cite one example, the Roman aqueducts, some of which exceeded in length 100 kilometres.

required would have to be, necessarily, distributed over a span of many years (say about two decades as an order of magnitude), a time span anyway required by the lengthy construction process, and as a consequence the yearly allocations would be of the same order of magnitude as the economic losses – not to mention the loss of human lives on which no monetary evaluation can be attempted- now incurred every year (on the average) as a consequence of extreme hydrological events.

2 THE CONCEPT OF LONG-DISTANCE HYDRAULIC INTERCONNECTIONS: PROS AND CONS

In the first instance it has to be remarked that the 'long-range hydraulic interconnection systems' considered in the present paper should be set apart, in principle as well as in functions, from the 'changes of river watershed' plans that in some cases were put in effect in the past, with great costs and efforts, and that quite often caused unexpected drawbacks, especially in the form of harmful environmental side effects.

The difference between the two kinds of operation can be simply put as follows:

- A change in watershed usually diverts to a different final destination (a side valley, a different geographic watershed) the bulk of the river inflows, depriving the natural stream course of the inflows coming from its hydrographic basin down to the point of diversion.
- A long-range hydraulic interconnection puts the hydraulic resources of distant regions in mutual communication, so that the surplus of each of the connected regions may be shared with other regions suffering from a paucity of inflows, without depriving to the first ones of the availability of the useful contributions of the local rivers. In principle, no harmful environmental side effects should derive from this kind of infrastructure; quite otherwise, insofar as local problems tied to large floods in the hydrologically richer regions would be made less dangerous thanks to the draining of excessive natural inflows provided by the interconnection. (Figure 5 illustrates the worldwide experiences mixing the two types of projects).

In the second place, relevant consideration should be given to the important question of the probable public opinion reactions to any large-scale plan intended to modify the natural distribution of the water cycle inflows by means of artificial interventions on the territory. While in countries ruled by strongly centralised authoritarian regimes the planning and the actual building of such infrastructures are carried out overriding without regard a public opinion that may, to all practical effect, be either absent or muted, in democratic countries the mind of the citizens does not fail to find vocal expression, and to exert influence, thanks to the media and through the free actions of interest or political group associations. In the latter type of societies, indeed, past experience has repeatedly shown that enterprises of the kind herein considered are almost invariably viewed with heavy suspicion, and excite lively outbursts of fervent opposition, by large sections of the public opinion. This attitude is not to be easily changed, hence – for the topic in hand- the necessity of a serious effort on the part of the interconnection proponents in order to demonstrate in effective ways the usefulness (or indeed the necessity, as the case may be) of undertaking the realisation of such systems.

A further consideration of weight concerns the difficulties implied by the financial requirements, insofar as enterprises of this kind demand large capital outlays over a substantial period of time. Once again, these difficulties can be overcome with apparent ease by a strong central authority wielding absolute planning and allocation power, but in a democratic context, with strong parliamentary control over expenses to be met by a tight public budget, such financial necessities would come into open conflict with other needs, which would often appear, in the citizens' eye, to be of more pressing importance and thus would rank higher in their perceived scale of priorities. A rational objection with which it would be reasonable, and indeed advisable, to counter this opposition is of course the hard fact that in the absence of such infrastructures huge sums need anyway to be allocated, practically each year, in order either to repair the hydro-geological damages or to compensate the losses incurred by economic actors by way of halted or lost production brought about by floods or droughts; these allocations, which inevitably go to the detriment of more productive initiatives, would be largely avoided by the coming into service of the interconnection. On the other hand, it is only fair to admit a point which would not fail to be made by the opponents of the project, namely the undeniable fact that until system completion is reached the yearly allocations of public funds would have to meet not only the costs of construction of the infrastructure, but also the costs of the above cited damages. As the construction period is likely to cover -even in the most optimistic view- a range of at least one or two decades, all of these considerations are of relevant weight and should be taken into careful account in the financial planning of the enterprise as well as in the communication strategies.

Last but not least, in times such as the present ones, of volatile geo-political and economic climate as well as of unfavourable macro-economic expectations under governments of ephemerous duration, it may appear hazardous for national Authorities (or other corporate investors) to embark into important public works programs implying long-duration financial engagements[2]. In the even worse event of the occurrence of a serious economic crisis in the middle of building activities it could become necessary to decide the interruption of the project, thus bringing about the undermining of the entire underlying premises on which the planning of the interconnection had been based. Such an unfortunate event would not only entail passive interest costs, but would burden the public budget also with the hidden costs of hydrologic adversities and lost production ensuing from the want of the system benefits; hence these costs of the alternative consisting in 'not constructing (or not completing) the interconnection' should in any case be realistically appraised.

To complete a realistic critical appraisal, it has to be said that the mere physical existence of an interconnection system would not, by itself, magically eliminate the adverse effects of regional and temporal hydrologic imbalances. Rather, a specialised technical organisation should be set up from the beginning to ensure the effective management of such a huge, complex system. The central headquarters of this 'interconnection management authority' should monitor in real time the hydrologic situation, effect real-time projections for the near future (with the help of sophisticated mathematical models of the system dynamics), decide the best strategy to be adopted to the users' advantage and issue the relevant directives to the local operational units. The central staff of this governing authority would have to interface itself – and coordinate its actions- with other local and central powers (economic actors, power production companies, agricultural end users, civil protection organisations. . .). As a not inconsequential corollary, not negligible yearly management costs should be anticipated and of course included in the cost/benefit analyses. To these should be added, of course, the operating, maintenance and system updating costs.

3 SOME PRESENT EXAMPLES OF LONG-RANGE INTERCONNECTION PROJECTS

Three long-range interconnection projects will be herein briefly illustrated (in decreasing order of advancement of planning/realisation) prior to a short presentation of the European situation and a description of the preliminary Italian study. The three extra-European projects are being developed in the following countries:

Lybia, see § 3.1; China, see § 3.2; Brazil, see § 3.3.

3.1 *The Lybian project: 'The Great Man-made River Project' (GMRP)*

In the classical times (in particular during the Roman domination of the Mediterranean Sea) Lybia was a fertile land, often cited as 'the grain belt of Rome'. There followed centuries of desertification,

[2] The reasonable assumption is here made that such large-scale projects would necessarily be financed, in each country, within the framework of the national public budget; although privately funded schemes could not in principle be ruled out, experiences such as the Channel Railway Tunnel seem to make such a prospect a poorly appealing one. In particular cases, the financial contribution of the World Bank or of EU regional development funds could be sought for.

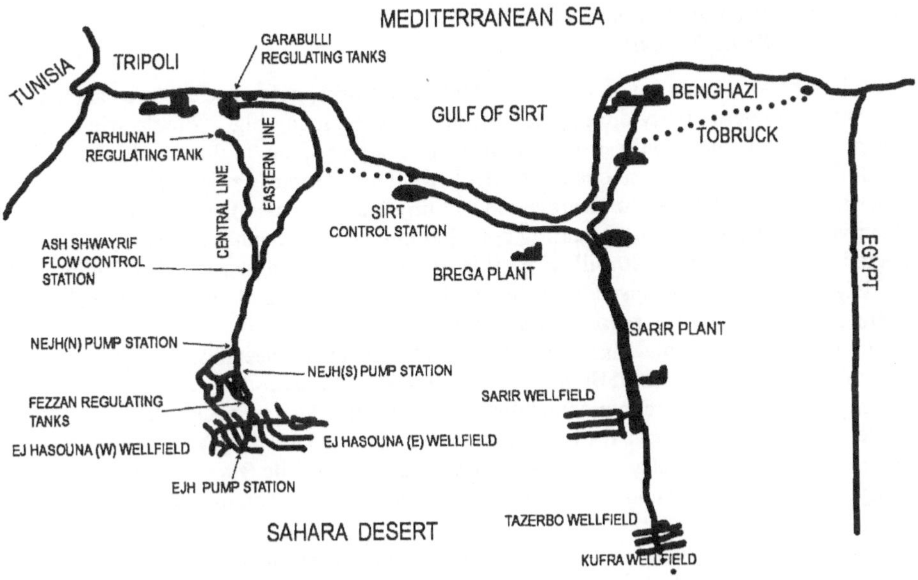

Figure 1. A schematic plan of the Libyan project.

intervened for a variety of causes (not last deforestation and destructive grazing of pastures by cattle) and the desiccation of water courses. Water is today a scarce commodity in Lybia both for urban consumption and for land irrigation, but in the last years a grand scheme of water interconnection facilities is taking shape.

This daring project, whose cost is estimated in 25 million $, is nearing the completion of the second phase of its realisation (out of the five phases anticipated for completion of the infrastructure, to be carried out over a span of 25 years). Its aim is the transfer of fresh water from underground fossil deposits, existing in the southern Lybian Sahara desert, to the northern coastal zones where the water will be put to civil, agricultural and industrial uses.

The fossil fresh water deposits, which were formed during the glacial era, lie at a depth of 500–600 m under the surface. Their discovery dates back to the Seventies of last century, when during the boring of exploratory wells intended to search for oil a huge 'subterranean lake' of fresh water was discovered instead.

The Lybian interconnection project actually envisages a twin system of water conveyance (see Figure 1), contemplating the pumping of fossil water from about 270 wells situated partly in the heart of the desert some 600 km south of Tripoli in the region of Jabal Hasawnah and partly some 700 km south of Benghazi in the region of Sarir-Tazerbo; the pumped water will then aliment a system of more than 4,000 km of pipelines and of two aqueducts some 1,000 km long; these lines will convey through the desert a discharge of more than 5,000,000 cubic metres a day.

The pipelines are built from pre-fabricated concrete units, 7.5 m long, of 4 m diameter, weighing some 80 tons each; they are buried in excavated trenches with the help of special giant trucks.

The short-term purpose of this project consists on one hand in substituting the use of the fresh water tables of the coastal zones, which are suffering from sea water intrusion caused by excessive exploitation, and on the other hand in increasing the surface of the arable lands by more than 500,000 hectares and in allowing the farming of about 3,000,000 sheep. Moreover, the system will provide the hydraulic supply to the inhabited regions as well as to other zones which have been earmarked for industrial development; the long-term strategic intent is to accompany the foreseen strong demographic growth and economic expansion not only of Lybia, but also of the adjoining countries.

3.2 *The Chinese interconnection scheme: The Western Route, Middle Route and Eastern Route Projects*

China can boast a long tradition of large-scale hydraulic works, going back to very ancient times; more recently, and particularly during the last 20–30 years, many big hydroelectric and mixed-purpose projects have been built in that country, with some of the highest and largest dams in the world. Foremost among these developments is the Three Gorges Project, directed not only to the power production, but also to improve navigability and to reduce the inundation risks of the Yang-tze river. [This project is now nearing completion which is scheduled for the year 2009, and is expected to cost more than 30 billion €; it is to be remarked in passing that this project has been widely and heatedly criticised, both for its feared environmental consequences and for the displacement and relocation of more than one million people].

A new interconnection project has been under study for several decades; its aim is the transfer of a part of the water resources of Southern China towards the semi-arid regions of the North. It will constitute a great hydraulic interconnection system destined in perspective to redress a serious imbalance in the water resources availability within the Chinese territory, thus fostering the long-term development of the country. Indeed, China suffers on the whole from a limited water supply, the average availability being about a quarter of the world average: it amounts in fact to only 2,700 cubic metres per capita per year.

Besides, there is a definite unbalance -both timewise and spacewise- in the water inflow distribution; the southern part enjoys an abundant hydraulic supply, accounting for about 80% of the overall national availability, whilst the northern part suffers from an arid or semi-arid hydrologic regime. After a prolonged period of studies and field investigations, the relevant technical/administrative departments proposed the building of three lines of interconnection, starting from the lower, middle and upper course of the Changjiang River, which will create among them a powerful South-North hydraulic interconnection structure: the Eastern, the Middle and the Western Route projects, respectively. The project was officially activated at the end of 2003; the Middle Route system, the first on which work is to be started, will achieve completion in the year 2010, while it is anticipated that the last of the three interconnection lines will enter service in 2020 or not much later. The total capital outlays will amount to about 8 billion €.

The experts estimate that, once the whole system is operational, from 38 to 48 billion cubic metres of water could be transferred annually from South to North [this amount is roughly equivalent to the whole discharge of the Yellow River (the second largest in China)], and the Chinese economists think that the removal of the drawback consisting in the scarcity of water resources will result in a strong impulse to the industrial and agricultural development of the North. At the same time, the coming into service of the project will also improve the ecological condition of Northern China, where the excessive reliance on pumping underground waters is causing the subsidence of vast surface areas as well as coastal erosion and other ecological degradation phenomena.

As above mentioned, the Chinese interconnection project is conceived around three distinct sub-systems: the Western Route Project (WRP), the Middle Route Project (MRP) and the Eastern Route Project (ERP), see Figure 2; a brief description of these three sub-systems is given in the following.

The *Western Route Project (WRP)* will take water from the upper course of the Changjiang River and bring it to the Huanghe River, these huge volumes flowing partly by gravity, partly by pumping; the annual amount of water thus transferred will be about 20 billion cubic metres. The infrastructures to be built will include – besides the pumping stations – a large dam (200 m high) and tunnels through the Bayankala mountains, some of which up to 100 km or more in length.

The *Middle Route Project (MRP)* will take water, from Haijiang River (a tributary of Changjiang River), to a system of canals to be built across the Funiu and Taihang mountains and eventually to supply facilities of the city of Beijing. In a successive stage, more water will be obtained from the Three Gorges impoundment or from a dam on the main course of Changjiang River; all these sources, including the basin of Danjiangkou, whose dam will be raised from 162 to 177 m of height, will allow about 14 billion cubic metres to be diverted annually to the North. Within the year 2020, some compensatory works will also be carried out, aiming at ensuring that enough water from

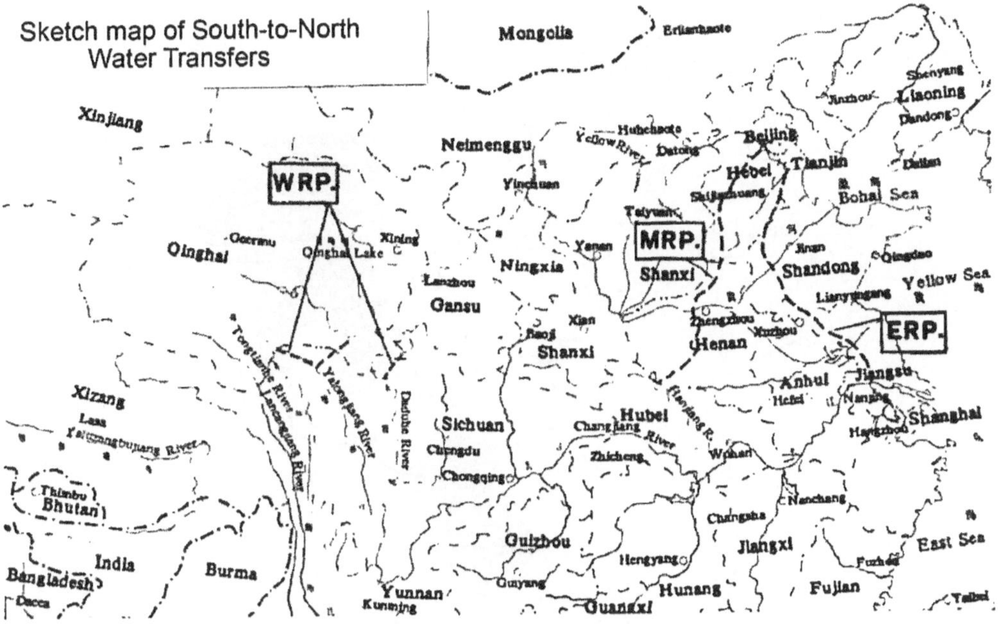

Figure 2. A schematic plan of the Chinese South-to-North hydraulic interconnection system.

the Haijiang River will continue to provide the volumes that are needed in order to sustain the agricultural and industrial development of the adjoining region.

The *Eastern Route Project (ERP)* will connect the lower course of Changjiang River, by means of the Great Canal Beijing- Hagzhou, to the eastern part of the Huang-Huai-Hai plain, the Northern terminus of the sub-system being the city of Tianjin. Thanks to this project, the annual water supply to the cities of Jiangsu, Anhui, Shandong, Hebei and Tianjin will be increased by more than 14 billion cubic metres. This sub-system will comprise reservoirs, conveyance works and power stations; the conveyance infrastructures will include, besides the main canal 1150 km long, some pumping stations and a crossing of Huanghe River through a 9 km tunnel. The availability of a total stored volume of about 8 billion cubic metres will be ensured by the impounding capacities of the lakes of Hongze, Luoma, Nansi and Dongping; the electric energy needed to operate the 30 pumping stations (890 MW of installed power) will amount to about 4 billion kW a year.

3.3 *The Brazilian project*

The impoverished populations living in the dry interior of Brazil have looked with thirsty envy, during more than a century, to the regions adjoining the São Francisco River (see Figure 3), the 3,200 km long, second-largest water course of the country[3]. While in the last 15 to 20 years the narrow strip of land (formerly considered a semi-arid zone) bordering the São Francisco stream has witnessed a vigorous economic development, thanks to extensive irrigation works, the arid regions to the west were not equally favoured and their inhabitants are now asking quite vocally for the realisation of interconnection schemes to partake of the São Francisco water resources. Quite recently, the newly elected Brazilian President Luiz Inacio Lula da Silva authorised the allocation of 1.4 billion € for the first stages of the building of a couple of canals many hundreds of km long,

[3] The low flow discharges at the mouth of São Francisco River are of the order of about 100 cubic metres per second, with high flow discharges up to 1,200 cubic metres per second.

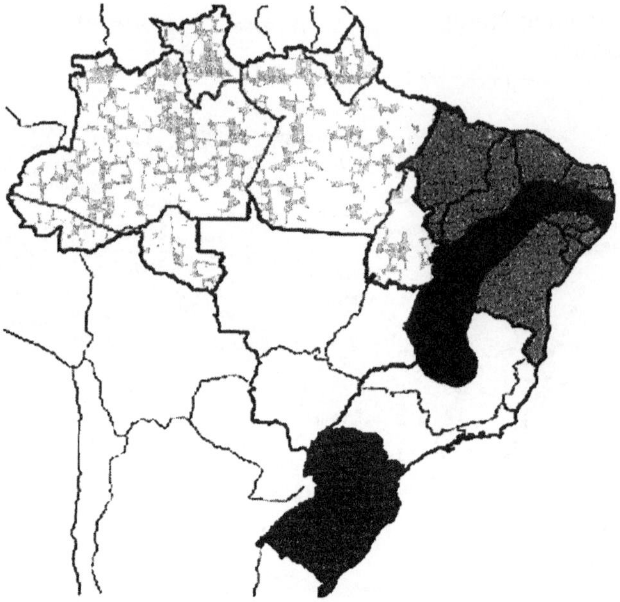

Figure 3. A schematic plan of the Sao Francisco river Brazilian region.

designed to bring a part of the São Francisco discharge to the arid regions of the interior, mainly to increase food production (but also to alleviate environmental problems of pollution, erosion, deforestation, and to counter the trend to the exodus of local people severely affected in the past by recurring droughts); this being intended as the first stage, started last year (2004), of a wider-scope project aiming at re-organising the distribution of the Amazon watershed outflows. This ambitious project involves four sub-projects, which will benefit about 16 million people in different States; its planning will be co-ordinated by the Brazilian Agricultural Research Corporation (EMBRAPA).

It is only fair to mention, anyway, that the Brazilian project is not meeting universal approval; it has spurred active opposition on the part, among others, of environmentalist groups which contend that such a large-scale project is not only too expensive, but also unnecessary insofar as the building of reservoirs, cisterns and wells could meet, in the opinion of these groups, the stated necessities in a more economical way. [Besides, these opponents of the project remark that many pharaonic projects attempted or carried out in the past (of which not a few were left unfinished) have failed to fulfil the original intentions aiming to 'transform Brazil'.]

4 THE SITUATION IN EUROPE AND A SHORT OUTLINE OF THE ITALIAN PRELIMINARY FEASIBILITY STUDY

4.1 *The European picture*

In the last decade, many European countries (France, UK, Germany, Poland, the Czech Republic...), and notably Italy, have experienced recurring, unusually heavy hydraulic and hydrogeological hazards. These are commonly attributed on one hand to an assumed trend towards climatic changes, on the other hand to the demographic expansion and the accompanying exponential development of environmental alterations inflicted on the territory by anthropic activities (deforestation, urbanisation, topsoil erosion...); both causes, indeed, bring about a change for the worse in the characteristics of the surface flow of meteoric waters. The crises in question (together with the ever-growing interactions of distant communities) pose with ever increasing urgency the problem of revisiting the traditional planning of water cycle management in the socio-economic context of developed countries. The hitherto prevailing localistic approach, indeed, shows its limitations as it becomes more and more evident that an integrated, regional or better inter-regional approach should instead be adopted. Only in this way the unbalances in the water availability of different regions – or the consequences of their different vulnerability to extreme hydrologic events-could be redressed. The basic issue to contend with is the recognition that within industrialised countries water is both a primary resource subjected to a rigid demand and a source of hazards; therefore the rational governance of the natural water cycle (including its artificial modulation and re-distribution) make for an essential part of a holistic approach to territorial planning. In this view, it appears 'right' not to wait for hydro-geological disasters to occur -and only afterwards go to great efforts and expenses in order to palliate their consequences-, but, rather, to ask the question: 'Would a different approach to water management be able to avoid (or strongly mitigate) the occurrence of these hazards in the first place?' It is the contention of the present paper that long-distance hydraulic interconnection infrastructures may constitute at least a part of the answer. In the following, the initial stages of the analysis made by CESI for the Italian peninsula will be shortly illustrated in order the better to expound the possible advantages of such an approach.

4.2 *The Italian case*

As concerns Italy, the perception of a turning point in the worsening of the hydrological hazards can be dated – albeit with some arbitrariness- in the autumn of 1994, which saw hitherto unheard of disastrous floods in the Piedmont region. Since then severe autumn and winter floods have occurred practically each year in the great plains of the North (Po valley), while at the same time damaging summer droughts have struck the Southern half of the peninsula (occasionally, hydro-geological catastrophes from exceptional floods do occur in the rainy season also in the geologically unstable mountainous regions of the South, and summer droughts tend now to appear with greater frequency also in the North). Actually, reflections on the necessity to undertake the realisation of an effective system of hydro-geologic protection predate these recent events: after the disastrous floods of Florence and Venice of Nov. 1966 an officially appointed study Commission, presided by the famous Professor Giulio De Marchi, issued a substantial Report (published in 1970) which analyses the situation and concludes advocating the necessity of building a number of new reservoirs and fluvial expansion zones. This plan was never implemented (exception made for a few reservoirs built in the last years); its total presumed cost can be estimated at about 15 billion € (in present value).

The technological progress intervened to date in the construction equipment makes now possible to conceive a somewhat different system, consisting in a complex of reservoirs situated at a uniform elevation and connected by pressure tunnels. The reservoirs would be located along the Apennines mountain range, which runs along the NW-SE central axis of the Italian peninsula for a length of more than 1,000 km. The function of such an interconnected system would be on one hand to compensate the time and space irregularities in the outflows of the many water courses of the Apennines valleys, on the other hand to store the excess winter discharges of Northern rivers floods and render them available to alleviate the summer droughts of the South. The reservoirs being situated at a uniform elevation, tentatively fixed at 500 m. a. s. l., makes it unnecessary to foresee the construction of pumping stations[4]; besides, the only gates to be installed in the tunnels would be the isolation gates to be operated in case of maintenance works. In this way the system would achieve the maximum simplicity and reduce the overall costs. The choice of a system elevation around the 500 m. a. s. l. contour line appears advisable on several more grounds too. In fact, a much lower elevation would put a not negligible part of the potential users at a disadvantage,

[4] Instead, new power stations tied to the proposed system could be created where the local conditions should prove favourable. The electric energy thus generated would be but one – albeit certainly not the main one- of the system benefits.

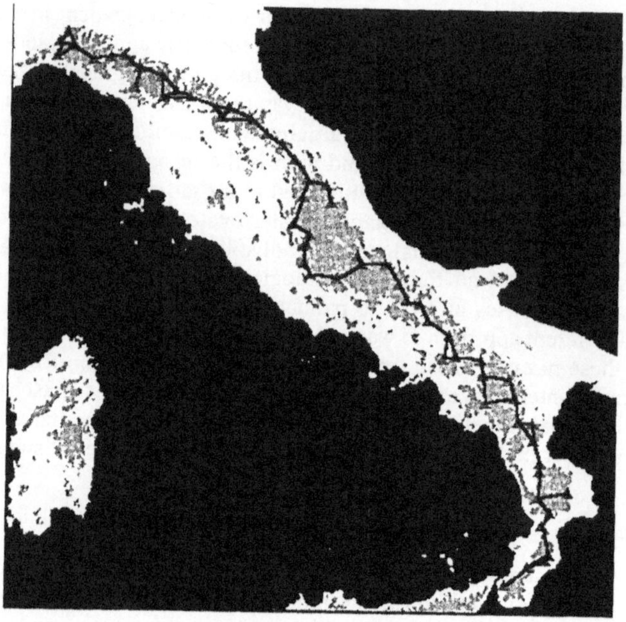

Figure 4. A schematic plan of a possible configuration of an interconnected hydraulic conveyance system along the Apennines mountain range.

while a much higher elevation would cause the reservoirs to fail to catch an appreciable part of the inflows of the relevant streams; besides, a lower elevation of the tunnel/reservoir lines would entail more interferences with the heavily populated valleys, while a higher elevation would entail more difficulties in crossing the rock masses separating the valleys to be dammed.

An analysis of the local possibilities has shown that the system could include about 60 medium-sized reservoirs, to be created by new dams not higher than 100 m (their average height being less than 50 m), for a total storage capacity of about 2 billion cubic metres; the total length of the connecting tunnels (of 4 m internal diameter, to be bored with 'mechanical moles' so as not to disturb the rocky masses and their water tables) would amount to about 1,500 km, extending from the Ligurian Apennines in the North to the Calabrian Apennines in the South (see Figure 4).

The location of the reservoirs was chosen by selecting the valleys where the local streams presented – statistically speaking- a greater value of the average yearly discharge; besides, the requirement that the connecting tunnels between each reservoir and the two adjacent ones should not exceed about 40 km in length was also given due weight in the selection. The relevant informations were desumed from official data of the National hydrologic service, crossed with the ample topographic informations available (also in graphic form) from the digital data bases of the GIS (geographic information system).

Some further technical features of the project conception, strictly tied to the 'freely communicating vessels' scheme underlying the whole idea, deserve to be briefly mentioned. Of course, at selected points along its route it will be possible to have outlet installations for the local users or for new power stations, and each of these outlets will profit from the sharing of the whole of reservoirs capacity, independently from local availabilities. This same feature of capacity sharing will provide an advantage in coping with local floods , which even in the case of nearly full reservoir will require spillway operation only for a fraction of the incoming discharge, because a part of it will flow from the reservoir directly concerned towards the adjoining ones. Speaking of spillways, it is good to remark that they will be of the 'fixed sill' type, without gates, not only in favour of simplicity and safety of operation, but also because, as well known, they assure that the spilled-over maximum

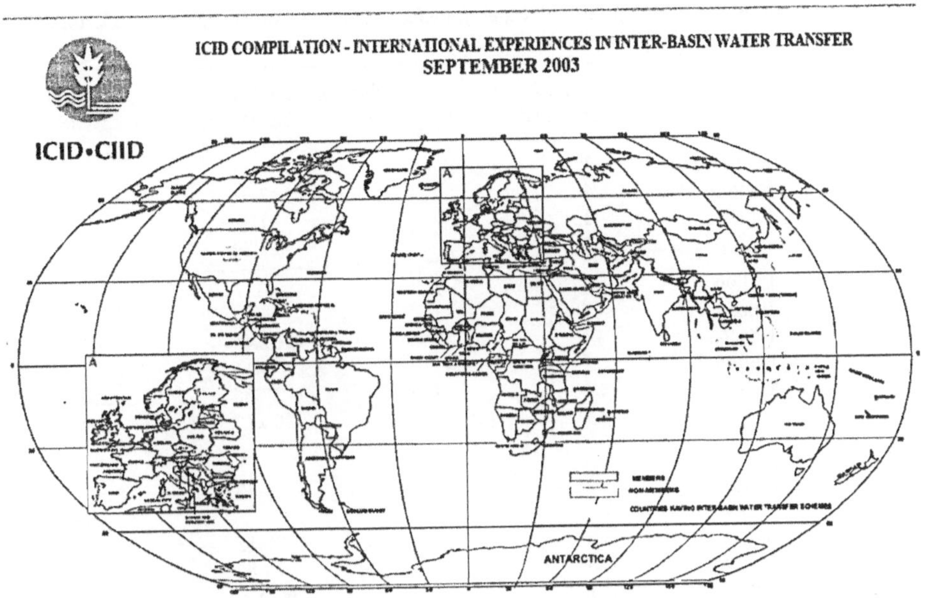

Figure 5. A scheme of the existing inter-basin water transfer projects, worldwide (from ICID report, 2003).

discharge will be less than the maximum of the incoming flood; this effect will be further increased by the already mentioned 'shared capacity' feature of the interconnection.

Such a large system, accepting irregular inflows and accommodating irregular demands of utilisation, would operate under conditions of transient hydraulic regime practically all the time. Thus, it becomes necessary to model the system behaviour in order to investigate its dynamic response features. A numerical model of a somewhat simplified version of the interconnection was therefore set up and run through a series of simulations, under a spectrum of different assumptions about flood inflows occurring at certain points, or peaks of demand concentrated at given sections. The conclusion of this investigation was that the transit times of the 'signal' of a perturbation from one end of the system to the other end is of the order of two-three days, while the time needed to reach a steady-state condition is of the order of several days. However, the maximum differences in the water levels occurring among the many reservoirs during such transients would remain limited to very few metres; the clearances between operating maximum levels and spillways crest[5] would have, of course, to be fixed keeping into account these results. Moreover, this study of the system dynamics highlights the necessity, already alluded to, of creating an effective management organisation endowed with all the necessary means of monitoring, real-time simulation and control of the state of the system.

The total cost of the project has been roughly estimated at about 18 billion €; the time span of construction activities could be of about 20 years, so that the yearly allocations needed would be of the same order of magnitude of the damage costs incurred each year as a consequence of extreme hydrologic hazards.

This preliminary survey has shown that the technical and economic feasibility of this kind of infrastructure does not appear out of the realm of present-day possibilities; of course, much more detailed study would be needed (the corresponding investigation time being of the order of about 5 years, during which not only the engineering side, but also the economic/financial practicability, and above all the environmental compatibility of the system, will have to be thoroughly researched)

[5] The spillways crest would be fixed at a uniform elevation for all dams, for evident reasons.

Table 1. Inter-basin existing and proposed water transfer schemes, worldwide (from ICI report, 2003) BCM = Billion Cubic Metres; N. A. = Not Available.

Scheme N°	Country	N° of schemes completed	N° of schemes proposed	Annual transfer, BCM Completed	Annual transfer, BCM Proposed
1	Morocco	1	–	1.51	–
2	South Africa	24	–	2.51	–
3	Sudan	1		7.3	
Total for **Africa**		**26**		**11.32**	–
4	Bolivia	–	1	–	0.2
5	Brazil	–	1	–	1.50
6	Canada	38	10	262.46	463.6
7	Chile	2	–	3.15	–
8	U. S. A.	19	7	37.56	381.6
Total for **America**		**59**		**303.17**	**846.9**
9	Australia	1	–	1.13	–
10	China	–	3	–	44.8
11	India	6	–	28.85	–
12	Iraq	6	–	15.8	–
13	Japan	1	–	N. A.	–
14	Malaysia	–	1	–	0.14
15	Pakistan	8	–	99.95	
Total for **Asia**		**22**	**4**	**145.73**	**44.94**
16	Czech Rep.	4	2	15.15	2.20
17	Finland	1		0.09	
18	France	5	–	2.35	–
19	Germany	2	–	0.47	–
20	Portugal	1	–	0.01	N. A.
21	Romania	3	N. A.	N. A.	N. A.
22	Russia & C. A	5	2	60	47
23	Spain	3	1	1.3	1
Total for **Europe**		**24**	**5**	**79.37**	**50.20**
Grand Total		**131**	**28**	**539.59**	**942.04**

before a definite proposal could be made and presented in the proper evaluation and decisional circles. Anyway, in Italy as in the rest of the Western world, the greatest uncertainty rests with the socio-political acceptability of such an ambitious program, which not only would meet severe opposition from the environmentalist circles, as already remarked, but would also quite surely – given the strictures of the national budget- enter in competition with other perceived 'public works' priorities (to mention just a few of them, in Italy the much discussed project of the Messina Straight suspension bridge, already approved by Parliament, would necessitate about 5 billion € by itself alone; another mammoth project scheduled for the near future is the system of gates designed to close the Venice lagoon in case of exceptionally high tides in order to protect the city from the recurring inundations it suffers more and more frequently. . .).

5 CONCLUSIONS, FUTURE PERSPECTIVES

The complex dynamics of demographic evolution, socio-economic progress, quality of life rights/expectancies, climatic changes impose an ever-growing strain on regional and national hydraulic infrastructures, implying the necessity to take a long-range view in the framework of an integrated approach to the planning of territorial development. In this framework, a possible useful component of infrastructural evolution could lie in the design and building of extended lines of hydraulic interconnection (not unlike the existing long-distance network facilities satisfying the needs of electric power, oil, gas, information...). Some examples of this kind of systems, chosen as most representative of the interconnection concept from among the many already realised, as well as others planned or under construction, for an imposing worldwide total of many hundreds of billion cubic metres transferred annually (see Figure 5 and Table 1, which albeit include also pure change-of-watershed schemes, not exactly answering the more restricted concept of long-distance interconnections envisaged in the present paper), have been summarily illustrated in the preceding sections of the present paper. A preliminary study concerning the possible example of an hydraulic interconnection system tailored to the particular conditions relating to Italy has also been presented (section 4.2). It is the opinion of the Author that the future will see more and more examples of similar large-scale endeavours, in spite of their requiring huge capital outlays and extended times of realisation. In Europe, besides the already existing water sharing schemes, possible candidates to further implementations of this kind of development could be countries like Spain, Portugal, Greece...

ACKNOWLEDGEMENTS

Grateful acknowledgements is here given to CESI S. p. A. and to CERSE (the evaluation panel of research proposals for the power industry) which endorsed the initial concept and financed the preliminary feasibility study for Italy illustrated under section 4.2 of the present report. The precious team work of Guido Mazzà, Renato Cadei (respectively a CESI executive and a CESI technical expert) and Pietro Picozzi (a hydraulic engineer from Pavia Engineering Faculty) has been an indispensable ingredient in the preparation of the preliminary report for the Italian study; their competent, enthusiastic contributions to the original Author's idea are here most gratefully acknowledged.

REFERENCES

Rome, E. (ed.) (1970) Conclusive Report (Proceedings) of the Inter-ministry Commission for the study of the hydraulic management and land defence in Italy ('Giulio DE MARCHI Commission Report'; in Italian).

Giulio, L. (2003) The deep differences in water resources availability in Italy: is there a possibility for compensation? (in Italian), *'L'Acqua'*, n°4.

Home page and linked pages of *'Integrated Management of Land-Based Activities in the São Francisco Basin'*: http://www.oas.org/usde/SAFUP/sf2.HTM.

ICID (International Commission on Irrigation and Drainage) compilation. (2003) *International Experiences in Inter-Basin Water Transfer*.

Jacques, B. (2004) *L'eau dans l'antiquité* (ed.) Eyrolles, Paris 1984 Wolff G. The risks and benefits of globalization and privatization of fresh water: Proceedings of the Seminar on Challenges of the new Water Policies for the 21st Century, 29–31 October 2002, Valencia (Spain), Balkema publishers.

Mario, F., Michele, F. & Carlo, N. (2000) The creation of hydraulic reserves through dams: environmental and financial implications (in Italian), *'Ingenierìa del Agua'*, 7(4), 375–390.

Ministry of Water Resources of the Popular Republic of China. *A brief introduction to the planning for South-to-North water transfers (on the Internet site:* http://www.mwr.gov.cn/english/projectintroduction/nsbd/index.htm).

CHAPTER 13

Water resources in developing countries: The millennium development goals in the 21st century

C. Fernández-Jauregui
Manager of Water Assessment & Advisory – Global Network (WASA-GN), Spain

ABSTRACT: A description of the water situation in the world is hereby afforded, with special emphasis being placed on developing countries; highlighting the fragile situation of water resources and the availability of water according to population, user demand, priority problems such as: water for human consumption and basic sanitary services; the challenges to meet the Millennium Development Goals; the reasons why the goals, particularly goal 7, will not be achieved except in one region; proposing the inclusion of the concept of the right to water as a human right and as a tool to achieve social harmony; and to detail the challenges that should be faced in the 21st century in social development, emphasising the role of water as support for life and improving global governance of water while recognising the fact that it is a responsibility shared by all of us.

1 INTRODUCTION

Water is a unique resource, transversal in all human activities and the management of water requires certain matters to be understood which make it different from other natural resources.

Water is a finite resource; the amount of water existing in nature has not changed significantly throughout the history of mankind. It is a fragile resource, susceptible to mineral and organic contamination, and when considering the human race's growing needs for water, it is a limited resource.

The ever increasing world population, consumer habits and the accelerated process of urbanisation have uncovered concerning situations about the capacity of countries to indefinitely maintain these rates and trends in terms of available water resources.

The vision of the current situation about water requirements that remain unfulfilled and the forecast for water demands in the very near future lead to the perception of a critical situation turning into a world water crisis. One third of the world's population suffer from some degree of water shortage which is either the supply of water itself or the quality of water and basic sanitary conditions and it is expected that two thirds of the world's population will be in a situation of water shortage by 2025.

It is assumed that water shortage is one of the symptoms of poverty, which is true to a certain extent. However, the crisis phenomenon seems to be ruled by other factors apart from poverty and threatens to reach medium and highly developed countries.

Recorded growth and forecasts for the world's population, at the three levels of the Human Development Index (HDI), superimposed by the forecast for water shortage are shown in Figure 1.

2 WATER RESOURCES AND DISTRIBUTION

Stemming from the research carried out at different universities around the world, and work by an endless number of national and international institutions about the availability of water in the

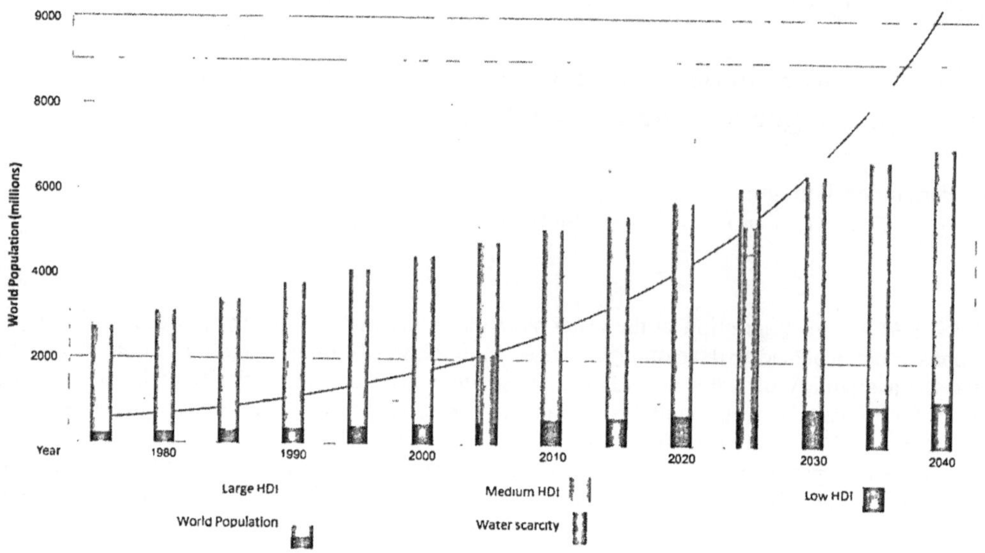

Figure 1. World population and water scarcity.

world, we are able to assert that the exact amount of water available in each hydrological cycle is not accurately known. In order to know how much water is available in one place, the hydrographic basin and the aquifers or hydro-geological basins first need to be defined. Afterwards, in the case of surface water, an annual and monthly balance needs to be made, and in the case of underground water, an assessment needs to be made of the aquifers leading to the production of hydro-geological maps.

The elaboration of water balances and hydro-geological maps, although it seems incredible, has still not been done in many places around the world owing to a number of reasons; although the first conclusion is that we do not accurately know how much water is available in all countries and international basins around the world.

Similarly, it can be asserted that no consolidated study about the quality of water in the different basins has been carried out, either owing to the high costs involved in assessment and monitoring or owing to the lack of knowledge and resources to do so.

Considering that the availability of water remains accurately undefined, we are able to say that the amount of water required per person / country / year is known. Its availability, nevertheless, continues to diminish at an ever accelerating rate, mainly because of three factors: population growth, economic development of countries and climatic change.

On the other hand, we are also able to confirm that the quantity of water on the planet is constant, and of the total amount, a small proportion is fresh water and therefore the economically exploitable proportion of this resource is also low. If, to this fact, we add the quality of the water in most countries being contaminated and untreated, we are also able to assert that in those countries the situation is even more serious, since the supply of water resources is not constant but is actually depleting.

Based on the information outlined above, a summary of the situation concerning water resources in the world can be shown, versus the population by continents and we can also conclude that the richest region in terms of water resources is South America and the poorest is Asia, as can be seen in Figure 2.

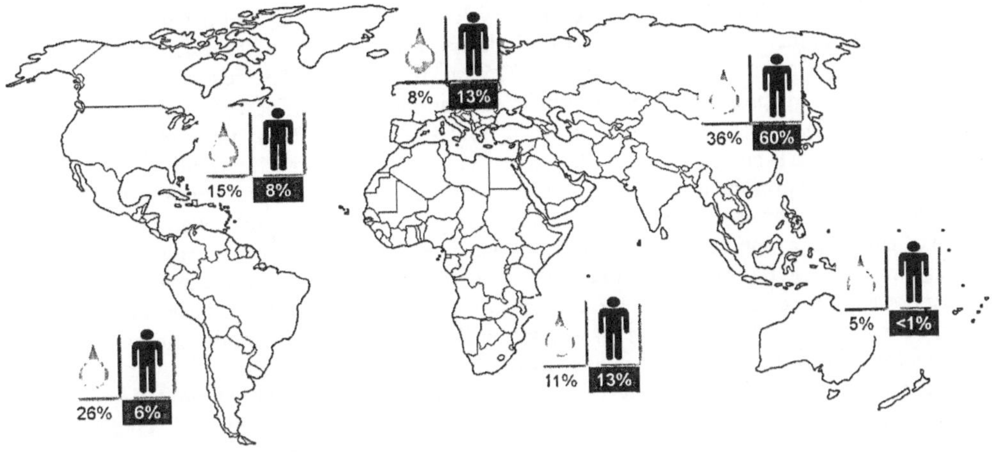

Figure 2. Water and population in the world.

3 WATER DEMANDS IN DEVELOPING COUNTRIES

As can be seen the water supply (resources) in the planet remain constant in terms of quantity whereas the quality of water is being systematically worsened as a result of human activity and the effects thereof. Today, the demand for water is subject to three "driving forces", which are as follows: The increase in the world's population, improvements in quality of life in developing and emerging countries, and climatic change which has lead to an increase in extreme events both in terms of their frequency and magnitude. Furthermore, users of water resources can be classified in the form of challenges for our society meeting short, medium and long term requirements.

Meeting basic human requirements. – There are currently 1200 million people who do not have access to safe water and 2600 million people who do not have basic sanitary systems, which is in itself a vicious circle where poverty and disease are concerned, resulting in one person dying every 20 seconds through causes related to the lack of basic water and sanitary services.

Guaranteeing food supply for the growing population. – There are currently 1000 million people suffering from undernourishment in developing countries which is a figure that has been steadily rising since 2003 from 777 million to 1000 million in less than eight years, as a result of the new demands for food in China and India, devastating droughts and the demand for biofuel in emerging and developed countries.

Cities and their diverging needs in urban environments. – Today most of the population live in urban areas, 51%, which means that the demands for public services increase and the water sources for city life are brought in from far off regions, in most cases from other hydrologic and/or hydro-geological basins, further accounting for water infrastructure problems through losses in distribution values that can account for up to 60% of the total distributed water entailing huge operational and maintenance costs.

Managing ecosystems for the welfare of the population and environment. – Ecosystems are delicate systems that demand constant amounts of water in suitable quality and quantities and as a result of the huge demands for water, particularly for agricultural purposes, there has been a reduction and alteration of large wetlands all over the world, in developed and developing countries, and in the 20th century 50% of wetlands were actually lost. One of the challenges for this century is to scientifically quantify the impact this activity has had to date on ecosystems, since today, this data is only managed empirically owing to the lack of knowledge about the water genome and its implications on the environment.

Promoting cleaner, less contaminating industry for all. – Emerging and developing countries forecast growth in their demands for water for industry, which could double water consumption in industry from the figure calculated for 1995 of 725 km^3 by 2020. It is also relevant to note that this sector is making huge progress in terms of efficiency and has notably increased efficiency values. Today, water consumption by industry at world level accounts for 22%, although in developing countries this figure is currently at 10% but does show a rapidly increasing trend.

Using energy to cover the needs for sustainable development. – There are currently 1000 million people in the world who do not have an electricity supply to their homes. Today water is not the main source of energy at a world level; however it is an indispensable source for energy transformation in many areas around the world. The two main uses are for electricity production and thermal and nuclear power plant cooling systems. It must also be mentioned here that tidal, wave and geothermal energy are other possible alternatives for generating electricity. Today, hydraulic energy production accounts for 19% of the total electrical energy produced around the world.

Managing risks in the light of uncertainty. – There are several types of natural disasters related to water, such as: Droughts 11%, Epidemics related to water 28%, famine 2%, landslides and avalanches 9% and flooding 50%. Flooding caused 15% of human life losses and droughts caused 42% of human losses caused by natural disasters. At regional levels, the distributions of natural disasters relating to water are: Africa 29%, Asia 35%, the Americas 20%, Europe 13% and Oceania 3%. These figures indicate how important it is to further knowledge about risk management at local, regional and world levels in order to prevent the loss of numerous lives and economic damage.

Sharing water in common interest. – Water is shared by different users and also by regions, countries and towns. There are 264 shared water basins in the world in which over 145 countries take part. Statistical data shows that in spite of being a potential for conflict, water is actually a source of cooperation, since cooperation resources prevail, and the same data shows that at world level, the number of cooperation activities is always higher that any possible conflicts, which in fact actually occur because of politics using water as an excuse, and not because of. There are also many examples in the world of how shared management can be carried out such as in the basins of the Danube, Lake Titicaca, Lake Peipsi, Senegal, etc.

Quantifying and evaluating the multiple facets of water. – One of the great challenges for social and natural sciences is to analyse and further knowledge in order to quantify the true value of water taking all components into account: social, cultural, economic and religious, since the fact that water is a social, cultural, economic and religious asset is well known; whereas to date only its economic value has been quantified, while leaving the other factors out of the equation. This shows us that when talking about values, costs and rates, we are not including all the quantifying parameters thereby leading to a number of conflicts and injustice particularly in the poorer countries where the poor pay a much higher price than in developed countries. From this analysis, we are able to conclude that water should always be state owned because of the type of resource it is and its role.

Guaranteeing dissemination of basic and applied knowledge. – Information and knowledge are the pillars of sustainable development in any society that is aware of the need for collective welfare. In recent years we have seen a wide choice of knowledge, particularly for underdeveloped countries, which sadly have not had the expected results owing to the fact that those offering knowledge do not pay attention to the reality of the situation and the needs of the community, thereby leading to great fiascos and loss of trust in the quality of the knowledge afforded. There are some notable cases where technological solutions for subject matter were not consulted with potential users and huge investments were made that did not prosper in water science. In 2001 an assessment was made of the offer of knowledge provided to a region in the developing world and its low demand leading to the conclusion that from what was actually offered only 23% was actually required, the rest was irrelevant. Training must be provided at the place where it is to put into practice, or at least with good background knowledge of the place. Another item that has been identified is the lack of training by water managers around the world, which is a type of training that is not yet exciting, but which is starting to be studied.

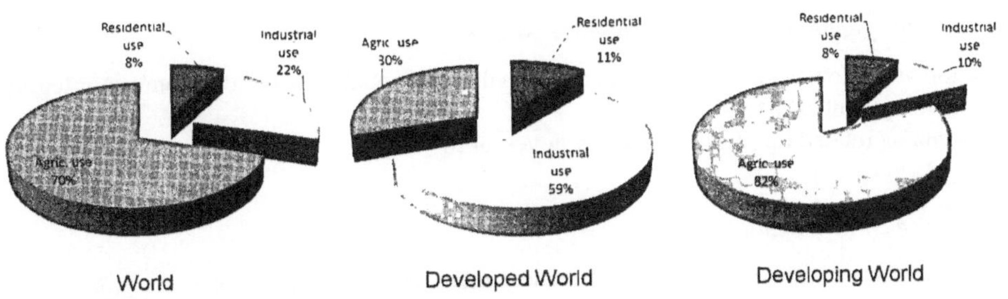

Figure 3. Water uses in developed and developing world.

Guaranteeing free circulation of reliable information. – If we agree that water management is a shared responsibility, we have to accept too, that the information must be reliable, freely available and free, since said information has been produced through taxes paid by society for its basic services. The decades of the 80's and 90's are called the lost decades owing to the fact that most of the developing countries were victims of the recommendation by the World Bank and International Monetary Fund concerning the need to reduce the size of the state, which did not happen in developed countries and the first victim of this erroneous recommendation was the water sector as can be seen from the cuts in budgets for activities relating to the collection and processing of hydro-meteorological data in most developing countries, and in many cases closing or privatising. During that time development of non renewable energy was promoted such as crude oil and thermal power plants that lead to drastic cut backs in water works. The huge development of the internet this century is now allowing popularising of information and the subject of water sciences is benefiting and beginning to recover from the damage caused in the last century leading to more visible dynamism in the sector.

4 THE MILLENNIUM DEVELOPMENT GOALS AND THEIR IMPLEMENTATION IN DIFFERENT REGIONS AROUND THE WORLD

All the Millennium Objectives are directly and indirectly related to water, since water is essential for life and is the single most important factor for sustainable development. We shall now move on to briefly describe each objective and point out the relationship with water. It is essential to point out here that all the objectives have the common goal of halving the number of people suffering the problem, consequently accepting that the other half will continue to suffer the problem and they will be formally condemned to suffer at an institutional level, which is ethically unacceptable, and what is more this is to be achieved by 2015.

4.1 *Eradicate extreme goal poverty and hunger*

Water is a production factor in nearly all businesses, including agriculture and the service sector. Better nutrition, along with food safety, reduces vulnerability to disease including HIV/AIDS and malaria, among others. In modern times, access to electricity is fundamental to improve quality of life. Competition between different sectors needs to be balanced, through policies that recognise the capacity and responsibility of all sectors to tackle matters relating to poverty and hunger.

This goal is related to: agriculture and life in rural areas.

4.2 *Achieve universal primary education*

Promoting a healthy school environment is fundamental to ensure universal access to education, schooling, attendance, permanence and results. Distribution of teachers has already improved. In order to achieve the above, access to drinking water and sanitary conditions are fundamental.

This goal is related to: improving knowledge and training of human resources.

4.3 *Promote gender equality and empower women*

Education will allow women and girls to develop their potential in order to become an integral part of the development effort.

This goal is related to: improving water governability.

4.4 *Reduce child mortality*

Improving access to drinking water and suitable sanitary conditions will help prevent diarrhoea and will set the baseline to control parasitic worms transmitted through the soil and schistosomiasis, among other diseases.

This goal is related to: improving human health through water and sanitation.

4.5 *Improve maternal health*

Improving health and nutrition reduces vulnerability to anaemia and other conditions that affect maternal mortality. A sufficient quantity of clean water for washing before and after childbirth reduces the possibility of contracting mortal infections.

This goal is related to: water and sanitary conditions for women.

4.6 *Combat hiv/aids, malaria and other diseases*

Improving water supply and sanitation reduces vulnerability to contracting aids and the seriousness of it, and other serious diseases.

This goal is related to: human hygiene and sanitation.

4.7 *Ensure environmental stability*

Healthy ecosystems are fundamental for maintaining biodiversity and human welfare. We depend on them to obtain the water we drink, to produce healthy, safe food and a number of other environmental assets and services.

This goal is related to: Water for human consumption and sanitation, agriculture, energy, ecosystems, cities and rural areas.

4.8 *Develop a global partnership for development*

Water has a number of values that need to be recognised when choosing governability strategies. The assessment techniques will guide decision making relating to water distribution, which in turn promotes sustainable social, environmental and economic development, further to transparency and accountability. Programmes and alliances for development should recognise the importance of drinking water and basic sanitation for economic and social development.

This goal is related to: the value of water, risks and governability of water.

5 WHY SOME REGIONS ARE NOT GOING TO ACHIEVE THE GOAL CONCERNING WATER FIT FOR HUMAN CONSUMPTION AND BASIC SANITATION?

In accordance with that described in the previous chapter, it could be said that from among the Millennium Development Goals, No. 7 is the most relevant since it specifically refers to the problem of water and focuses on safe water fit for human consumption and basic sanitation, and makes a more or less adequate assessment (2002) of the number of people who do not have access to both services. The first conclusion is that there are 1200 million people without access to safe water,

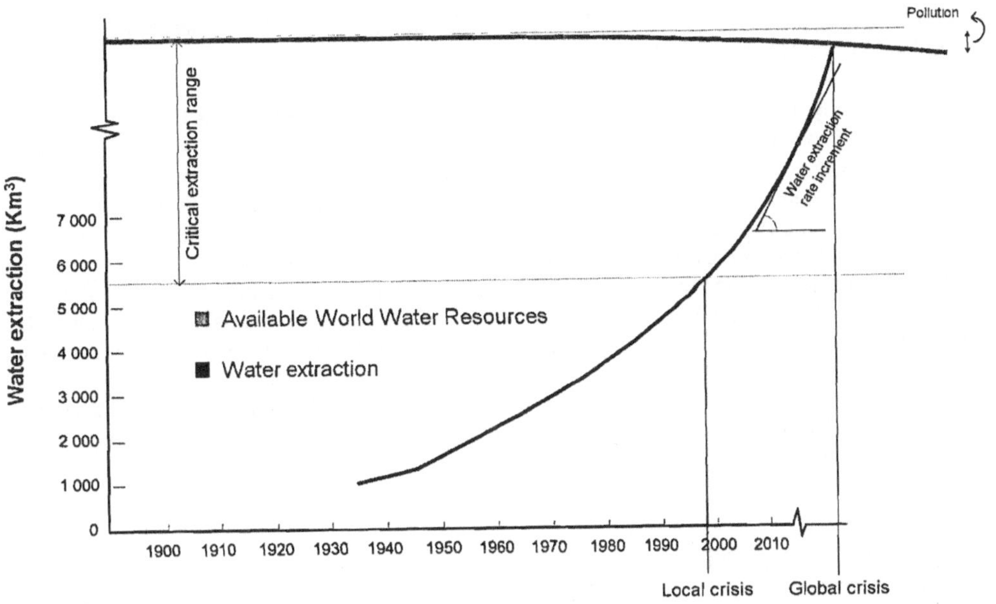

Figure 4. Water crises.

who are distributed as follows: Africa 27%, Asia 65%, Latin America and the Caribbean 6% and Europe 2%.

Concerning basic sanitation, the situation is as follows for the same year (2002): Africa 13%, Asia 80%, Latina America and the Caribbean 5% and Europe 2%.

What has the evolution of the situation in the world been? Population has continued to grow and progresses relating to this goal have been delayed to a large extent in Africa and Asia. Why has this happened? Why are more people without access to sanitation than safe water? Why is progress not made?

We have put forward several answers but no solutions: There are more people in the world without sanitation that without safe water because there are a lot of bad politicians who are only interested in activities that can be seen during their terms of office. Sanitation is not photogenic, and they therefore do not invest in things that do not lead to direct votes; which mean that every 20 seconds a child dies in the world owing to a lack of suitable sanitary conditions. Why is no progress made? Because many politicians only carry out short term activities and not long term action or other action in the general interest of the state.

In this year of 2010, we are able to assert the following: goal number 7 is a fiasco, since we are not going to achieve the objective of halving the number of people without access to both services, except in the case of Latin America where the goal will be achieved, but in the case of Africa, there will be a backward trend of around 30% in terms of safe water and 40% in the case of sanitation.

What is the cause of the fiasco? Lack of finance and political decisions to make this item a priority for states involved.

How are we possible able to understand that in 2009 funds from society were provided to bail out irresponsible private banks (for instance: 7,000,000 million US dollars and other equivalent amounts in Europe) and nobody in civil society objected to the corruption and immorality of the behaviour by the bankers, but 5000 million euros per year over 10 years is not available to achieve this goal? Human solidarity is going through a critical stage at world level, as well as an about turn in the moral values of our society.

6 WHAT SHOULD WE DO FOR THE MOST UNDERDEVELOPED COUNTRIES TO IMPROVE THEIR SITUATION?

There is currently a world water governability crisis which prevents progress being made in the development of society. It is said that a nation or country has governability of water when: A. – It has authority over water at the highest level within the structure of the state, which is neutral and does not manage demand. B. – When there is modern legislation that includes the basic principles of water resource management and when the experts and the community are taken into account in the political decision making processes, with special emphasis placed on the role of the community. C. – When there are suitable human resources in terms of quality and quantity using a holistic focus on training. D. – When there are enough long term financial resources available, and finally: D. – When there is enough transparent, reliable information in free circulation and of free access owing to the fact that it involves a public service for a common asset.

What has been done in recent years? The most underdeveloped and emerging countries have made a major effort in this subject showing some excellent results, which are being shared through horizontal cooperation between countries.

7 WHY THE IMPORTANCE OF WATER IS A HUMAN RIGHT AS A SUSTAINABLE DEVELOPMENT TOOL?

The right to water is a human right specifically or explicitly included in several international treaties and statements, in The Universal Declaration of Human Rights, adopted and proclaimed in December 1948, article 3 states:

"Everyone has the right to life, liberty and security of person" If we accept that without water there is no life, and that water is the basis for all life, the conclusion is categorical; water is a human right therefore making it a legal title and not a charity or merchandise.

It is important to highlight that as water is a human right, under no concept are we proposing it to be a free service, since the costs must be covered in order to guarantee the common asset.

Over recent years it could be seen that those countries which have improved the governability of their water, have included the concept of water being a human right in their laws and/or constitutions, and in this way have managed to convert some basic activities such as access to drinking water and sanitation in their policies as well as other types of collateral benefits.

Stemming from the above, it is deduced that the fact of including this concept in water management means that in many countries it becomes a tool for providing basic services and in this way their achievements go beyond the Millennium Development Goal in that 100% of society has access to safe water.

8 THE WATER RESOURCE MANAGEMENT CRISIS IS A CHALLENGE FOR THE INTERNATIONAL COMMUNITY

The management of water resources is a responsibility shared by all inhabitants of the planet where all local action taken can now be seen reflected somewhere else around the planet such as the case of climatic change in the management of water resources, which we are obviously not ready to affront either scientifically or effectively.

What has South America done in the Andes region to deserve not having now availability of water resources from snow and ice? Or in Asia? The source of drinking water supply in those countries and regions is snow and ice. What has happened with the wetlands, or exploitation of their water resources? From the above it can be deduced that the impact of water infrastructures is very important and delicate, particularly in the least developed countries that have done nothing until now to deserve being the victims of the development of other countries, and have had no benefits

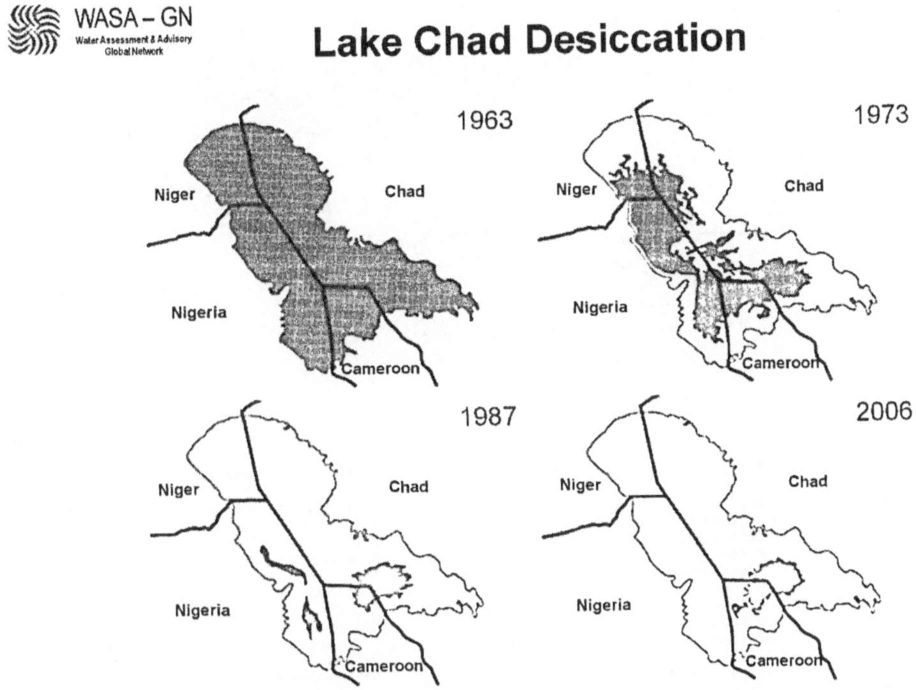

Figure 5. Lake Chad desiccation from 1963 to 2006.

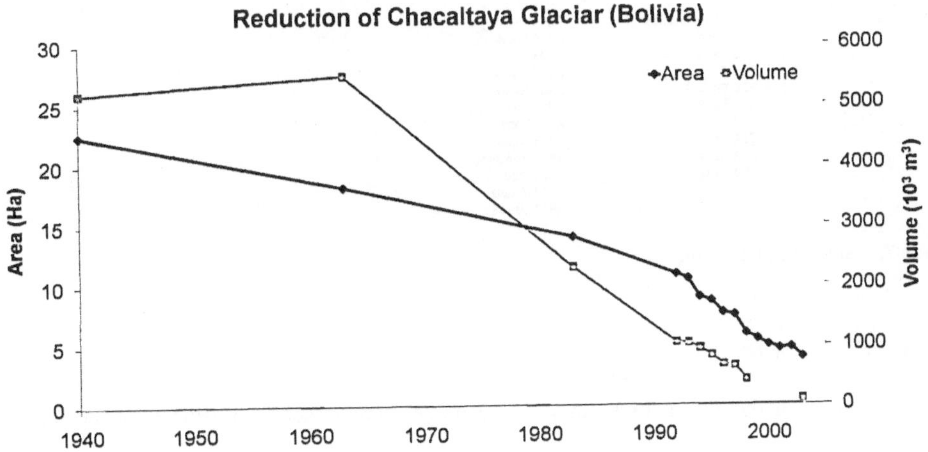

Figure 6. Reduction of Chacaltaya Glaciar (Bolivia).

of any kind from it. Consequently it must be very important for the international community to develop a new paradigm for fair development and water is not far removed from that challenge.

How to make the water governability crisis become a common challenge and for us to develop a new strategy to make progress in the implementation of global water management in terms of cross-border waters, national in terms of development and eradication of poverty.

Africa

A01 Lake Chad
A02 Congo
A03 Cuanza
A04 Cunene
A05 Jubba
A06 Limpopo
A07 Mangoky
A08 Mania
A09 Niger
A10 Nile
A11 Ogooue
A12 Okavango
A13 Orange
A14 Oued Draa
A15 Rufiji
A16 Senegal
A17 Shaballe
A18 Turkana
A19 Volta
A20 Zambezi

Europe

E01 Dalalven
E02 Danube
E03 Daugava
E04 Dnieper
E05 Dniester (Nistru)
E06 Don
E07 Duero
E08 Ebro
E09 Elbe
E10 Garonne
E11 Glomma-Laagen
E12 Guadalquivir
E13 Kemijoki
E14 Kizilirmak
E15 Kura-Araks
E16 Lake Ladoga
E17 Loire
E18 North Dvina
E19 Oder
E20 Po
E21 Rhine & Maas
E22 Rhone
E23 Seine
E24 Tagus
E25 Tigris & Euphrates
E26 Ural
E27 Vistula
E28 Volga
E29 Weser

Asia

AS01 Amu Darya
AS02 Amur
AS03 Lake Balkhash
AS04 Brahmaputra
AS05 Chao Phraya
AS06 Ganges
AS07 Godavari
AS08 Hong (Red River)
AS09 Huang He (Yellow River)
AS10 Indigirka
AS11 Indus
AS12 Irrawaddy
AS13 Kapuas
AS14 Kolyma
AS15 Krishna
AS16 Lena
AS17 Mahakam
AS18 Mahanadi
AS19 Mekong
AS20 Narmada
AS21 Ob
AS22 Pechora
AS23 Salween
AS24 Syr Darya
AS25 Tapti
AS26 Tarim
AS27 Xun Jiang
AS28 Yalu Jiang
AS29 Yangtze
AS30 Yenisey

North & Central America

NA01 Alabama & Tombigbee
NA02 Balsas
NA03 Brazos
NA04 Colorado
NA05 Columbia
NA06 Fraser
NA07 Hudson
NA08 Mackenzie
NA09 Mississippi
NA10 Nelson
NA11 Rio Grande
NA12 Rio Grande de Santiago
NA13 Sacramento
NA14 Saint Lawrence
NA15 San Pedro & Usumacinta
NA16 Susquehanna
NA17 Thelon
NA18 Yaqui
NA19 Yukon

South America

SA01 Amazon
SA02 Chubut
SA03 Magdalena
SA04 Orinoco
SA05 Parana
SA06 Parnaiba
SA07 Rio Colorado
SA08 Sao Francisco
SA09 Lake Titicaca & Salar de Uyuni
SA10 Tocantins
SA11 Uruguay

Oceania

OC01 Burdekin-Belyando
OC02 Dawson
OC03 Fly
OC04 Murray-Darling
OC05 Sepik

Figure 7. Shared water basins.

9 CONCLUSIONS

The challenges for the 21st century in water resource management are centred on the following areas:

- Guaranteeing safe access to drinking water and basic sanitation for all society, not only half as defined in the Millennium Development Goal.
- Guaranteeing safe food for the 1000 million undernourished people by developing more efficient irrigation technology.
- Strengthening the development of hydro-energy to produce electricity in conjunction with other sources of renewable energy.
- Strengthening shared management of resources between all users and nations, taking advantage of the role of water as a source of cooperation.

- Bigger investments in water sciences in order to provide the right tools to guarantee sustainable development that is fair for all.
- Eradication of corruption in the sector.

ACKNOWLEDGEMENTS

I would like to acknowledge the contribution by Engineer Alberto Crespo Milliet for the interesting discussions about the water crisis problem and its governability in developed and developing countries. I would also like to thank the members of WASA-GN for their contribution with ideas and proposals as to how to make progress in the subject of the world's water.

REFERENCES

Andreani, J.-L. & Orange, M (1997) *Le Monde/Dossiers*. Documents 258, Octubre 1997.

Bogardi, J. & Nachtnebel, H.-P. (1994) *Multicriteria Decision Analysis in Water Resources Management*. UNESCO-PHI.

Braverman, A. (1997) *Le Monde/Dossiers*. Documents 258, Octubre 1997. Israel, Univ. Ben-Gourion du Néguev.

Bretschneider, H. (1993) Taschenbuch der Wasserwirtschaft. Berlín, Alemania, Paul Parey, 7. Auflage.

Candela, L. & Fernández-Jáuregui, C.A. (Coordinadores) (1999) Agua y desarrollo. *Revista CIDOB d'Afers Internacionals*, 45–46. España, abril 1999. ISSN: 1133-6595.

Fernández-Jáuregui, C.A. (1994) Management von Wasserressourcen in Lateinamerika. Diskussionsforum, Technische Universität Berlin, TU-International 24/25.

Fernández-Jáuregui, C.A. (1994–1999) Sistema de información del ciclo hidrológico y las actividades en recursos hídricos de América Latina y el Caribe (LACHYCIS). UNESCO/PHI.

Fernández-Jauregui, C. (1995) *Conociendo la cuenca*. UNESCO (PHI-EPD)/UNEP.

Fernández-Jáuregui, C.A. (1997a) *Desarrollo de escenarios futuros del agua en América Latina*. Vol. 2. UNESCO.

Fernández-Jáuregui, C.A. (1997b) Proyecto GLOBESIGHT UNESCO/PHI. Los recursos hídricos superficiales y subterráneos: Elementos base para la modelización y generación de escenarios futuros del agua. UNESCO/PHI.

Fernández-Jáuregui, C.A. (1999) El agua como fuente de conflictos: Repaso de los focos de conflictos en el mundo. *Revista UNESCO-CIDOB d'Afers Internacionals*, 45–46.

Fernández-Jáuregui, C.A. & Johnson, I. (eds) (1998) Hydrology in the humid tropic environment. IAHS/UNESCO, IAHS Publication 253. ISSN: 0144-7815.

Fernández-Jauregui, C. & Milliet, A.C. (2008) *Agua, Recurso Único*, Expoagua Zaragoza 2008 S.A. Edición Prensa Diaria Aragonesa S.A. ISBN: 978-84-95490-82-7.

Fernández-Jauregui, C. & Pochat, V. (2008) *Agua Compartida*, ExpoZaragoza 2008 S.A. Edición Prensa Diaria Aragonesa S.A. ISBN: 978-84-95490-87-2.

Fernández-Jauregui, C. & Milliet, A.C. & Agbar, F. (2009) *Las Aguas Transfronterizas en el marco de la Crisis Mundial del Agua*. Editorial Ormoprint.

Fernández-Jáuregui, C.A., Fleming, G., Folkard, A., Larsen, L.C. & Pupolin, S. (1999) Assessment for the development of a master plan and certification of environmental monitoring networks. Padova, Italia, Centro Sperimentale per l'Idrología e la Meteorología (CSIM/ARPAV).

Fleischer, G. (1994) *Abfallvermeidung*. TU-Berlín, Alemania, Ökobilanzen.

Fleischer, G. (ed.) (1994) Prozeß und Bilanzprinzip. TU-Berlin, Ökobilanzen.

Fundación CIDOB (1998) El agua. *Fundación CIDOB*, 62. España. ISSN: 1132-6107.

Garduño, H. & Arreguín-Cortés, F. (eds) (1994) *Uso Eficiente del Agua*. UNESCO-PHI.

Gioda, A. (1997) *Historia del Agua*. ORSTOM, Francia, Archivo y Biblioteca Nacionales de Bolivia, SENAMHI-Bolivia, CONAPHI-Bolivia. UNESCO-PHI.

GWP/SAMTAC, UNESCO, CATHALAC, CEPAL y consultores (1999) La Visión de América del Sur. Documento de trabajo. Agosto 1999.

GWP/SAMTAC, UNESCO, CATHALAC, OEA, CEPAL e instituciones nacionales de cada país (1999) La Visión de Centroamérica, Caribe, USA y Canadá. Documento de trabajo. Julio–agosto 1999.;IMTA y Comisión Nacional del Agua (México) (1994) *Uso Eficiente del Agua*. UNESCO/PHI, IWRA.

Jiménez, J.M.M. (1997) Priorización y toma de decisiones ambientales. España.
Klemes, V. (1996) Risk analysis: The unbearable cleverness of bluffing. Canada.
Lundqvist, J. & Gleick, P. (1997) *Comprehensive Assessment of Freshwater Resources of the World. Sustaining Our Waters into the 21st Century*. New York, UNESCO. pp. 1–51.
Moreira, A.M.M. (1997) O modelo multicriterio de decisão em grupo. Brasil.
Postel, S. (1996) Dividing water: Food security, ecosystem health and the new politics of scarcity. *WorldWatch*, 132, September 1996.
RSC (1994) Catalogue of rivers for Southeast Asia and the Pacific, Regional Steering Committee on IHP-IV Project H-5-1 for Southeast Asia and the Pacific, November 1994.
Saaty, T. (1997) Toma de decisión para líderes.USA.
Shiklomanov, I. (1998) *World Water Resources – A New Appraisal and Assessment for the 21st Century*. UNESCO.
The politics of scarcity of water in the Middle East. London, 1988–96.
UNESCO (1994) Agua, vida y desarrollo. Manual de uso y conservación del agua en zonas rurales de América Latina y el Caribe. Proyecto Regional Mayor para la utilización y conservación de los recursos hídricos en áreas rurales de América Latina y el Caribe (PRM).
UNESCO (2008) *El derecho humano al agua, situación actual y retos del futuro: UNESCO Etxea y Oficina de Naciones Unidas de apoyo al Decenio Internacional para la Acción "el agua, fuente de vida" 2005–2015*. Editorial Icaria. ISBN 978-84-9888-020-5.
UNESCO/OMM (1993) Evaluación de los recursos hídricos – Manual para un estudio de apreciación de las actividades nacionales.
UNESCO–WWAP (2003) Water for People, Water for Life. 1er Informe sobre el Desarrollo de los Recursos Hidricos en le Mundo. ISBN: UNESCO: 92-3-103881-8.
UNESCO–WWAP (2006) *El agua: Una responsabilidad compartida. 2° Informe sobre el Desarrollo de los Recursos Hídricos en el Mundo*. UNESCO-WWAP. ISBN: UNESCO: 92-3-104006.
Vickers, A. (1995) Technical issues and recommendations on the implementation of the U.S. Energy Policy Act, report prepared for the American Water Works Association, September 1995. Amherst, MA, Amy Vickers & Associates, Inc.
World Water Vision project. (2007) Messages to initiate consultation for the World Water Vision. First and second draft for internal use only. GWP/SAMTAC working document, other documents from World Water Vision Management Unit.

CHAPTER 14

Water challenges in the 21st century

Philip H. Burgi
Water Resources Consultant, Wheat Ridge, Colorado, USA

ABSTRACT: The global water challenges we face in the 21st century will focus more on management of the resource than its development which was the focus of the 20th century. If this globe is to serve as the home for all peoples with provision for basic living conditions including adequate supply and quality of fresh water, then our policies and technologies must be adaptable to the challenges of the 21st century. As water scientists, engineers and managers we must be cognizant of the new global transitions in demographics, economics, and the environment. This paper will present a brief summary of present conditions and provide insight to the challenges we will face in 21st century in water engineering and management as our global water resources are further stressed.

1 INTRODUCTION

A challenge is a summons to answer and explain; a call to action, often the challenge is made in response to an identified need. Keen (1987) credits the advancements in 19th century hydraulics on the simple truth that "Necessity is the mother of invention." Our heroes in hydraulics developed innovative solutions as they met the challenges they faced: Chezy's need to channel a water supply to Paris; Darcy's need to devise a piped water distribution system for Dijon; Reynold's need to determine the influence of waves and currents on the Mersey Estuary; Simpson's need to suppress cholera in urban areas and John Stevens and Dr Gorgas' need to control malaria to build the Panama Canal. "No man is an island" reflects a second truism Keen credits for such unprecedented progress in the 21st century. The importance of collaborative investigations is often under-valued in the importance it plays in innovation. As we progressed into the 20th century, water engineers developed hydraulic laboratories where physical models were used to resolve complex flow patterns in structures and rivers. However, by the last quarter of the 20th century many developed nations moved away from a focus on water development to a focus on water management. A new paradigm appeared as public values changed and concerns developed over environmental and societal consequences of water development. This was certainly true in the United States where water development agencies such as the Corps of Engineers and Bureau of Reclamation lost much of the political power they enjoyed earlier in the century. These concerns lead to the advancement of water resources planning and integrated water resources management (IWRM) as professional disciplines. As we enter the 21st century, it is appropriate to pause, and consider what challenges we will face as individuals and national and international agencies to deal with the future water needs of our globe.

Predicting the future is a very speculative venture and open to great risk of humiliation. Some would say one should not make predictions for the period of one's own lifespan for fear of being proven wrong. However, in spite of these warnings, let's look back at recent human history to see if we can gain a perspective on where we have been and then take a look through the window to the future and see what water challenges we might face in the 21st century. In a special issue of *Scientific American* entitled "Crossroads for Planet Earth", several scientists and economists gave their insight into what lies ahead for humanity in the 21st century. George Musser reviews what he refers to as the three great transitions set in motion by the Industrial Revolution and coming to their culmination in the next few decades: *demographic*, *economic*, and *environmental*. We are about to pass through the "bottleneck" – a period of maximum stress on natural resources and human ingenuity. Our future will be greatly influenced by our success with how well we can bring our ingenuity to bear on the challenges of these three transitions (Scientific American, 2005). To face the water challenges of the 21st century we must appreciate and consider how water is associated with all three of these transitions and bring our water engineering and management ingenuity to bear on these three transitions.

Latest predictions show world population will reach a plateau of about nine billion by mid-21st century. This is in stark contrast to a sense of uncontrollable population explosion as recent as thirty years ago. Another observation is that even with continued population growth that will occur over the next fifty years, extreme poverty is receding both as a percentage of population and in absolute numbers. This is especially true in China and India. However, as population and wealth grow they increasingly press against the limits of our planet's natural resources. Environmental degradation is most noticeable as we look at CO_2 emission predictions. We currently pump out carbon dioxide three times as fast as the land and oceans can absorb it. Figure 1 illustrates these trends and, although we may have a sense of comfort with the demographics and economics, there are serious repercussions if mankind cannot meet the water and environmental challenges we will face in the 21st century.

Although these broad-brush challenges may be difficult to grasp, we have the responsibility as water engineers and managers to effectively engage them in a professional and equitable manner. This is especially important as we look at regional differences in our extremely interdependent global economy and environment. How do we respond to the increased global population characterized by hyper-urbanization and the fact the virtually all of the future population growth will occur in today's economically less developed and water short regions? Can we develop water resources in a sustainable manner that will sustain life and bring prosperity to people in developing nations who now have no firm water supply? Will our role in expanding hydroelectric development and

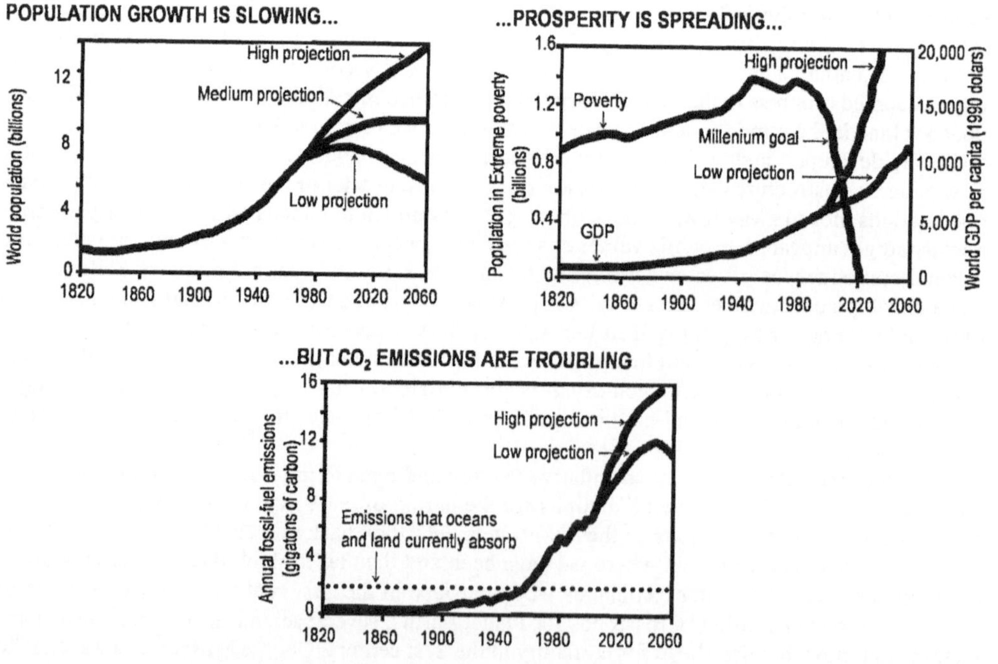

Figure 1. Three World-Changing Transitions (Scientific American, 2005).

utilizing other renewable resources as power alternatives have a positive impact in decreasing CO_2 emissions? These are the broad-brush issues we will face as we cross into the next millennium.

2 WATER ISSUES AT THE TURN OF THE CENTURY

Recent major water-related disasters include: the 2004 Indian Ocean tsunami; the 2004 and 2005 hurricanes in the Caribbean, west Pacific and the United States; the 2005 floods in central and eastern Europe; and the extensive droughts in Niger, Mali, Spain and Portugal reminds us of both the destructive power of water and the misery caused when risks are not properly addressed. The lack of water after such disasters also causes extreme misery and death. The occurrence of these disasters illustrates the fragile nature of life and the fundamental changes that are affecting water resources worldwide and add to the challenges we face as water engineers and managers.

Major migration patterns are seriously affecting the quality and quantity of available freshwater on the planet. Today China is experiencing the largest population migration in history as 120 million people leave rural villages and move to industrial zones in search of employment. More than 400 million Chinese will move to cities in the next 25 years. In many rapidly growing urban areas, it is proving impossible for local governments to build the infrastructure necessary to deliver water supply and sanitation facilities to service the growing population, leading to poor health, low quality of life and in some cases social unrest. These urban demands for water only exacerbate the increasing demands on water for food production, energy creation and industrial uses. They also are a reminder of poor management decision made by previous generations. To understand where we are today, let's look at *water supply* and the *global differences in water development.*

2.1 *Water supply*

Figure 2 illustrates the growth of global population, water withdrawal and irrigated land use in the 20th century. Water consumption rate increased six fold between 1900 and 1995 – more than double the rate of population growth – and continues to grow as agricultural, industrial and domestic demands all increase. Although there is more than enough water available on our globe, it is not always equitably distributed nor of the quality required. Some of the inequity results from

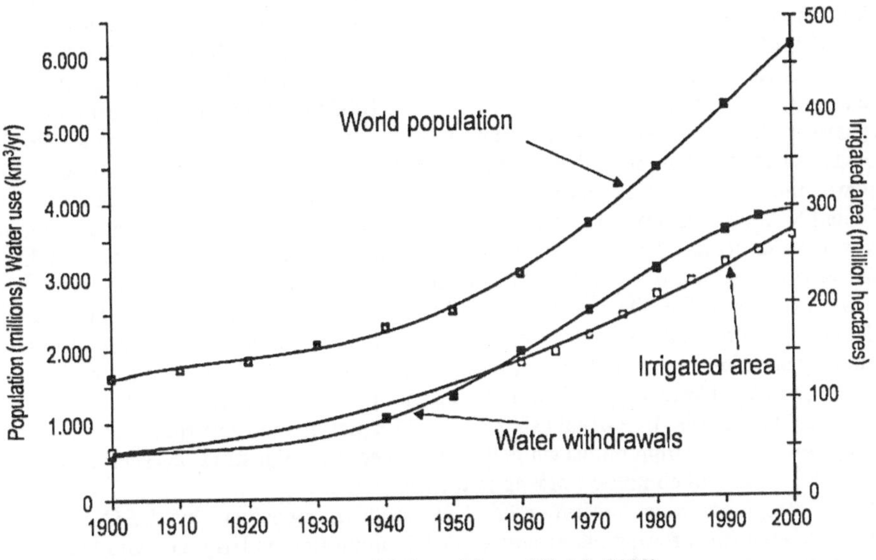

Figure 2. World Populations, Water Use, and Irrigated Area. (Gleick, 2000).

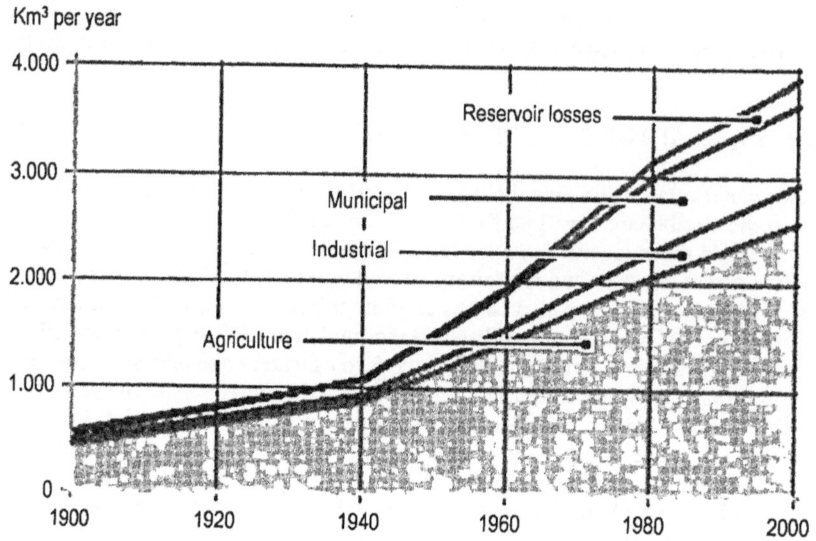

Figure 3. Estimated Allocation of the World Water Use (From FAO Aquastat, 2005).

nature's diversity in distribution of the water and some are caused by poor water management. As illustrated in Figure 3, seventy percent of the water used worldwide is used for agriculture. Water for agricultural purposes will continue to increase if we are to feed the world's growing population. Some predict that the rate of water consumption for agriculture will soar further as more people expect Western-style lifestyles and diets. For instances, production of one kilogram of grain-fed beef needs at least 15 m^3 of water, while a kilo of cereals needs only 3 m^3. The good news is that in the last quarter of the 20th century there has been a decrease in the rate of water used for agriculture as shown in both Figures 2 and 3.

Keizrul Abdullah, President of the International Commission on Irrigation and Drainage states "At present, only 12% of the global land area, that is, about 1.5 billion ha, is cultivated for food production. The irrigated area, which is of the order of 270 million ha (18% of cultivated land), contributes 40% of crop output, providing food for people and livestock. More than 1 billion people, including one third of the population of China and India and 350 million people in sub-Saharan Africa, currently face severe *water scarcity*. If we use 1000 m^3 or less as a per capita threshold for water availability, in 1990 eighteen countries in the world were classified as scarce (12 of these countries had less than 500 m^3 per capita), and this number could increase to 30 by 2025" (Hydropower and Dams, 2005). Figure 4 illustrates the water stress by river basin in 2002.

The United Nation's World Water Assessment Program recommends that people need a minimum of 50 litres of water per day for drinking, washing, cooking and sanitation (UNWWAP, 2003). The developed world uses on the average 350 l/d for these basic necessities, but many countries are not able to provide even the 50 l/d recommended by the UN. In 1990, over a billion people, 1/3 of the global population, did not have access to that basic need. Providing universal access to this minimum daily per capita consumption worldwide by 2015 would take less than 1% of the amount of water we use today. However, at this time we are not prepared institutionally to deliver on this right. Many feel that a good part of this shortage will be met by non-governmental organizations (NGOs) working in rural communities while the large international institutions such as, the United Nations, US AID, etc. will continue working in the urban areas.

The future for global water resources has been studied extensively by the United Nations World Water Assessment Program, the International Commission on Irrigation and Drainage, the International Water Resources Association, and many other governmental and non-governmental

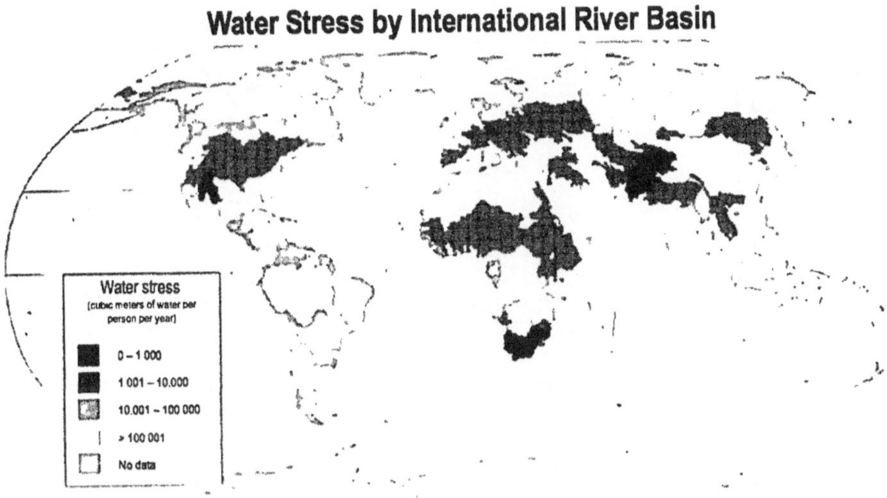

Figure 4. Water availability per capita by basin (UN Environmental Program and Oregon State University; 2002).

organizations. The United Nations through its Division of Sustainable Development created a plan of action first approved at the 1992 Rio Earth Summit. Eleven challenges were developed and adopted as the primary focus of the First World Water Development Report (WWDR), issued at the occasion of the 3d World Water Forum in Kyoto in March 2003 (UNWWAP, 2003). These global challenges cover the issues related to the ways we humans use water and the increasing demands we place on the water resource. Signs of stress and strain are apparent across every sector: health, ecosystems, cities, food, industry and energy. With population growth and continuing pollution, these pressures are likely to increase.

2.1.1 *Eleven water challenges – UN*

1. *Meeting Basic Needs* – Having access to safe and sufficient water and sanitation are now recognized as basic human rights. Being able to wash one's hands and drink clean water has a major impact on family hygiene and health. Because people who are poor are most likely to get sick, and ill health perpetuates poverty, it triggers a vicious cycle that hampers economic and social development.
2. *Protecting Ecosystems* – The possible negative impact of human activity on the environment must be considered when managing water resources in a sustainable way. Human beings must learn to respect the resource base on which life ultimately depends and to see land and water as two sides of the same coin. For this reason, decisions should be taken at river basin level, when possible.
3. *Water and Cities* – By 2030, over 60% (nearly 5 billion people) of the world's population will be living in urban areas. As a result, competing demands from domestic, commercial, industrial and peri-urban agriculture are putting enormous pressure on freshwater resources.
4. *Securing the Food Supply* – The challenge here is to increase food production and security by getting 'more crop per drop', while also devising ways to ensure a more equitable allocation of water for food production.
5. *Water and Industry* – Industry is both a major user of water and a major contributor to economic and social development. To move towards sustainability, industries must be assured of having an adequate supply of water. Industries should see that water used in industry is used efficiently and not returned to nature as untreated waste that polluting the environment.

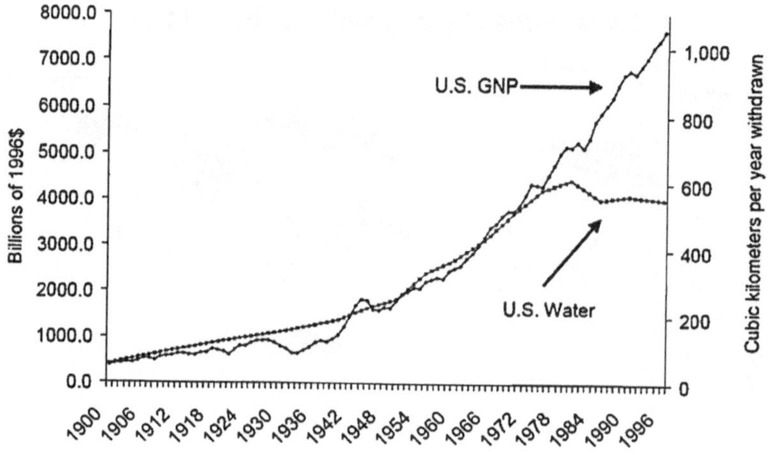

Figure 5. The Link between Economic Growth and Water in the USA (Gleick, 2002).

6. *Water and Energy* – Tremendous increases of energy and water will be required in the near future as the world's population increases from 6 to over 9 billion. Even now, some 2 billion people do not have access to a reliable supply of electricity.
7. *Managing Risks* – Water related hazards, such as floods, droughts, tropical storms, erosion and various kinds of pollution should be factored into any integrated approach to water resource management and policy.
8. *Sharing Water Resources* – Competition over scarce or poorly allocated resources can lead to tension and insecurity. Therefore decision-makers, communities, governments and regions must strive to develop policies that allow for sharing among all stakeholders.
9. *Valuing Water* – This whole question is among the most controversial of all the challenges identified in the Ministerial Declaration emerging from the Second World Water Forum in the Hague. In many societies the whole notion of putting a price tag on something as intrinsically valuable as water is unacceptable.
10. *Ensuring the Knowledge Base* – This target takes account of the whole range of technical and non-technical information and knowledge, and seeks ways for all societies to benefit from their development, exchange and dissemination.
11. *Governing Water Wisely* – This challenge area is particularly complex and sensitive. It moves the debate about sustainability beyond water management issues and into processes of political, social and institutional change.

One encouraging note, as we look to the future, is Peter Gleick's assessment that we appear to be breaking away from the previous relationship where economic growth was tied inexorably with water withdrawal trends. In the United States this was a fixed trend through much of the 20th century. However, around 1980 the connection between water demand and economic growth separated. As shown in Figure 5, the trend in the United States for total withdrawals of water flattened or even lessened in subsequent years. In the period from 1980 to 2000, California's water usage actually dropped. Some of the decrease can be attributed to changing industries (computer chips take less water than steel), but most resulted from improved efficiency in delivery systems and use, especially in the agricultural sector. This encouraging trend has caused a number of planners around the world to reconsider their long-range demand projections (Gleick, 2002).

The issue of water supply has been identified and is fairly well understood as we enter the 21st century. A more unsettling issue might be the global water development differences and perceptions, as well as expectations we have as people, communities, and nations.

2.2 *Global differences in water resources development*

The major global driving force in water use in the 20th century focused on water development. It was generated and sustained in an era when public values were founded on expansion by development of natural resources. This was especially true for developing nations at the start of the 20th century including the United States. There is a global philosophical difference today over how nations should develop in a sustainable manner. Political battles between developed and underdeveloped regions continue within countries and between countries. For example, in Spain an argument is raging over how the drought should be handled. One side calls for a revival of a multibillion dollar transfer program that called for the water-richer northern areas to supply the central and southern areas. However, the other side lead by the present government and ecological groups has scrapped that plan. They argue for a total rethink of water resources, based on desalination plants and water banks, with an emphasis on conservation and protection of aquifers and forests.

Many Spanish agriculturists are urging the government to return to the water transfer program. Some have said "The problem is not the water but the politicians and their plans. Transfers worked for the Romans and there wasn't much they didn't know, referring to the ancient aqueducts that still carry water across parts of Spain today. But statistics indicate the government has a point about the need to reform water policy. Spain loses more than 60 percent of its water before it reaches the tap and only 1.5 percent is recycled. The country is tops in Europe for using up to 80 percent of its water in irrigation systems, of which only a fifth could be considered modern" (US Water News, June, 2005).

These comments reflect the difficulty in trying to change and adapt national water policies to fit changing public values as regional water resources become developed. The water stakeholders have different expectations and needs depending on their region's water development. There is also a growing political opinion among those nations with developed water resources to impose their values on others who are still developing their resources. The point raised by many at the 1994 I COLD Congress held in Durban, South Africa, was that developed nations have no right to dictate their new found environmental values on developing countries that are still in the early stages of development. In fact, their statements reflect an arrogance of spirit by those who have already reached a comfortable level of development (Saavedra, 1995).

The international complexities in water engineering and management are even more challenging than those within the borders of any one nation as efforts are made to balance inequities in water availability. As an example, global hydroelectric development has reached about 19% of feasible potential. However, there are extreme differences on where that development has occurred. Developed nations have reached levels of utilization as high as 45%, where developing nations in Africa may be as low as 5%, Figure 6. These increasingly complex differences in development philosophy will need to be reconciled on a global level if we are to safely pass through the "bottle neck" suggested by Musser.

These are the circumstances and realities we face at the start the 21st century. The challenges they present require a new look at how we will respond in the 21st century to provide the basic water needs necessary for quality of life and adequate economic growth for all. It is imperative that we understand this changing water paradigm and actively seek new collaborative and technical approaches to meet the challenges. The purpose of this paper is to focus our attention on what are the perceived needs and therefore challenges we will face as water engineers and managers in the 21st century.

3 WATER CHALLENGES IN THE 21st CENTURY

"It is one thing to find fault with an existing system. It is another thing altogether, a more difficult task, to replace it with another approach that is better." Nelson Mandela, speaking of global water resources management, World Commission on Dams, November 2000.

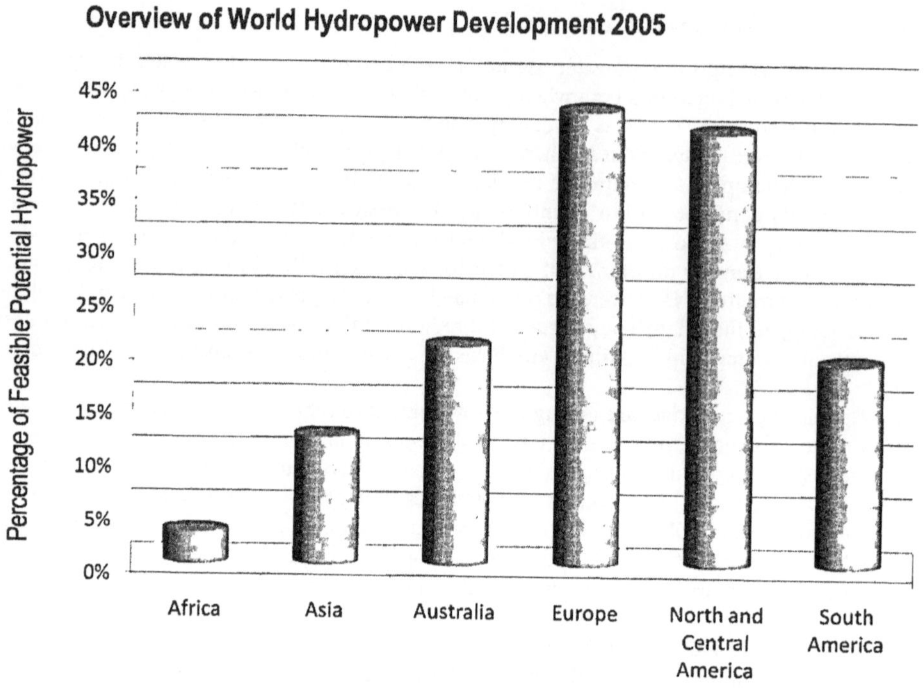

Figure 6. Overview of World Hydropower Development in 2004/5 (World Atlas, 2005).

Today we are witnessing very rapid and significant changes on our globe. These changes are most evident as they relate to *population, prosperity* and the *environment.* Water is less visible, but just as important. We are not making any new water, nor have we lost any. As, Avinash Tyagi, Director, Hydrology and Water Resources Department of the World Meteorological Organization recently stated at the World Water Forum held in Mexico City, "There's plenty of water. Over millions of years, the Earth hasn't lost a drop. The problem is management-keeping water clean and available" (Enriquez, 2006).

Total global water withdrawals are presently estimated at approximately 4000 km^3/yr. As shown by the dashed lines in Figure 7, there is a considerable range of prediction as to what will be the future withdrawal rates as we move through the 21st century. The present encouraging change in the water withdrawal rate most likely results from improved conservation and efficiency in water use both in agriculture and urban areas world-wide. Will that encouraging rate continue?

The future water challenges facing the global community are quiet variable. Keizrul Abdullah, President of the International Commission on Irrigation and Drainage (Hydropower and Dams, 2005) characterizes them under two broad categories:

For developed countries:

Water quality management,
High technology in irrigation,
Conservation through water pricing,
Public awareness through water saving campaigns.

For developing countries:

Poor infrastructural set-up,
Rehabilitation of old systems,
Low irrigation efficiency,

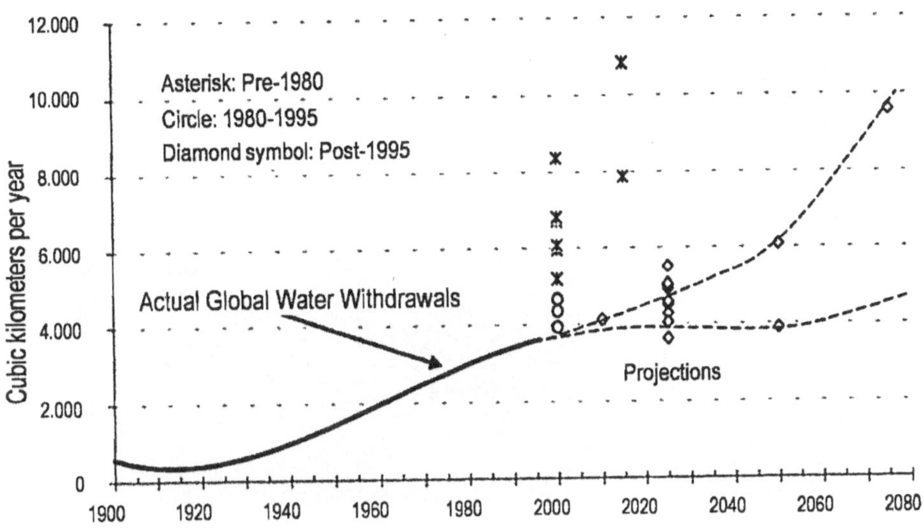

Figure 7. Water Scenarios; projected and actual global water withdrawals (Gleick, 2000).

Negligible re-use of water,
The use of low quality water, and
The continued use of conventional irrigation methods yielding low crop outputs.

There are those who would say rehabilitation of aging systems belongs in both lists.

During the 1960s and 70s, the prominent assumption was that the answer to global water and sanitation issues lay in engineering solutions. When such approaches did not lead to an appreciable impact on the problem it began to become clear that the issue was not merely a technical problem, but a complex social, economic and political problem. In the 1990s, increasing emphasis was placed on the development of *policy and legislation*, and the development of *viable institutions*. This was in response to the recognition that water and sanitation services cannot be sustainably developed in isolation from other sectors and without an enabling legal framework. Closely related to these issues is the need for *good governance* leading to sound public administration and effective governmental planning. Unfortunately, over the past 30 years the practice of good water and sanitation principles has not been emphasized nearly as much as the provision for water supply. Due to this lack of emphasis on developing good practices of water use as well as water supply, the potential health and development of the water supply sector have not been fully achieved (Abrams, 2001).

As discussed previously, the UN's *World Water Development Report* has studied future needs and identified eleven water challenges for the 21st century. We in the global community of water engineers and managers would do well to fully embrace these challenges as we consider our roles in finding solutions to today's challenges just as our heroes of the past fulfilled their duties. In the discussion that follows, a few of the *water management* and *water engineering or technology* challenges will be presented.

3.1 *Challenges for water managers*

There is a lack of comprehensive strategies at the national and international levels to meet the water challenges we face. This is true for the developed or underdeveloped world. Attitudes and behavioral issues are responsible for part of the failure but a lack of leadership and a world population not fully aware of the scale of the problem also exacerbates finding solutions. Innovative water management tools will play a critical role in the next thirty years as new technologies are developed, but strong

Figure 8. Community "Sweat equity" in Potable Water System in Santo Tomas, Ecuador.

leadership from water managers will be required to put innovation into practice. Here are a few observations:

3.1.1 *Meeting the basic water right of the global citizenry*

Access to basic water supply and sanitation services are increasingly acknowledged as fundamental for health and development, and are being increasingly accepted as a fundamental human right. This water right has to be free of politics and inclusive of all people. This is a huge challenge and one that will not easily be met in a sustainable manner.

Certainly large international institutions such as the United Nations cooperating with developed nations will have the major responsibility and leadership in meeting the challenge. However, there is a significant and increasing role for NGOs to help with development of basic water and sanitation in many rural communities world-wide. One challenge the NGOs face is the standardization of basic criteria and definition for the construction, ownership, and sustainable maintenance of small water systems. Burgi and Rydbeck (2001) discuss an example of a sustainable water system policy and the spirit of self-determination it has provided rural communities in Ecuador, Figure 8. The requisite list of commitments by the local community must make before assistance is offered includes: forming a community water council, commitment to community, "sweat equity" to build the project, identification of a community dump site and basic sanitation including latrines, implementing a minimal water usage fee for all who use the system, community health training, and providing a community water person – "aguatero" to maintain the water system. NGOs such as Lifewater, Water for People, Engineers Without Borders, are but a few working worldwide to develop potable water supplies in rural areas.

Although there are some engineering challenges, the real challenge is institutional in nature. Is it possible that we in the engineering community might step into more of a leadership role in encouraging international and national government agencies to practice good governance and ethical standards in addressing this identified right for all?

3.1.2 *Integrated Water Resources Management (IWRM)*

IWRM will become an increasingly important tool in the successful management of water resources. Rahaman and Varis (2005) in an article summarizing the evolution of the concept of integrated water resources management over the past thirty years make an excellent argument for using this tool. They quote the Technical Advisory Committee of the Global Water Partnership from the

2002 Johannesburg World Summit on Sustainable Development (WSSD), by defining IWRM as, "..a process, which promotes the coordinated development and management of water, land and related resources in order to maximize the resultant economic and social welfare in an equitable manner without compromising the sustainability of vital ecosystems." Water should be managed in a basin-wide context, under the principles of good governance and public participation. Examples of older versions of IWRM are included from the 10th century Valencia water court up to the 20th century Tennessee Valley Authority (TVA).

Rahaman and Varis identify seven challenges to the effective future implementation of IWRM principles. These are issues not adequately addressed in the present IWRM models.

3.1.2.1 Challenges for implementation of IWRM

Privatization – Where basic infrastructure is not yet complete, the issue of privatization versus subsidies needs to be ethically and practically addressed.

Water as an Economic Good – The application of economic principles to the allocation of water is acceptable; however, water should not be treated as a market-oriented commodity when it comes to domestic use for very basic needs. The UN has set this basic right at 50 liters/day/person.

Transboundary River Basin Management – The enforcement of river basin plans across political boundaries needs to be addressed. Formal agreements on international water allocations based on local participation are needed. They must be accepted and respected regardless of economic and military power imbalances between shared states on the river basin.

Restoration and Ecology – IWRM principles do not clearly focus on river restoration needed for sustainable water resources management in areas where water resources have been substantially developed such as the United States and Europe.

Fisheries and Aquaculture – Fisheries are generally undervalued in water and management models. Aquaculture is one of the fastest growing industries for human consumption of protein and IWRM and other such tools need to include the interplay of the impact of sediment and nutrient loads from fresh waters on coastal and estuarine zones.

Integrating Lesson Learned from Past IWRM Experience – Lessons from past initiatives are vital to the implementation of future IWRM principles and policies.

Spiritual and Cultural Aspects of Water – Our appreciation for the spiritual and cultural aspects of water and its impact on local acceptance of water management plans is poorly understood. This is a challenge that must be addressed if we expect to develop sustainable water management plans globally.

3.1.3 *Water transfers*

In developing countries such as China, water transfer is one solution to regional differences in water availability. China has 21% of the world's population and only 7% of the world's total water resources. Exacerbating the problem is the uneven distribution of water in China, particularly in the drought-prone north. This region is home to over 30% of the country's people but has only 6 percent of its water. China has launched an ambitious south-north water diversion scheme. Nearly 45 billion cubic meters of water from the Yellow, Yangtze and other rivers will be sent north every year when the project is finished in 2050, at an estimated total cost of almost 500 billion yuan ($60.42 billion), twice that of the Three Gorges Dam, the world's biggest hydroelectric project. Measured against its economy, China consumes five times more water than the global average (Water Conserve, 2005).

As mentioned previously, Spain has also looked at a large water transfer program that would move water from the water-richer northern areas to supply the agricultural and tourist areas in the central and southern areas of the country. Water transfer schemes carry great political burdens and will only be successful when all people affected by the transfers have opportunity to participate in the decision making process.

In the United States, the recent voluntary transfer of Colorado River water from Imperial Irrigation District to southern California municipal water agencies is a good example of effective water transfer. Work remains to ensure that potential adverse affects on farm workers, the local economy and the nearby Salton Sea are addressed. Another recent example is the legal settlement between an environmental group interested in saving the silvery minnow on the Rio Grande River and the City of Albuquerque, New Mexico these examples show the potential for voluntary transfers to be part of the solution to complex disputes.

As water demand in developed countries continues to grow in the face of limited supply, water transfers are an increasingly attractive alternative to developing new water supplies. The use of "water banks" in major river basins is gaining acceptance as a water transfer alternative. Some water bank accounts can be arranged through short-term leasing, "interruptible supply contracts" or may be permanent water rights transfer. The interruptible supply contract can be arranged between agricultural interests and cities where agriculturists are paid over longer periods of time to let cities take part of their water during droughts. One major challenge to the effective use of water transfer concepts is the development and application of water law. In many countries, water laws are not clearly written or understood.

3.1.4 *Water conflicts*

The increasing scarcity of clean fresh water impedes development, undercuts human health, and provides the potential for conflict between and within countries. Water scarcity can assume an increasingly contentious and violent role when, for example, water-dependent sectors such as irrigated agriculture can no longer sustain farming livelihoods, leading to political conflict.

However, water can also provide a path for dialogue in otherwise heated conflicts. In politically unsettled regions where water is essential to regional development, water use negotiations can serve as de facto conflict-prevention strategies. Although it is rare that wars start over water, water allocation is often a key sticking point in ending conflict and undertaking national and regional reconstruction and development. Wolf (2005) is quoted as saying, "Water has also proven to be a productive pathway to confidence building, cooperation, and arguably conflict prevention. Cooperative incidents outnumbered conflicts by more than two to one from 1945–1999. The key variable is not absolute water scarcity, but the resilience of the institutions that manage water and its associated tensions."

In some cases, such as in the Middle East, water provides one of the few paths for dialogue in otherwise heated bilateral conflicts. Over the past 50 years, Turkey has been at the center of controversy surrounding the rivers which rise in Turkey and flow into neighboring countries. Turkey has moved forward with significant river basin development projects, while its neighbors have accomplished little in terms of development, see Figure 9. According to current consumption targets put forward by the three riparian states (Turkey, Iraq, and Syria), they intend to use a combined, impossible total of 149% of available water from the Euphrates and 112% from the Tigris (Ünver, 2002). Ünver supports a new paradigm of thinking suggesting the possibilities of meeting the region's water need by better regional collaboration with water management at the core. As the upstream country, he feels Turkey's development successes must flow downstream, metaphorically and literally.

To that end, Ünver has launched the Euphrates-Tigris Initiative for Cooperation (ETIC). ETIC is a diplomatic project that aims to bring together Iraqi, Syrian, and Turkish water experts and graduate students to build group identity, promote networking and an exchange of ideas, and develop a trilateral institutional framework removed from major political pressures. It is hoped that the group can open dialogue that avoids hydro-nationalist sensitivities, and which collects and shares multi-stakeholder data. This group can also pass along its recommendations to politicians with the benefit of scientific reason and agreement.

3.1.5 *Virtual water*

Virtual water may be a significant future solution to global imbalances in water distribution. The challenges of effectively using virtual water lie in a good understanding of its benefits and trying to balance these benefits with security and other concerns perceived by the recipient country. Since water is an important variable in crop production, countries should consider how much water is needed to produce the food they require and if there are other means for providing their food needs.

Figure 9. Euphrates-Tigris Rivers.

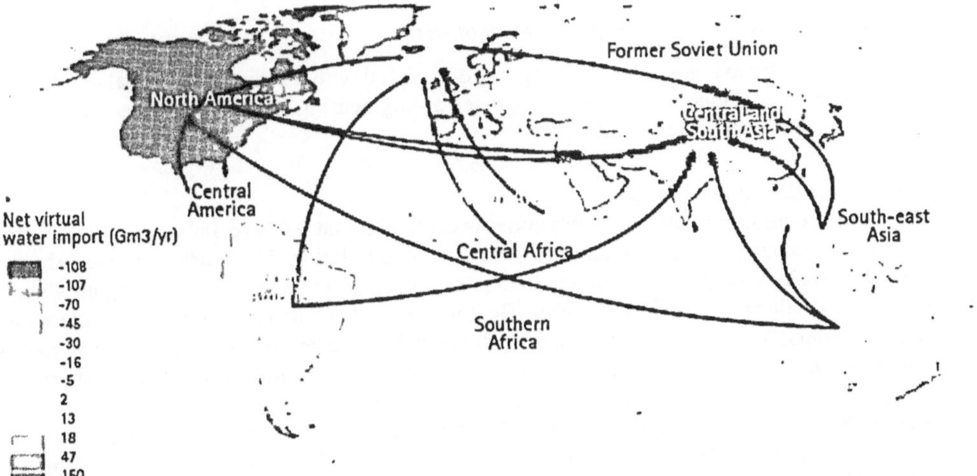

Figure 10. Net Imports of Virtual Water Around the World (Chapagain & Hockstra, 2004).

When a country imports a ton of wheat or maize, it is in effect, also importing "virtual water", i.e. the water required to produce that crop. Trade in virtual water generates water savings for importing countries – it is estimated that Egypt's maize imports in the year 2000 generated a global saving of about 2 700 million m^3 of water. The global real water saving is significant: a first estimate shows that water savings from virtual water transfer through food trade amounts to 385 000 million m^3 (Chapagain & Hockstra, 2004). Virtual water is becoming an ever increasing reality of our global economy. Figure 10 illustrates the net imports of virtual water around the globe.

Table 1. Loss of Storage Capacity in Large Dams in Pakistan Due to Sedimentation (nipalahore.gov.pk).

Water Storage	Storage capacity (km^3)	Loss up to 2002 (km^3)	Loss by 2013 (km^3)
Tarbella	14.35	3.74 (26%)	5.21 (36%)
Mangla	7.26	1.46 (20%)	1.91 (27%)
Chashma	1.07	0.46 (43%)	0.62 (58%)
Total	22.68	5.66 (25%)	7.74 (34%)

3.1.6 *Water efficiency labeling requirements*

The water conservation concept of *efficiency labeling* in the use of appliances is growing in acceptance in the developed world. The U.S. government-supported Energy Star program has been very successful in promoting the development, sale, and use of high energy efficient appliances. While the primary focus has been on energy, most high energy efficient appliances also use less water. At some point in the future, the manufacture of high water use appliances will be prohibited. Efforts in the United States are now underway to develop a program for *water efficiency labeling* requirements.

Technological advances in plumbing fixtures, appliances, irrigation equipment, and landscaping techniques have led the way in the water conservation effort. However, focusing on management of water use, which requires education and understanding, is equally important.

3.2 *Challenges for water engineering and technologies*

In addition to the water management challenges, we are faced with numerous technical challenges, many of which are ill defined or not even recognized as engineering challenges at this time. Listed below are a few of the identified challenges:

3.2.1 *Sedimentation of reservoirs*

Sedimentation is common to all reservoirs and it is estimated that 0.5% to 1% of the world's total reservoir storage capacity is lost annually to sediment accumulation. There are a number of ways to address the problem, ranging from watershed management as a means to minimizing erosion, to flushing sediments through storage reservoirs. But there is no standardized remedy that will remove sediment and extend the useful life of storage at all dams. Even reservoir flushing is seen as having a potential negative effect by releasing accumulated contaminants into downstream fisheries or water supplies.

Presently there is a heated political battle in Pakistan over a proposal to build two large dams on the Indus River: Kalabagh Dam (To replace the storage lost to sedimentation in Tarbella dam) and Basha (To collect upstream sediments before they enter Tarbella). Since Tarbella's completion in 1974, sedimentation in the reservoir to the extent of 3.74 km^3 has reduced its gross storage capacity from 14.35 km^3 to less than 11 km^3. Mangla dam, which originally had a gross storage capacity of 7.26 km^3, is also filling with sediments; though not as fast as Tarbella, thanks to watershed management and a silt-trap storage project in the catchment area above the dam.

These large dams which have brought irrigation and agricultural development to Pakistan are quickly filling with sediments, see Table 1. Siltation is a major problem affecting the efficiency of Mangala and Tarbella. The Indus is the fifth largest silt carrying river in the world bringing with it a load of 200 million tons of silt a year. Half of the new storage capacity that Pakistan hopes to gain if it builds two more large dams on the Indus – Kalabagh and Basha – will go towards replacing what is lost due to sedimentation from the existing dams (Sharma, 1998).

Sedimentation also impacts long term storage in thousands of smaller reservoirs across the globe. Duncan with the US Department of Agriculture's Natural Resources Conservation Service, reports

Figure 11. Gap weir built in 1902 (Chanson, 1998).

that since 1948 over 11,000 small flood control dams have been built in 2,000 watersheds in the United States. Many of the earlier constructed dams were designed for a 50-year life expectancy and have silted up (Dunbar et al., 2001). Figure 11 clearly presents an example of the total loss of storage behind a small dam and reservoir in Australia.

Reservoir sedimentation challenges include: Improved watershed management, effective flushing techniques, advancements in dredging technologies, and improved understanding of the environmental impacts on downstream users resulting from sediment flushing.

3.2.2 *Minimum environmental flows below dams*

The setting of environmental flows below water control structures such as dams and diversion structures has been the subject of considerable interest internationally and several countries are currently addressing this problem. There has been a change from the concept of minimum flows below large dams developed in the 1950s and 1960s, followed by instream flows in the 1970s, hydrologic and habitat based methods in the 1980s to today's multidisciplinary watershed based criteria. For instance, in Norway the idea of "minimum flows" is still prevalent and there is a clear need to think more in terms of environmental flows. A certain degree of flexibility by setting different flows at different times of the year has been instigated in many instances, although there is a need to incorporate year to year variations. For example, it may be possible to allocate more water in wet years compared to dry ones (Brittain, 2002).

3.2.3 *Climate change*

Climate change will have a major impact on future water allocations. Water resource planners as well as, dam designers work on the assumption that historic hydrological variables such as average annual flow, annual variability of flow, and seasonal distribution of flow are a reliable guide to the future. As global warming takes hold, however, there are likely to be significant changes in seasonal and annual rainfall patterns and other factors affecting streamflow such as the rate and timing of snowpack melting, and the nature of watershed vegetation. Historical and geological evidence for floods in past millennia indicate that even small changes in climate can cause major changes in the size of floods. Reservoir sedimentation will also likely be significantly affected. For example, in arid areas, an increase in average annual precipitation of only 10 percent can double the volume of sediment washed into rivers. Dams and other large-scale water development projects will play an equalizing role in the variability of water distribution, but there is a lot of uncertainty in how effective they will be.

McCully, in his book entitled The Ecology and Politics of Large Dams, states, "Calculations of the amounts of water available to turn turbines, the maximum flood which spillways will have to discharge, and the rate at which reservoirs fill with sediment, will thus become increasingly unreliable as global warming takes hold and as, inexorably, year by year, decade by decade, the earth's climate changes. Insurers are increasingly convinced that global warming is to blame for the increased frequency and severity of violent and expensive storms, floods and droughts since the late 1980s, weather events which have already resulted in burst dams, increased sedimentation and reduced hydropower capacity. A 1991 report from the UN Intergovernmental Panel on Climate Change noted that: 'Increased run–off due to climate change could potentially pose a severe threat to the safety of existing dams with design deficiencies. Design criteria for dams may require re–evaluation to incorporate the effects of climate change. Thus, not only is global warming not the godsend to save an ailing industry which many hydro backers hope, but it is going to render dams less safe and less likely to perform as their builders' claim" (McCully, 2005).

Watts (1997), discussing the engineering responses to global warming warns, "If currently prevailing human activities do not change in the next few decades we shall likely be faced with a climate substantially different from that of today." He goes on to predict that the rate of climate change will in all likelihood be unprecedented in recent history and may well prove to be large enough to make it very difficult for many species of plant and animal life to adapt to the new climate as they sometimes have in the past. In Chapter 5 of the same reference, McAnally, Burgi, et al. describe the engineering responses needed by water resource systems to effectively deal with climate change. Projected changes in the magnitude, timing, and distribution of hydrometeorological parameters, particularly if coupled with demographic shifts and changes in industrial and agricultural activity, will impact the safety of hydraulic structures, as well as the ability of water resource systems to effectively balance available supplies against competing water uses.

Prieto (2005), in a paper discussing droughts and water stress situations in Spain states, "A major increase in water demands in Spain is not foreseen because improved water use efficiency neutralizes the effect of population or irrigation growth. However, the effects of global warming on water resources availability could break the equilibrium between water supply and demands. Moreover, although Spain on average has sufficient water to meet demands, its uneven distribution in time and space produces local and temporal water stress situations" (Prieto, 2005).

Water engineers' responsibility and responses to natural disasters will play an increasingly important role as climate change impacts become a major concern on our globe. In the newly released report, *Water a Shared Responsibility*, the authors state; "In the last decade, 90 percent of natural disasters have been water-related events. Tsunamis, floods, droughts, pollution and storm surges are just a few examples of hazards that can constitute a risk for societies and communities. These are likely to increase in the changing environmental context projected for the future. Such hazards become disasters when risks are not managed with the objective of reducing human vulnerability. Floods and droughts are the most deadly freshwater disasters, disrupting socio-economic development in particular in developing countries. Efforts to reduce disaster risks must be systematically integrated into policies, plans and programs for sustainable development and poverty reduction" (Water a Shared Responsibility, 2006).

At the 2005 World Conference on Disaster Reduction held in Kobe, Japan, Jan Egeland, the UN Under-Secretary General for Humanitarian Affairs stated in response to the recent tsunami event in Indonesia, "Technology is not a cure-all. Experience shows us that people, not hardware, must be at the centre of any successful disaster warning and preparedness measure……All disaster prone countries should adopt clear goal-oriented disaster reduction policies and action plans, underpinned by detailed structures and resources….a minimum of 10% of the billions spent on disaster relief should be spent for *disaster risk reduction*" (Egeland, 2005).

In the future, water engineers must work collaboratively with water management and environmental professionals. We will continue to face challenges of hydrologic modeling, new and innovative risk protection schemes and an increased emphasis and more scrutiny toward risk reduction and safety of the public from water related disasters.

3.2.4 *Safe drinking water*

The future for global safe drinking water will challenge our engineering and institutional capacities requiring much more interdisciplinary interaction and collaboration. A few of the challenges are listed below:

Water Quality – As we move into the 21st century, it is important to realize that as scientists continue to improve their tools for detecting and identifying substances that can be measured in the environment, the number and type of chemicals and organisms known to exist in drinking water will increase. This knowledge will generate new debates in the medical and public health arenas regarding chronic health issues.

Waterborne Disease – "Many infectious agents have been categorized as emerging diseases and have not been recognized until recently, or at least not in association with water, including legionella pneumophila, Cryptosporidum parvum, ..., hepatitis E, and Helicobacter pylori." ". . we believe that the new evidence of carcinogenic and other health effects from exposure to disinfected water cannot be ignored and will likely challenge the public health and water communities in the 21st century" (Levin, 2002).

Groundwater – Contamination of groundwater supplies will continue to be a challenge in the 21st century. Regulation of groundwater quality and quantity will need to be institutionalized globally with accepted water quality standards to have any chance of managing this critical water supply intrinsically tied to surface water supplies.

Emerging contaminants – Although today's water treatment technology can remove metals and several pollutants, many are not technically capable of removing hormones, pharmaceuticals and other chemicals flushed down toilets or rinsed down drains. There is growing suspicion that increased use of anti-bacterial agents in human medicines, household cleaners and veterinarian medicines has encouraged the development of germs that are resistant to antibiotics. This could be the next 'big unknown" on the horizon of water quality threats in the developed world.

There are over one hundred new chemical compounds introduced annually. Personal care and pharmaceuticals, as well as per-chlorate compounds are an increasing issue in developed countries. Endocrine disruption is another recently identified problem seen in some waterways affecting the fish and wildlife populations. These new, complex chemical compounds are showing up in minute quantities in treated water returned to rivers, as well as from feedlot operations in the United States and are an emerging challenge in the waste water treatment community. Unfortunately people have few choices as how to dispose of unused and unwanted pharmaceuticals. Some guidelines need to be developed related to disposal of these unwanted drugs to prevent them from entering our vital waterways.

Alternate Urban Water Supplies – The technological challenges in the areas of *water supply*, will focus on conservation, reuse, desalination and other new technologies yet to be developed.

Desalinization – Desalination appears to be the wave of the future for many areas of the globe. Newer and better technologies continue to be developed that improve energy efficiency and lower production costs.

Figure 12 illustrates the growth in production of desalinated water since 1959. Today, it is estimated that the global production of potable water using desalination processes has risen to over 35 million m^3/d. Systems with capacities of up to 300 000 m^3/d are being constructed. Additionally, desalted seawater costs have decreased from \$2.00/m^3 in 1998 to about \$0.50/m^3 last year, according to Mark Wilf (2004).

Desalination systems will continue to gain popularity as their capacity and performance improve. High Efficiency Electrodialysis (HHED) is one of the latest technologies developed using up to 40% less membrane area than previous process. Many people in the water community believe that membranes will emerge as the first choice for treatment because they are readily available and affordable, they provide disinfection, and they solve multi-contaminant problems and provide a

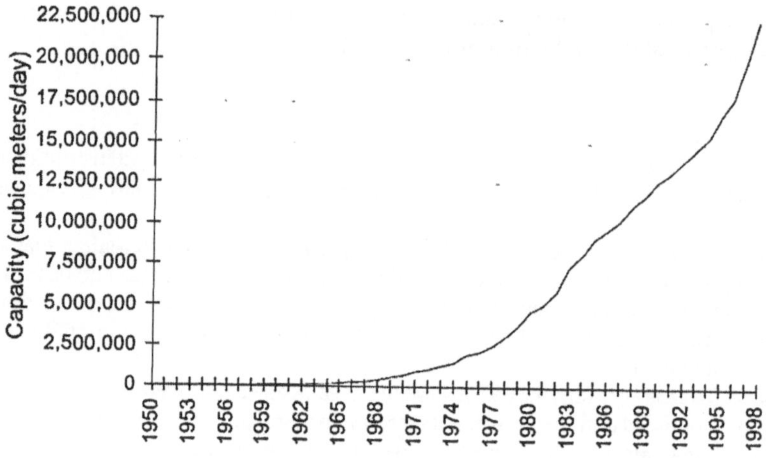

Figure 12. Cumulative desalination capacity worldwide, 1950 to 1998 (Gleick, 2000).

Figure 13. The City of Sarasota, Florida USA. Reverse Osmosis Water Treatment Plant Processes 17,000 m^3/day (CENews.com, May 2005).

physical barrier. Figure 13 depicts a relatively small desalination systems recently put in use in Sarasota, Florida.

For decades, Singapore has relied on Malaysia to supply the major portion of their water needs, but the two neighbors sometimes disagree and Singapore wanted to be less reliant on others. Therefore, they are building a 136,000 m^3/day desalination plant which is the world's largest facility making potable water from sea water (Hawlader, 2000). The new Ashkelon desalination plant in Israel will ultimately produce 274,000 m^3/d. The Bureau of Reclamation in cooperation with Sandia National Laboratories in New Mexico, USA is looking at ways to desalinate brackish or salt saturated groundwater for drinking water purposes in land-locked areas of the globe. These new technologies could turn present day contaminated water into potable water.

The negative aspects of desalinated water include: the future energy costs, release of concentrated brines to waterways or the oceans, as well as some concerns over the health risks of chemical compounds produced in the desalination process. Water authorities in the Emirate of Abu Dhabi are working to rid drinking water of elevated levels of bromate, a chemical compound thought to cause cancer in humans. Bromate can appear when drinking water is produced from salty sources, and in the United Arab Emirates and other desert countries, desalination of sea water is the dominant process for producing potable water (Times Online, 2005).

River-Bank Filtration (RBF) and Aquifer Recharge and Recovery (ARR) Technologies – In this century there are few locations with access to "first use" water. Even those with such luxury are looking at augmenting their water supply with reuse technologies. Cities such as Aurora Colorado, a neighbor city to Denver, is moving toward indirect water reuse using RBF and ARR technologies to provide potable water for its citizenry. The project will use reusable water rights on the South Platte River to extract water a second time from the river. The water will be pumped from the river banks through vertical wells and high-capacity horizontal collector devices. The water will then be reintroduce through an aquifer recharge and recovery scheme thus providing additional purification as the water moves through the river alluvium. RBF and ARR technologies are proving to be a cost-effective, reliable, and a sustainable barrier for the majority of organic micropollutants. After further purification, the reuse water will be mixed with "first use" water to supply the city.

Recycled Water for Non-potable uses – Denver, Colorado will recycle up to 113,600 m^3/d at its new recycling plant located on the site of the city's wastewater treatment plant on the South Platte River. The recycle plant uses a process similar to the one used to treat drinking water – coagulation, sedimentation, filtration and disinfection.

Although this recycled water doesn't meet drinking water standards and shouldn't be consumed, Denver Water's process will make it safe for fish, wildlife and, of course, plants. The system serves users such as: power plant use it for cooling purposes, parks, school lawns, the Denver Zoo, and municipal golf courses, see Figure 14 (Denver Water, 2004). Although recycled water will play a critical role in future global water supply, the challenges of water quality and water rights remain to be conquered.

3.2.5 *Agricultural water supply*

Rain-fed agriculture which currently produces more food than irrigated agriculture has benefited from the practice of collection of rainwater. *Water harvesting* – collecting water in structures ranging from small furrows to dams – allows farmers to conserve rainwater and direct it to crops. Water harvesting can boost yield two to three times over conventional rain-fed agriculture. Introducing improved crop varieties, better cropping patterns, correct crops for the climate, and using minimum tillage methods which conserve water, further increase yields. An example of the spectacular results that can be achieved by making these improvements is found in the Keita Valley in Niger. The valley was transformed from a barren desert to a garden for crops, livestock and trees (Water Spotlight, 2002).

Though developing countries depend on both irrigated and rain-fed crops to feed their people, much of the future increase in food production will need to come from irrigated land. FAO expects that irrigated areas in developing countries could expand by 20% by 2030. Coupled with increased cropping intensity, the effective harvested area will grow by about one third, from 250 million to 320 million hectares. Finding enough water to support such an increase requires producing more "crop per drop". The most common forms of irrigation: surface irrigation, in which water floods fields; and sprinkler irrigation, which mimics rainfall – can waste water. More efficiency is realized with localized application methods, such as drip irrigation, which puts water only where it is needed. FAO believes that the efficiency with which irrigation water is used can be increased over the coming 30 years – from an average 38 percent to about 42 percent. An FAO analysis of

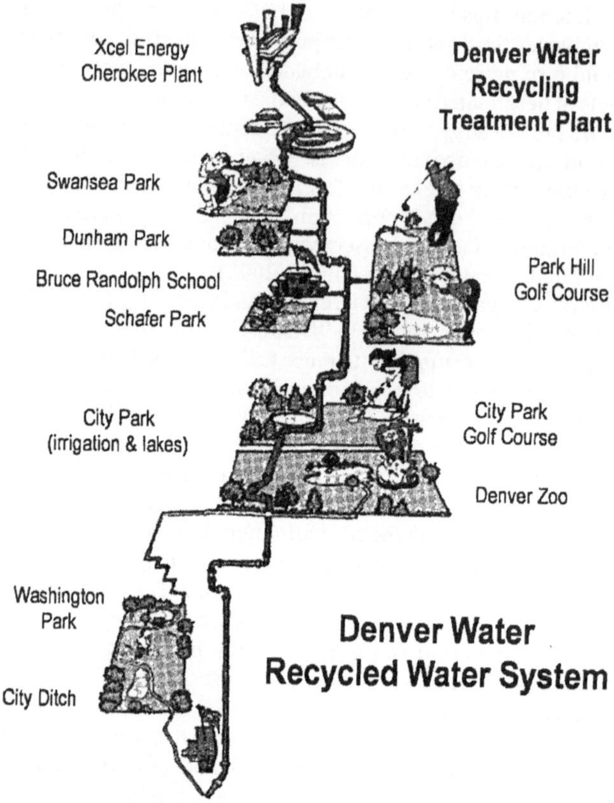

Figure 14. Denver Water Recycled Water System.

93 selected developing countries shows their water abstraction for agriculture in 1998 was about 2128 km^3 a year. They calculate that by 2030 only 2 420 km^3 of water will need to be abstracted to irrigate a net harvested area more than one-third larger than it is today. These countries and these regions will need special attention in the years to come, and they will need to increase their irrigation efficiencies by much more than just 4 percent" (Crops and Drops-FAO, 2000). One way to improve access to water is to treat it as an economic good as well as a social right. A pricing policy that makes the wasting of water expensive is one of the best incentives to save water.

To adequately confront the challenges of the 21st century, public policy as well as ingenuity in water engineering and management will be needed. In a summary report called *Addressing our Global Water Future* prepared by Sandia National Laboratories in New Mexico, the authors' state, "...... it is clear that institutional capacities in governance systems across the world—varied as they are—must all be strengthened to adequately address the magnitude of future challenges involving water. Improving governance will enable and facilitate the development of strategies and responses engaging the full range of available water-related technologies—from high-tech, high expense to low-tech, low expense" (CSIS, 2005). New water solutions are appearing daily and must be applied at new and greater scales to reduce the impacts on public health, economic development, environmental degradation, and political stability. Continual effort and investment is needed to develop new technologies and initiate policy approaches and synergies that could jumpstart new solutions in the decades to come. Policy, management and technology must evolve together to effectively link innovative strategies with innovative technologies.

4 CHANGING EMPHASIS IN WATER RESOURCES ENGINEERING IN THE 21st CENTURY. A CASE STUDY – THE U.S. BUREAU OF RECLAMATION

We have looked at some of the water challenges the global community will face in the next century. I would now like to narrow the focus to one water agency in the western United States – The Bureau of Reclamation. The same demographic, economic, and environmental transitions occurring on a global level are causing major impacts in the Bureau of Reclamation as it faces a future that looks a lot different than its first 100 years. A look at these changes may provide contemporary insight into the emerging issues of water engineering and management in the developed and developing world.

4.1 *The Bureau of Reclamation's Past*

A century ago the Bureau of Reclamation was established by the U.S. Congress to develop the water resources of the arid western United States and thus promote the settlement and economic development of that region. The focus was on development of the much needed water resource flowing in untamed western rivers such as the Colombia, Colorado, Missouri, Sacramento, American, Green, Rio Grande, etc. This focus on development was generated in an era where public values were founded on expansion by development of natural resources. Since 1902, the Bureau of Reclamation has built hundreds of projects that served as the foundation for settlement and economic development in the western United States. This development has resulted in Reclamation being the largest wholesale supplier of water in the United States, the sixth largest electric power generator and manager of over 45% of the surface water of the western United States. The successful water development model used by the Bureau of Reclamation in the 20th century, has been considered a model for replication by many countries around the world. However, is it a viable model for the 21st century?

An example of this climate of development in the 20th century is best illustrated in the excerpt from an article dealing with Reclamation's Newlands Project written 100 years ago: "...... *June 17, 1905, was an occasion of great moment to the interest of the State of Nevada, for then it was that the immense government irrigation canal known as the Main Truckee Canal received its first water from the Truckee River. The Indians will not suffer, even though Pyramid Lake does dry up on account of its main artery, the Truckee River, being diverted from it into the irrigation canals. When the lake runs dry, if it ever does, the Indians will have water from the irrigation system. Therefore there is no cause for alarm over the care of the Indians in the event of the Pyramid Lake drying up.*" Figure 15 shows the geological feature for which the lake was named. [Editor's note; In 1968 the Paiute Indians, alarmed by Pyramid Lake drying up, initiated litigation to reverse the process] (Scientific American, 2005).

Some 89 years later, at the 18th Congress of the International Commission on Large Dams held in Durban, South Africa, Daniel P. Beard, Commissioner of the Bureau of Reclamation at the time, declared that the dam building era in the United States was now over and indicated that others at this international venue might want to join Reclamation in setting a new direction in water resources (Beard, 1994). A number of reasons for this change in direction were given including:

- Water resources policies were originally conceived and implemented to meet the needs of agriculture and mining
- Federal funding is limited
- Environmental policies are more stringent and demanding
- Public support in favor of agriculture subsidies has diminished
- Indigenous peoples and environmentalists have a critical voice in political and legal proceedings

Beard's comments were received with some suspicion by the international community of water engineers and managers. The Spanish Committee on Large Dams took exception to Beard's comments as being politically motivated and not well representing the United States and more

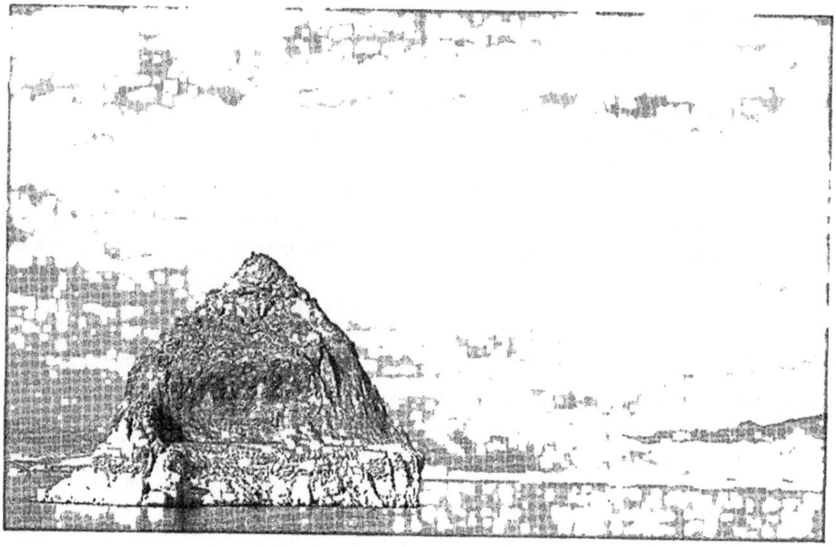

Figure 15. Pyramid Lake, Nevada.

specifically the Bureau of Reclamation on such an international platform as an ICOLD Congress (Saavedra, 1995). Many water managers regarded his comments as self serving and not worthy of the leadership role in large dam construction which the United States and the Bureau of Reclamation had assumed in previous decades in the international water community.

However, to its credit, the Bureau of Reclamation continues to change its emphasis from a federal agency focused on water development to one focused on confronting the reality of managing water in the western United States under new rules of engagement. Marc Reisner, author of *Cadillac Desert*, and a good friend of Daniel Beard quoted Beard as saying: "*The greatest challenge facing state and federal water leaders, is how we can effectuate transfers from agriculture to urban and environmental uses in a politically acceptable fashion. I don't have any specifics on how we can guide these transfers. I just know they will happen. Los Angeles, Las Vegas, San Diego, Tucson – urban regions in the desert West will not run out of water. If agricultural interests say, 'Sorry, but we need to keep using all this water to raise hay and alfalfa,' well, that's just not going to happen. As some sage said, water does run uphill to power and money.*" Reisner said, "*Beard is short on specifics because, in a region where frontier thinking still prevails, redistributing water is, to many, a sacrilegious idea. As Mark Twain noted, in the West, 'Whiskey is for drinking, water is for fighting over.' Also, water transfers are sanctioned mainly through state law, and every state has written different laws*" (Reisner, 1995).

4.2 *Role of the Bureau of Reclamation in the 21st century*

The Bureau of Reclamation still faces major challenges as it continues to define its role in the 21st century. Its new role as a water broker to various water stakeholders such as agricultural, urban, environmental and recreational interests will be an exercise in developing collaboration.

Water allocations continue to be a contentious issue in the Klamath Basin, where the Bureau of Reclamation is charged with providing suitable habitat for endangered suckers in Upper Klamath Lake, environmentally threatened coho salmon in the Klamath River, and more than 1,000 farms in the Klamath Reclamation District which straddles the Oregon-California border.

The Bureau of Reclamation operates a "water bank" in the Klamath Basin as a way to hold water in reserve to supplement Klamath River flows. The water bank was instituted after the severe drought conditions in 2000 and 2001. A Congressional report found that the Bureau of Reclamation has delivered the required amount of water. But it failed to provide stakeholders with clear information about the water bank's management and status.

In recent hearings before the U.S. Senate Conference Committee on Energy and Natural Resources, Reclamation's clients generally see a future role for the agency as a water broker in the next century. Reclamation's history is entwined with the development of the West (U.S. Senate, 2005). That development continues today at an unprecedented rate, and is placing significant pressure on a finite water supply. There are several possible directions the Reclamation program can move in the immediate future. Some of these challenges align closely with the global challenges mention above. Others are specific to the arid west of the United States. The following are a few of its most critical challenges:

Dam safety must be a priority – Reclamation manages over 350 high dams in the west. While the construction of large new federal dams and reservoirs is unlikely for the foreseeable future, Reclamation faces an enormous challenge related to its portfolio of aging dams and related infrastructure.

Agricultural Water Conservation – Reclamation is actively pursuing programs to help irrigation districts and other water users make the most efficient use of available water supplies. Some government agency needs to be administratively responsible for the operation and maintenance of these facilities.

Fishery Protection – Endangered species and western water management are and will continue to be intertwined. Finding water for fish and farmers, as well as the growing municipal and industrial needs, within the parameters of state water law and federal environmental law is a challenge that must be successfully met.

Dam Removal Issues – With respect to the issue of dam removal, the engineering issues and legal and socioeconomic issues, as well as functional alternatives to small and large dams need to be carefully considered. Reclamation has developed experience and expertise in these areas.

Research into Alternative Water Supplies – Reclamation has been and should continue to be a leader in the development of a number of alternatives and technologies that promise to help meet future water needs: (1) ground water recharge, storage and recovery projects; (2) water reclamation and reuse projects; (3) desalination; and (4) eradication of salt cedar (Tamarisk). There may be other opportunities to increase water storage and yields from wetlands/streambanks through better management of state and federal lands and riparian zones.

Research on aquifer storage and recovery or artificial recharge – This is an important item in the portfolio of future water in the west.

Research on emerging contaminants – This challenge is directly related to water quality issues. Work is needed on the development of remediation technologies that can be used to address new and current pollutants.

4.3 *New directions for Reclamation's hydraulic research*

An even more specific focus can be placed on Reclamations water resources research. In the period from 1930–1980's Reclamation produced world class technological advancements in pursuit of water development. These advancements are documented in publications such as Reclamation's Water Measurement Manual (1953, 1997), Engineering Monograph #25 Hydraulic Design of Stilling Basins and Energy Dissipators (Peterka, A. J., 1958), Design of Small Dams (1960), Design of Small Canal Structures (1974), and Engineering Monograph #41 Air-Water Flow in Hydraulic Structures (Falvey, 1980).

However, as public values shifted from an emphasis on water resource development to management of western waters, Reclamation's contemporary water resources research program has also changed. This evolution from *water development* to *water management* has led to an emphasis on technological innovation for public safety associated with existing infrastructure, encouraging water conservation, and emphasizing environmental restoration on regulated river systems.

Figure 16. Ute Dam 14-Cycle.

4.4 *Protecting infrastructure Dam Safety*

The Bureau of Reclamation administers the U.S. Department of the Interior Dam Safety Program, which involves, among other goals, the development of new technologies to cost-effectively solve dam safety problems. Inadequate spillway capacity is a primary reason for dam failure; therefore, Reclamation's dam-safety research includes hydraulic investigation of alternative spillway designs, fuse plug concepts, and overtopping protection concepts. The safety improvements, innovative design concepts, and construction cost savings realized from this research have been significant.

4.5 *Increased spillway capacity: Labyrinth spillways*

Reclamation has used labyrinth spillways on existing dams where the discharge capacity of a spillway is insufficient or where a reservoir must be enlarged. Research on the labyrinth spillway concept produced design criteria that were applied to augment the spillway capacity at Ute Dam on the Canadian River, New Mexico, and generated significant savings in field construction cost, Figure 16. Ute Dam labyrinth spillway was constructed for $10 million, a $24 million cost savings over the estimated $34 million cost for a traditional gated structure (Houston, 1982).

4.6 *Emergency spillway concepts: Fuse plugs*

Pugh (1984) defines a fuse plug as "an embankment designed to wash out in a predictable and controlled manner when the capacity in excess of the normal capacity of the service spillway and outlet works is needed." A number of laboratory embankments at scales of 1:10 and 1:25, were tested in the laboratory to develop fuse plug spillway design criteria. Fuse plug designs have been selected for the dam safety corrective action plan for the Horseshoe and Bartlett Dams on the Verde River in Arizona. The fuse plug for Bartlett Dam is designed with an erosion-resistant invert and abutment structure and will pass 10,100 m^3/s, Figure 17. Three erodible embankment sections will operate in a step-wise sequence. The Horseshoe Dam fuse plug is designed to pass 6,850 m^3/s through three 44–52 m long openings that vary from 6.0–7.9 m high. The documented construction cost savings of $150 to 300 million on the recently upgraded Verde River dams are an example of the significant benefits resulting from this hydraulic research of innovative alternatives to traditional spillway design.

Figure 17. Bartlett Dam, Verde River Arizona, 9-m high labyrinth Spillway Note fuse plug at the right side of the photograph.

4.7 *Concepts for embankment overtopping protection*

Flood flow overtopping an embankment is considered unacceptable. However, hydraulic protection systems are available, not only for low dams (under 15 m high), but now for high dams as well. Frizell et al. (1994) have reported on cooperative research, funded by the U.S. Department of the Interior, the Electric Power Research Institute, and Colorado State University, that has resulted in design criteria development for concrete step overlay protection for embankment dams. After initial laboratory studies, near-prototype tests were performed in a large-scale, outdoor overtopping facility at Colorado State University, Figure 18.

The 1.5 m wide, 15 m high outdoor test facility subjected wedge-shaped blocks to unit discharges as high as 3.2 $m^3/s/m$. The 35 cm long, 5 cm high, and 60 cm wide blocks were placed in an overlapping pattern on filter material. The blocks are designed to aspirate water from the filter layer through small drainage slots formed in each block. The block shapes developed through these studies have been patented and are effective for a range of embankment slopes. Armortec of Bowling Green, Kentucky has exclusively licensed this product and is developing a block system trademarked as Armorwedge®.

Reclamation is working with a design team from Spain to assist in the design and first application of the Armorwedge block on a rockfill dam in Spain. Barriga Dam, located in Burgos, Spain, northeast of Madrid, is 18.5-m-high with a trapezoidal spillway section. The spillway crest is 26 m wide and has a unit design flow of 6.5 $m^3/s/m$. The flow drops 11 m ending in a bucket toe block above the tailwater. It is anticipated that the Barriga Dam project will be completed by the end of summer 2006.

4.8 *Environmental restoration*

In the 21st century, water development and environmental interests must coexist at a new level of collaboration. In recent years, hydropower production and agricultural water supply have been cut back substantially in the U.S. to meet regulatory environmental requirements. Rivers regulated for hydropower development, urban and agriculture water supply, and flood control are complicated ecological systems; operational decisions must consider fishery issues and other environmental resources as well as engineering design. A bioengineering (biological and engineering) focus has led to new, innovative concepts for using hydraulic structures to manage regulated aquatic ecosystems

Figure 18. Outdoor test facility at Colorado State University. Flow on 2:1 slope with a unit discharge of 3.2 $m^3/s/m$.

in the west. A look at fishery and stream restoration issues in the western U.S. illustrates these new technological approaches.

4.9 *Reservoir selective withdrawal*

All runs of chinook salmon in the Sacramento River have declined because of several factors, including warm water temperatures in the upper reaches of the river. From 1987 to 1997, Reclamation used the river outlet works at Shasta Dam, which bypassed the powerplant, to provide cooler water for the salmon. The cost of replacing power lost by bypassing the powerplant was more than $35 million over seven years. Despite these efforts, the winter-run Chinook salmon were listed as a protected species in 1989. Three years later Reclamation was directed to install and operate a Temperature Control Device (TCD) at Shasta Dam to reduce the loss of salmon, Figure 19. Reclamation engineers began working on preliminary designs in 1988. After assessing several alternatives, engineers recommended a shutter-type device developed in Reclamation's hydraulics laboratory. As tall as the Statue of Liberty and as wide a football field, this steel structure is one of the largest man-made mechanisms ever constructed for fish preservation. Constructing the Shasta TCD is comparable to building a 28 story steel building under water and contains approximately 9,000 tons of steel. High level withdrawal from the reservoir is controlled by the 76-m-wide by 91-m-high shutter structure that projects about 15 m upstream. The structure is open between shutter units to permit cross-flow in front of the existing trashrack frames. Three openings with hoist operated gates and trashracks on the front of each shutter unit allow selection of the reservoir withdrawal level. To the left of the shutter is the low-level intake structure, which is 38-m-wide by 52-m-high and also projects about 15 m upstream. It acts as a conduit extension to access the deeper, colder water near the center of the dam. Similarly, a reservoir selective withdrawal structures was also designed and installed on the upstream face of Flaming Gorge dam in 1978. Reclamation is currently designing selective withdrawal structures for 2 of the 8 penstock intakes at Glen Canyon Dam on the Colorado River in an effort to restore suitable habitat for native fishes.

Figure 19. Installation of the Shasta Dam. Temperature Control Device (General Construction Company).

The 1988 through1992 drought in northern California resulted in limited volumes of stored cold water deep in reservoirs. Because of the urgent need to reduce reservoir outflow temperatures, the Reclamation initiated an active research program to develop and install temperature-control curtains in shallower reservoirs, such as Lewiston and Whiskeytown Lakes. Two curtains were designed and installed in Lewiston Lake in August 1992. The primary reservoir curtain was designed to hold back the warm surface water while colder water traveled under the curtain and was released through Clear Creek Tunnel into the Sacramento River (Vermeyen and Johnson, 1993).

In a continuing multiagency effort, two additional flexible curtains were laboratory tested, designed, and installed in Whiskeytown Lake in 1993. The use of these new temperature control technologies will ensure continued hydropower production at Shasta Dam, increase the selective withdrawal capability within the Sacramento River basin, and provide improved management by selective withdrawal of the limited cold water storage in Shasta Lake.

4.10 *Fish Protection at Water Diversions*

More recently, the Bureau of Reclamation recently published a new manual entitled *Fish Protection at Water Diversions* (Burgi, et al, 2006). The 450 page manual summarizes the various fish screen designs currently used on western rivers in the United States. Figure 20 illustrates one of the new drum screen installations at a water diversion in the State of Oregon.

4.11 *Fish passage*

Considerable effort has been placed on improving fish passage technologies in recent years, including new designs for fishways, improved spawning facilities, fish barriers with associated bypass designs for canal headworks, and various screening and fish behavioral control concepts. Most recently, efforts have centered on returning the Sacramento River near Red Bluff Diversion Dam to a run-of-river condition by raising the dam gates for much of the year. Several alternatives are being studied to improve the fishery. One alternative, proposed by Liston and Johnson (1992), is to evaluate the feasibility of replacing the diversion dam with a pumping station utilizing fish-friendly pumps. The full-scale plant would deliver 76.5 m^3/s, with a lift of 4.3 m to the Tehama Colusa Canal, while incurring minimal fish mortality. Every effort has been made to minimize fish entrainment

Figure 20. Drum Screen Installation.

at the pump intakes. Construction of a pilot pumping plant, which pumps up to 9.5 m^3/s, was completed in the spring of 1995. It is designed to evaluate and monitor the mechanical performance of two fish-friendly pump concepts as well as evaluate fishery issues associated with pumping. A screw-centrifugal (helical) pump and an Archimedes screw pump are being evaluated. Two 3.0 m diameter, 8.0 m long Archimedes pumps, placed on a 38° angle, deliver a total of 4.5 m^3/s at a rotational speed of 28 rpm. One 1.2 m helical pump delivers 5.0 m^3/s at 400 to 600 rpm. Fishery and mechanical issues will be evaluated over several years at the research pumping plant before construction of a larger, permanent pumping plant.

4.12 *Stream restoration*

Water development projects have altered the character of rivers and watersheds. Restoring a watershed or ecosystem damaged by physical alterations to the natural flow regime requires multidisciplinary research involving engineers, biologists, geomorphologists, landowners, and the public.

Muddy Creek, near Great Falls, Montana, is an example of the Bureau of Reclamation's stream restoration efforts. This creek has been drastically altered by irrigation return flows. Muddy Creek historical flows before irrigation were on the order of 12.3×10^6 m^3/year. The creek now sustains runoff of 98.7×10^6 m^3/year, eight times greater than historical flows. A 15 m incised channel has been carved in the glacially deposited silty soil since the early 1900s. Active measures to restore stream gradient and reduce erosion are now under way. The erosion control demonstration project includes 19 rock ramps and three barbs placed along a 6.6 km reach of the creek in 1994 to control 8.2 m of stream gradient. The demonstration project will look at long-term performance of the in-place technology as monitoring programs track stream response in the short and long term.

4.13 *River water-quality improvement: Dissolved gases*

Hydraulic structures often dramatically affect water quality and aquatic life by changing the spatial and temporal distribution of dissolved gases within reservoirs and regulated streams. Reclamation is conducting research to address problems of both nitrogen supersaturation and reduced dissolved oxygen concentrations downstream of energy dissipation structures and power plants.

Spillways and outlet works associated with hydraulic structures affect the dissolved gas content of the released flow. Depending on the structure and local conditions, there may be positive or negative effects on water quality. Releases may aerate flows depleted in dissolved gas, create supersaturated dissolved gas levels, or reduce supersaturated levels in the flow. Johnson (1975) presents an analysis to predict the effect of a wide variety of hydraulic structures on the dissolved gas content of the flow.

In the 1990's, the Tennessee Valley Authority (TVA) led efforts to improve dissolved oxygen conditions in reservoirs and downstream of power-plants (Bohac and Ruane, 1990). The Bureau of Reclamation has cooperated with the TVA and the U.S. Army Corps of Engineers to develop autoventing turbine technologies that use aeration of power plant flows to improve dissolved oxygen concentrations of power plant releases. Reclamation has also retrofitted turbines at Deer Creek Power Plant on the Provo River, Utah, to improved dissolved oxygen concentrations through turbine aeration (Wahl, 1995).

4.14 *Ongoing efforts*

There are numerous practical problems and environmental impacts associated with hydraulic structures on water resource projects that will motivate applied hydraulic research in the future. Efficient water use, of necessity, requires continued use of hydraulic structures to effectively manage water resources. Some of Reclamation's past successes in water resource development have produced new problems in water management as societal values have changed. The challenge is to keep hydraulic research contemporary by clearly identifying the problems and working as partners with the social and scientific communities to develop holistic solutions.

5 CONCLUSIONS

Water challenges in the 21st century are intimately tied to the global transitions in *demographics*, *economics*, and the *environment*. As water engineers and managers, we have the responsibility in our generation to effectively engage and solve these challenges so we might successfully pass to the next generation a sustainable approach to water development and management world-wide. Future generations will look back at our generation, the first to face these 21st century water challenges, and evaluate our effectiveness to pass on to them viable solutions and a road-map to the last half of the century. Viable institutions at the local, national and international levels will be the key to future success. "... it is clear that institutional capacities in governance systems across the world—varied as they are—must all be strengthened to adequately address the magnitude of future challenges involving water. Improving governance will enable and facilitate the development of strategies and responses engaging the full range of available water-related technologies—from high-tech, high expense to low-tech, low expense." (CSIS, 2005) As water engineers and managers, we see many of these global challenges as beyond our level of influence and therefore are tempted to abrogate our responsibility. That being said, we have two courses of action; 1) expand our level of influence to intentionally impact our world in areas beyond our historic role, and 2) fully engage our ingenuity to tackle the challenges of the 21st century. These challenges will require an engineering philosophy much broader than the days of Hoover Dam's design and construction. Because global freshwater is finite and in ever greater demand, 21st century engineering will be; more about managing water demand than expanding water supply; more collaborative and multi-disciplined than the specialized activities of the past; more micro than macro; more watershed level projects than multi-regional; more consensus seeking that authoritarian; and more about fish, soils and trees than steel and concrete.

Water engineers and managers today are measuring salinity, modeling watershed runoff, removing or preventing the spread of contaminated sediments, managing stormwater, restoring riparian zones or working with conjunctive use groundwater modeling. The 21st century calls for a continued reconciliation between the engineering and ecological communities.

As we are challenged to be more influential in the formulation of public policy, we also need to look closely at our own ethics and integrity to do the best of our abilities as articulated in the Order of the Engineer- Code of Ethics "... to serve humanity by making the best use of the Earth's precious natural wealth." (Order of the Ring, 1970).

Shinn (2006), in an article entitled *Living Upstream*, suggests we look beyond what he calls the "information stalemate" of bickering over environmental issues and shift the paradigm toward asking, 'What would it mean if we saw ourselves as living upstream instead of viewing ourselves as living downstream?' He states that, "When we view ourselves as living downstream from others' decisions and actions, we usually focus on the inherited elements of our current situation and debate our current options and actions...But to understand ourselves as living upstream from future generations requires much more focus on our intentions and actions rather than a debate on what our inheritance from the river is and what we want to make of it. ... we are responsible for what we put into the river of life (literally and metaphorically), and thus our care and compassion for future generations requires caution as we act." Viewing ourselves as living upstream instead of downstream calls for an "equity of generations" that commits to the best interests of, not only our grandchildren but, our great-great grandchildren (fifth generation). The American Iroquois tribes have a decision making principle – *Each decision should be made with seven generations in mind.* Such a mindset would produce a permanent paradigm shift in our planning and thinking about the future.

Are there water challenges that we face today and our children and grand children will face in the future? Of course, just as there has always been. However, water engineers and managers today and in the future face more daunting issues than in the past that will challenge our ingenuity and skills to the limit. Future generations are counting on us to act responsibly with our challenges today. We, as well as future generations of water engineers and managers, will need to work in a more collaborative world that will involve input from water users, environmental interests, political and policy interests.

REFERENCES

Abrams, L. (2001) *Water for Basic Needs*, 18 July (Commissioned by the World Health Organization as input to the 1st World Water Development Report).

Beard, D. (1994) International Commission on Large Dams, Durban, South Africa, November 9.

Bohac, C.E. & Ruane, R.J. (1990) *Solving the Dissolved Oxygen Problem*, Hydro Review, Feb.

Bureau of Reclamation. (1960) *Design of Small Dams*, Water Resource Technical Publications.

Bureau of Reclamation. (1974) *Design of Small Canal Structures*, Water Resources Technical Publication.

Bureau of Reclamation. (1997) *Water Measurement Manual*, Water Resources Technical Publication.

Burgi, P.H. et al. (2006) *Fish Protection at Water Diversions,* Water Resources Technical Publication, Bureau of Reclamation, http://www.usbr.gov/pmts/hydraulics_lab/pubs/manuals/fishprotection.html.

Burgi, P.H. & Rydbeck, B.V. (2001) *Sustainable Potable Water Systems Strengthen Rural Communities in Developing Nations.* EWRI World Congress, ASCE, Orlando, FL, August.

Brittain, J.E. (2002) *The Norwegian Research Program For Environmental Flows.* Enviro flows 2002 4th Ecohydraulics, Norwegian Water Resources and Energy Directorate (NVE).

Cabrera, E. & Cobacho, R. (2004) *Challenges of the New Water Policies for The XXI Century.* A.A. Balkema Publishers.

Chanson, H. & James, P. (1998) Rapid reservoir sedimentation of four historic thin arch dams in Australia. *Journal of Performance of Constructed Facilities, ASCE,* 12(2).

Chapagain, A.K. & Hockstra, A.Y. (2004) *Saving Water Through Global Trade*, UNESCO-IHE.

CSIS (2005) *Addressing Our Global Water Future.* Center for Strategic and International Studies, Sandia National Laboratories, New Mexico, September 30.

Crops and Drops-Making the Best Use of Water for Agriculture (2000) FAO, Rome.

Denver, W. (2004) (http://www.water.denver.co.gov/recycle/recycleframe.html)

Dunbar, J.A., Allen, P.M. & Bennet, S.J. (2001) Acoustic Imaging of Sediment Impounded by a USDA-NRCS Flood Control Dam, Oklahoma, USDA National Sedimentation Laboratory, Oxford, MS, Research Report No. 22.

Egeland, J. (2005) Keynote, World Conference on Disaster Reduction, Kobe, Japan January 18. unisdr.org/wcdr/.../WCDR-proceedings-of-the-Conference.pdf.

Enriquez, S. (2006) Children share ideas at world water forum. *Los Angeles Times*, March 19.

Falvey, H.T. (1980) Air–Water Flow in Hydraulic Structures. *Engineering Monograph* #41, Bureau of Reclamation.

FAO (2003) *Raising Water Productivity*, Spotlight, Agriculture 21, March (http://www.fao.org/ag/magazine/0511sp2.htm).

FAO Aquastat (2005) http://www.fao.org/ag/agl/aglw/aquastat/main/index.stm.

Frizell, K.H., Smith, D.H. & Ruff, J.F. (1994) Stepped overlays proven for use in protecting overtopped embankment dams, *Proceedings, 11th Annual*, ASDSO Conference, Boston, Mass.

Gleick, P.H. (2002) *Scientific Challenges in Watershed Hydrology*, Seminar Series on U.C. Davis, Pacific Institute, May 9.

Gleick, P.H. (2000) The changing water paradigm a look at twenty-first century water resources development. *International Water Resources Association, Water International*, 25(1), 127–138.

Hawlader, M.N.A., Ho, J.C. & Teng, C.K. (2000) Desalination of seawater: an experiment with RO membranes, *Proceedings of the Conference on Membranes in Drinking and Industrial Water Production*, 2, 537–542. Desalination Publications, L'Aquila, Italy, October.

Houston, K. (1982) *Hydraulic Model Study of Ute Dam Labyrinth Spillway*, Report GR-82-7, Bureau of Reclamation.

Hydropower and Dams (2005) *World Atlas and Industrial Guide*. The Friary Press.

Johnson, P. (1975) *Prediction of Dissolved Gas at Hydraulic Structures*, Report GR-8-75. Bureau of Reclamation.

Khaleej Times Online. (2005) *AP/Khaleej _ 12/31/05*.

Keen, M.J., 1987, *Advances in Hydraulics and Fluid Mechanics in the 19th Century*, pp. 159–172. Hydraulics and Hydraulic Research A Historical Review, A.A. Balkema.

Levine, R.B., Epestein, P.R., Ford, T.E., Harrington, W., Olson, E. & Reichard, E.G. (2002) U.S. drinking water challenges in the twenty-first century. *Environmental Health Perspectives*, 110 (Suppl 1), February.

Liston, C.R. & Johnson, P.L. (1992) Design criteria and environmental setting for a pilot pumping installation at Red Bluff diversion dam on the Sacramento River, California. Internal Report Bureau of Reclamation.

McCully, P. (2005) Silenced Rivers, The Ecology and Politics of Large Dams. Zed Books.

Order of The Engineer (1970) http://www.order-of-the-engineer.org/.

Peterka, A.J. (1958) Hydraulic design of stilling basins and energy dissipators, *Engineering. Monograph* No. 25, Bureau of Reclamation.

Prieto, M.M. (2005) *Droughts and Water Stress Situations in Spain*. Chapter 14, Drought and Water Crisis: Science, Technology, and Management Issues. Taylor and Francis.

Pugh, C.A. & Gray, E.W. (1984) Fuse plug embankments in auxiliary spillways: developing design guidelines and parameters, U.S. Committee on Large Dams.

Rahaman, M.M. & Varis, O. Spring (2005) *Integrated Water Resources Management: Evolution, Prospects and Future Challenges*. Sustainability: Science, Practice, & Policy, ejournal.nbii.org.

Reisner, M. (1995) The Fight for Reclamation, *High Country News*, 27(5), Paonia, Colorado.

Saavedra, F.M. (1995) *El Ocaso Del Bureau of Reclamation De Los Estados Unidos De America*, Revista De Obras Publicas, September, Madrid, Spain.

Scientific American (2005) *Crossroads for Planet Earth*. pp. 18 and 44–115, September.

Sharma, K. (1998) The Hindu – 12 December.

Shinn, L.D. (2006) *Living Upstream*. On Tap, National Environmental Services Center, Winter.

Times Online (2005) United Arab Emirates working to rid drinking water of chemical linked to cancer, 12/31/2005, http://www.waterwebster.com/Desalination.htm.

Ünver, O. (2002) Southeastern Anatolia development. *IIEEE Power Engineering Review*, March.

UN/WWAP (United Nations/World Water Assessment Programme). 2003. *1st UN World Water Development Report: Water for People, Water for Life*. Paris, New York and Oxford. UNESCO (United Nations Educational, Scientific and Cultural Organization) and Berghahn Books.

U.S. Senate Committee on Energy and Natural Resources (2005) Role of the Bureau of Reclamation in the 21st Century, April 5. http://energy.senate.gov/conference/waterconference.cfm

US Water News (2005) June.

Vermeyen, T.B. & Johnson, P.L. (1993) Hydraulic performance of a flexible curtain used for selective withdrawal: a physical model and prototype comparison. *Proceedings ASCE National Hydraulic Engineering. Conference*, ASCE, Reston, Virginia.

Wahl, T.L. & Young, D. (1995) Dissolved Oxygen Enhancement on the Provo River. *Waterpower '95*, San Francisco, California.

Water a Shared Responsibility (2002) The United Nations World Water Development Report 2 (WWDR 2) March.

Water: Precious and Finite Resource. (2002) Spotlight, Agriculture 21, October FAO (http://www.fao.org/ag/magazine/0210sp1.htm).

Water Conserve (2005) *Drought-plagued China sees water crisis peak in 2030*, 08 Jun 04:15:00 GMT Source: Reuters, http://www.waterconserve.info.

Watts, R.G. (1997) *Engineering Response to Global Climate Change – Planning a Research and Development Agenda.* Chapter 5 – Water Resources. CRC Press LLC.

Wilf, M. & Bartels, C. (2004) *Optimization of Seawater RO Systems Design.* Elsevier.

Wolf, A.T., Kramer, A., Carius, A. & Dabelko, G.D. (2005) *State of the World 2005 Redefining Global Security.* Chapter 5: Managing Water Conflict and Cooperation.

WSSD. (2002) Report of the World Summit on Sustainable Development, A/Conf. 199/20. http://www.johannesburgsummit.org. February 21, 2005.

Part D
Conclusions

Conclusions

There is no doubt: we are going through a changing and searching moment, in all aspects of our life and, fortunately or not, water management does not escape to this situation. Internationally, climate change is becoming better known and we are trying to analyze and estimate its consequences in daily water management. Nationally speaking, we are trying to adapt our infrastructures and our management service tools to keep an adequate quality level according with current situation.

Under these circumstances, a calm reflection exercise can let us to deal with this situation in a more constructive and positive way, and maybe a good approach could be to have a look at the past and check where we come from, what we did and how humanity has evolved throughout history in two important aspects such as "Hydraulic Engineering" and "Water Management". In any case, it is not enough. We need, after a calm look at the past, give out our know-how to the future, transmitting all good things coming from previous experiences. Because of that, I think now is the right moment to think about *Water engineering and management throughout history*, joining past – Water throughout time-, with future – Big water challenges in 21st century – and drawing as many conclusions as possible from this reflexive exercise we have launched in this project.

At the beginning of history, without doubts, agriculture initiated the need of development of hydraulic management, in Central and East Asia, long time before the diverse location of urban settlements became necessary a specific development for water supply. Till then, placing them close to watercourses was enough. The big Mesopotamia civilizations started with water management in the two large rivers Tigris and Euphrates.

If there is a civilization throughout history which has joined its own development to its fate in a natural watercourse, this is the Egyptian. Around 5,000 years, this civilization started depended entirely on the Nile River and its annual inundations; by this time, the cause remained a mystery. Until the 19th century it was not possible to make an engineering use of this natural periodic episode, flooding lands near the river for improving their fertility as much as possible.

During the "classical age", between the Archaic and Roman epoch, the political situation was characterized by the numerous wars between the various big city states, the Persian wars and the fall of the Alexander Magnus Empire. This policy absolutely conditioned the planning and construction of Mediterranean hydraulic infrastructures, becoming compulsory to build peculiar aqueducts when locally available water resources were not enough to guaranty water supply to all users in urban settlements. At the beginning, opened aqueducts were built, where water was in permanent contact with open air. Later on, because of health problems, they were converted into closed pipe-lines. In any case, we can conclude in this period huge structures for transport water were built, with more than a hundred kilometers long.

In Al-Andalus, water engineering and management were almost limited to irrigation, its system for legal regulation included. The extraction of water by means of mill dams, waterwheels and qanats constitutes one of the most significant contributions of Islam culture. This is the starting point of the later Spanish agricultural knowledge. Although some people think Spanish irrigation tradition comes, however, from Roman period, even in regions with later Islam influence like Valence and Murcia. In any case, as we show here, the culture of any place, even in such as specific issues like hydraulic ones, it is always consequence of different civilizations crossed its ground, being complicated to identify what comes from whom. Really, every period and every culture had to adapt their hydraulic technologies to the peculiar Mediterranean climate in our country, what made a special contribution to its spectacular development.

Hydraulics is as old as humanity, because water is a basic need for all life. However, urban supply, as it has already been said, promoted a hydraulic engineering development in a more educational way, along the 18th and 19th centuries, mainly in the European schools and universities, motivating and creating the hydraulic and urban service engineering specialties.

Leaving the importance of water in ancient cultures and going into the new challenges of water management in the 21st century, we can start saying agriculture represents a small percentage of the economic activity in modern societies but, at the same time, remains the biggest consumer sector. It does not mean now, in the current scarcity moment when an equitable water distribution is more difficult, any investment for improving water efficiency must not be done. The EU Water Framework Directive from 2000 becomes it mandatory and has promoted the Spanish modern irrigation plans, from a technical point of view, and the cost recovering in water pricing, from the financial one, even though consequences in such as sensitive sector to water pricing and used to be subsidized.

The passage of time demonstrates the growing importance of having such a precious natural resource, water, especially in crowded urban areas where water is strongly demanded. This basic premise for human surveillance is more important nowadays, if possible, due to the dramatic migration of the rural population to urban centers over the last decades in many countries. In current circumstances of demographic explosion and water stress, water sustainability concept become important among developed societies and what kind of legacy we are leaving to next generations, assessing solidarity concept for today and future.

On this hand, there is an increasing recognition that the "water crisis" is mainly a crisis of governance. I would be necessary to define correctly what "governance" means when we talk about water management and we could go into dialectic discussions, sometimes not so dialectic ones, depending on the different ways to interpret management in each country and each culture, in this current complex and global world. Undoubtedly, this word is related to the "political" treatment of all aspects connected and involved with tasks, being a paradox because in many places it would be convenient just to depoliticize the management of this basic resource. And then, another important issue comes up: social participation and how it is articulated inside all components of the process of real democratic governance. Maybe, the key point is a good coordination and a role definition for each participant in this activity: water engineers, hydrologists, sociologists, politicians, economists, consumers, journalists, etc., with the aim of turning such as a basic resource into a universal one, accessible for everybody and unchangeable with not related matters.

Link to this last concept, the universal water access, a new necessity is created which put on the table, one more time, the convenience of collaborate and go into the governance concept, just in this moment, when a demographic boom, climate change effects and urban settlement size are demanding it. In many cases, the only possibility would be to build extended lines of hydraulic interconnection for getting closer water to demanding places, like the existing long-distance network facilities satisfying the needs of electric power, oil, gas or information. Curiously, however, water conditions and determining factors are quite different. I will only mention the Spanish problem with Ebro basin river transfer. In any case, whatever we say, resources must be transfer from generation places to wherever they are needed, although the transferring element was water.

For concluding, we must say, water challenges in the 21st century are linked to concepts like demography, economy and environment. As water engineers and managers, we have the responsibility to effectively plan a water future use in similar conditions to current one, and better if possible. So, we have to be conscious a local, national and international coordination and adaptation of water management is needed, in such as way none jeopardizes another.

Moreover, we must not forget in such as global world like ours, any of our decisions or actions, even the littlest or the most local ones, have consequences for everybody and, even more, when we are taking decisions on a natural resource management. Therefore, we have to start working technically for solving current problems and avoiding creating new ones for future: no increase of salinity in water resources, reusing water as much as possible in low quality uses for preserving bigger quantities for the highest quality uses, restoring damaged zones, taking care of natural watercourses, preserving flowing waters, avoiding aquifer overexploitation, . . . but, mainly, we need to work in a more collaborative world that will involved input from water users, environmental, political and policy interests.

In short, as professor Cabrera writes in one of his documents, for a better adaptation to the quick changes we are suffering in water management terms, it is needed to:

- Depoliticize water management
- Educate citizens
- Reform Administration
- Implement legal reforms to improve water efficiency
- Establish control and surveillance mechanisms

Only in this way we will be able to answer, efficiently, to challenges of water management in 21st century.

Enrique Hernández Moreno
Civil Engineer
Director Services Aqualia

Author index